# Gaia in Turmoil

# Gaia in Turmoil

Climate Change, Biodepletion, and Earth Ethics in an Age of Crisis

edited by Eileen Crist and H. Bruce Rinker
foreword by Bill McKibben

The MIT Press
Cambridge, Massachusetts
London, England

For information about special quantity discounts, please email special_sales@mitpress.mit.edu

This book was set in Sabon by Toppan Best-set Premedia Limited, Hong Kong. Printed and bound in the United States of America.

Library of Congress Cataloging-in-Publication Data

Gaia in turmoil: climate change, biodepletion, and earth ethics in an age of crisis / edited by Eileen Crist and H. Bruce Rinker; foreword by Bill McKibben.
    p. cm.
Includes bibliographical references and index.
ISBN 978-0-262-03375-6 (hardcover : alk. paper)—ISBN 978-0-262-51352-4 (pbk. : alk. paper) 1. Gaia hypothesis. 2. Climatic changes–Environmental aspects. 3. Environmental ethics. I. Crist, Eileen, 1961– II. Rinker, H. Bruce.
QH331.G224 2010
577-dc22

2009007502

10 9 8 7 6 5 4 3 2

We dedicate this collection
to Rob Patzig
and to the next generation of Earth stewards,
especially
Jordan Chase Littrell and
Hunter Marshall Littrell

# Contents

# Foreword

Bill McKibben

To say that this book is timely would be an understatement. It literally couldn't be more of the moment, more crucial, more necessary.

The Gaian idea—or really, the very framework of the Gaian idea, the notion that we could think about the planet as a functional unit—seemed largely abstract until the question of global warming emerged in the late 1980s. With it came the thought that we might have introduced a real disequilibrium into the system whose remarkable stability for the duration of human civilization had masked the very existence of the Earth system. But even global warming seemed a little abstract, hard for most to picture, until the fall of 2007 when satellite pictures started appearing of the rapid melt of sea ice in the Arctic. All of a sudden we could see in virtually real time something that looked an awful lot like the experiments first modeled in Daisyworld. The ice began to melt; as it disappeared, and albedo shifted, the melt seemed to accelerate. By the end of September, the *New York Times* was describing scientists as "shaken" by the pace. Instead of the long, slow problem many had imagined climate change to be, we seemed to be staring at a dynamic system bent on flipping into some new state. In early December, America's foremost climatologist, NASA's James Hansen, gave a paper at the annual meeting of the American Geophysical Union laying out the latest Arctic data, and the most recent paleoclimatic interpretations: only if we somehow got back beneath 350 parts per million $CO_2$ in the atmosphere was there some hope of avoiding massive tipping points—a process that would require leaving most of the carbon still underground safely in place. Which would require, in turn, massive changes in human desire and appetite, in societal trajectory and organization. This was, at least in the loose sense, a Gaian diagnosis.

And it wasn't just scientists who were shaken. Any aware human shudders at the pace of change, at the inability to weigh ourselves on the old and reliable scales. At the same moment we seem very small against the backdrop of the great tempest now underway, and very large—for we have unleashed that tempest with our combustion. It is as disconcerting a moment philosophically as it scientifically, Copernican and anti-Copernican all at once. And we have no real clue what to make of it.

This remarkable collection will help. Its essays cover the gamut from physics to metaphysics, from biology to phenomenology. They offer something we badly need—a way to think about who we really are and what place we actually occupy in the scheme of things. And in that calibration there is some hope. James Lovelock notes that we are the species who, at the least, managed to leave the Earth far enough to look back in wonder. And David Abram, in a sparkling and compassionate essay, reminds us that even the horrors of a warming world carry some graces. "Only through the extremity of the weather are we brought to notice the uncanny power and presence of the unseen medium, and so compelled to remember our thorough immersion within the life of this breathing planet," he writes. "Only thus are we brought to realize that our vaunted human intelligence is as nothing unless it's allied with the round intelligence of the animate Earth."

This moment, in some sense, tests whether our brains and, more important, our hearts have evolved enough to deal with the troubles those same brains and hearts have created. Comprehending correctly the place where this battle will be fought, and the rules that govern its operation, is the first step in figuring out what to do, and what not to do. Physics, chemistry, biology—Gaia—do not bargain. They won't meet us halfway, allow us the luxury of an easy and slow change. We will see in the very near future if we're up to meeting the challenge of our moment, which calls us not to act heroically but humbly, to figure out how to shrink ourselves and our impact, how to fit in instead of dominate. It is the great challenge of our career as a species and hence, of course, for each of us as individuals once we understand the stakes. In every sense of the phrase, it is our moment.

# Preface

With its new and at the same time ancient understanding of the Earth as a living whole, the Gaian perspective has inspired awe, generated controversy, and stimulated research collaboration. Since its inception in the early 1970s as "the Gaia hypothesis," the science has morphed into a fruitful theoretical and research tradition. Gaian scientific inquiry, and the related field of Earth system science, have become progressively more established within the scientific community, yet understanding the biosphere continues to be a work-in-progress. For example, both organic and mechanical metaphors for the Earth abound in the Gaia literature; Gaian scientists grapple with whether biotic interactions are powerful enough to have tipped the Earth system into a habitable zone for billions of years; and Gaian theorists strive to articulate compelling formulations of the congruence between Darwinian natural selection and the evolution of a planetary living-cum-nonliving system.

Almost from its inception, Gaian inquiry took interdisciplinary form. For forty years Gaian science—itself a fusion of life and Earth sciences—has interfaced strongly with politics and culture: from the contested naming of the planet after a goddess and the lively clash with the selfish-gene view of life to the culturally celebrated (on the one hand) and scientifically censured (on the other) animistic intimations of the Gaian worldview. Gaian thinking has also had a stormy affair with green politics, given the tension between understanding the biosphere as a robust system versus an environmentalist view of our planet as fragile.

Since the turn of the twenty-first century, the ecological and political implications of Gaia have intensified with the deepening awareness of the effects human beings are having on Earth's atmosphere, hydrosphere, and land surface. The originator of the Gaia hypothesis, James Lovelock, has galvanized scientific and environmental communities with his latest prognosis of a looming ecological-climatic crisis. Indeed Gaia theory has led Lovelock to predict that unless arrested, the continued outpouring of greenhouse gases will lead to disastrous consequences by overstepping climatic thresholds. His ideas are now commanding increased attention, finding their way into the broader scientific and political discourse on global climate change.

This interdisciplinary volume forms a continuity with previous works about Gaia. The edited collection of papers has long been an important venue for Gaian science and scholarship—being an ideal medium for diverse threads of an unfolding understanding of the biosphere and allowing readers to experience the multidisciplinary nature of Gaian knowledge. At the same time this volume breaks new ground by focusing on global ecological problems in connection with Gaian knowledge. Analyses from a diversity of perspectives—by natural scientists, social scientists, philosophers, technologists, educators, and conservationists—center on two immense challenges facing biosphere and humanity: climate change and biodiversity destruction. We have concentrated especially on these two interrelated problems because of their daunting spatial and temporal scopes. Climate change and biodiversity destruction are occurring on a planetary scale, at an accelerating tempo, and their devastating repercussions will endure for millennia or longer. A correspondingly broad context of understanding can contribute to meeting and preempting these catastrophes underway. A whole Earth view presents such a context.

Gaian science investigates how the biosphere works as a whole. Today such pragmatic knowledge offers both theoretical and practical tools for navigating toward a future that seems increasingly uncertain. For example, the Gaian understanding of the global carbon cycle provides a biospheric baseline for the scale of its anthropogenic disruption and the consequence of climate change, in particular. Exploring the ties among biodiversity, nutrient and element cycling, and climatic patterns similarly serves to highlight the enormous risks of dismantling the planet's biological wealth.

The Gaian worldview has often been criticized as presenting a rosy picture of a biosphere inhabited by an integrated, interdependent, cooperative, and mutually supporting biota. It is true that Gaian science has resisted joining the bandwagon of a recent "paradigm shift" in ecological science: from the view of ecosystem as an integrated community of organisms to that of ecosystem as a dynamic aggregate of individuals.

The former, balance-of-nature perspective (currently out of favor) accents ecological stability and cooperation, while the latter perspective (now in vogue) highlights the pervasiveness of disturbance and competition. We believe that to its credit, Gaian science has eschewed the wholesale embrace of either paradigm (both with respect to smaller-scale ecosystems and the biosphere as a whole). Indeed it is highly unlikely that anything as magnificently complex as the Earth's global ecosystem is strictly describable by the lights of either model.

It is part of the richness of Gaian inquiry that it is equipped to perceive an intricate biogeochemical unity on planet Earth, while resisting the temptation to turn such a vision of exquisite integration into a spiritual (or academic) metaphysic. In fact small and large catastrophes, instigated from within or outside the planet, can and have happened to the biosphere. Whether by sheer luck, or by virtue of the biota's strong steering hand, Earth has remained unbrokenly inhabited despite such dangerous episodes as the first flooding of the atmosphere with poisonous biogenic oxygen, planetesimal collisions, runway heating episodes, and other extreme planetary events. Moreover the fact that fewer and fewer people are today willing to contest is that humanity's catastrophic impact has become comparable in scale to that of an asteroid strike. Gaian science, without unnecessary moralizing, illuminates why and how, from the perspective of the biosphere, we can count our net effect as a calamity that sooner or later will shift the Earth into a qualitatively different state. The chapters in this book converge in making this urgent point.

Past catastrophic episodes have instigated major reshufflings of life forms, climatic regimes, and proportions and cycles of chemical constituents within the biosphere. Thus far in Earth's 3.8 billion-year living history, the postcalamity biosphere has each time resettled into new states, stable enough for the evolutionary birthing and branching of novel life forms that have both altered and reinforced, through their interconnected adaptations, the biogeochemical regimes favoring them. The renewal of Gaia after global catastrophes—via the eventual generation of yet another taxonomic, biome, and genetic biota (i.e., the regeneration of biodiversity)—has always created a new chapter in the biosphere's wondrous natural history.

Even so, if humanity continues on the current path of excessive disturbance of the Earth system, the eventuality of the biosphere's self-restoration should offer us little consolation. The restoration of Gaia, as a whole, occurs within the range of millions of years—a time span that holds no meaning for human scales. Meaning of such timescales is certainly bountiful when we peer backward into deep time, using the powers of our imagination and technologies, but such meaning is virtually nonexistent when we attempt to project our minds into the future. Gaian science warns that without changing our way of life—without averting the depletion of life forms and ecosystems as well as halting climate change to at least current levels—we are heading into a biospheric period of profound indigence and instability for all future human generations as well as for much of our Holocene nonhuman cohort.

The role of technology in mediating the relationship between human beings and the natural world has been both excoriated and exulted. Sophisticated critics of modern civilization, from Lewis Mumford to John Zerzan, have deplored the denaturing reach of the machine through which wilderness is turned into playpen and artifact and the natural world is substituted by virtual reality. On the other hand, technocrats and planetary managers have put all their eggs in the solution-basket of technological fixes as though sheer technical innovation could reset our course toward a harmonious existence within the biosphere. The contributors of this work stake a different ground about the role of technologies—be they information, communication, or energy technologies—now and in the future. While technological approaches are necessary, they are not by themselves sufficient to heal the relationship between humanity and natural world. For example, while a drastic technological shift (and speedy technological transfer) can go a long way toward preempting the worst consequences of global warming, there are no technological solutions for habitat destruction, mass extinction, the depletion of marine life, desertification, and the diminution of freshwater sources. To address these problems, which are as pressing as global climate change, demands not simply forgoing destructive technologies and adopting green ones but embodying an Earth ethic in how we live. The chapters of this book thus support Lovelock's call for retreat—meaning an obligation to shrink our ecological footprint and relinquish the malady of the growth imperative.

Many of the chapters discuss the importance of life in shaping the global ecosystem called Gaia, accenting that long periods of relative stability within the biosphere are created through networks of biogeochemical cycles, the interaction between large-scale natural systems, the exchange of nutrients, and the complementary metabolizing of wastes. Yet there is neither scientific nor metaphysical guarantee that life forms always behave according to some etiquette of biospheric unity. Gaian science also explores (via interpreting the geological record or through computer modeling) what happens when life forms take the rogue path of massively disrupting the relatively stable parameters of the biosphere or of their local environments. It is sobering to contemplate that *Homo sapiens* may be describable as what scientists Tim Lenton and Hywel Williams in this volume call a "rebel" species—one that by overexploiting the nutrient stocks of the biosphere may end up decimating both itself and the global ecosystem.

This is admittedly the negative message of Gaian analysis: that we are overexploiting the biosphere we are embedded within and dependent on. The chapters of this book warn against such disturbance of the Earth system, for it can have unpredictable, large-scale, and irreversible effects. In this respect the politics of Gaian thinking are, for better or for worse, alarmist—agitating for change in order to avert disasters and worst-case scenarios. There is, however, also a deeply positive message coming from the Gaian platform. Gaia, or an integrated conception of the biosphere, can be a unifying idea of our time beyond national, ethnic, religious, spiritual, or ideological boundaries. A Gaian perspective can connect people, for it is as much an ancient and indigenous view as it is modern and scientific: Earth as an integrated, living entity consisting of its air, soils, rocks, waters, and all living beings.

The personification of the Earth as "living entity" is healthily broad enough to accommodate a diversity of ways to enliven the image: from mere linguistic metaphor and heuristic research concept to autopoietic system or superorganism. As Aldo Leopold speculated in a 1923 paper (first published in 1979) that foreshadowed the Gaia idea, the regard of the Earth as "organism" rather than "dead object" can inspire the respect and consideration we reserve, at least potentially, for living beings. What Leopold described in that paper as the Earth's indivisibility is coextensive with the Gaian matrix of interconnected ecosystems, nutrient flows, mutual consumption of waste by-products, recycling of elements, and tuning of macro- and microclimates.

Perhaps more than anything else today we need such a universal and fundamental vision of the global ecosystem to contextualize and solve our ecological and social quandaries. For we all live within this self-creating biosphere—one that is *living* in whatever sense we bring to that idea as individuals or cultures, a biosphere exquisitely sensitive to the impact of abundant life and fragile in the face of massive disruption, and one that deserves our respect as much for its dangerous unpredictability as for its magnificent beauty. We know that Gaia is robust as far as its lifespan is concerned—it is already measurable in the order of eons. But as far as our window of opportunity, not just to survive within the biosphere but to live the grace of our fullest potential as a species, time is fleeting and apparently running out.

# Acknowledgments

Some two dozen contributors—natural scientists, social scientists, philosophers, theorists, technologists, and educators among them—helped to shape this book, inspired by the conference "Gaia Theory: Model and Metaphor for the 21$^{st}$ Century" held in October 2006 outside Washington, DC. Hosted by the Northern Virginia Regional Park Authority and George Mason University, the conference brought together professionals and the public in an interdisciplinary dialogue to examine Gaian ways of understanding life on Earth. The initial idea of a conference précis morphed into a more precisely defined project: examining the intersections of Gaian science, Gaian analyses of the foremost ecological crises we face, and Gaian ways of rethinking political culture, environmental ethics, human experience, technological power, and education.

Much appreciation is extended to our contributors whose collective thought provides a hopeful way of knowing—knowing ourselves, knowing the biosphere and its dynamic history, knowing our options for the way forward. We also acknowledge and thank our editor, Clay Morgan, at MIT Press; Virginia Tech, particularly the Department of Science and Technology in Society; Pinellas County (FL) Department of Environment Management's Environmental Lands Division; and Martin Ogle, chief naturalist at Potomac Overlook Regional Park (Arlington, VA) and architect of the Gaia conference. We are keenly indebted to Bill McKibben for his heartening foreword. We are ever thankful for the support of our families, extending our gratitude especially to Rob Patzig and Ruth and Harry B. Rinker Jr. We also wish to recognize Lynn Margulis and Thomas E. Lovejoy for their long-term mentorships. We hope our readers will view this volume not so much a search for but a finding of Gaia as "one grand organic whole" that richly imbues our lives.

# I

## Introductory Essays

# 1

# One Grand Organic Whole

Eileen Crist and H. Bruce Rinker

In 1876 Alfred Russel Wallace, co-progenitor with Charles Darwin of the theory of evolution by natural selection, wrote in his classic book, *The Geographical Distribution of Animals*, that naturalists "who are disposed to turn aside from the beaten track of research may find...a study which will surely lead them to an increased appreciation of the complex relations and mutual interdependence." These, he continued, "link together every animal and vegetable form, with the ever-changing Earth which supports them, into one grand organic whole" (1876: vol. 2, 553).

One could hardly find a more succinct description of Gaia than "one grand organic whole." It submits that biota and their environments have been integral since the early eons of our ancient water world. It provides for feedback on multiple scales—from global processes like climate change and biogeochemical cycles to the minutiae of local environments. It highlights the primary impact of living beings and processes on the physiognomy of that world that even observers from the outer reaches of the galaxy would recognize as a life-bearing planet. It describes Gaia in a language of consilience that both scientists and religious thinkers can understand.[1] It underscores the unity and grandeur of the Earth by choosing the capital "E" spelling over the lowercase alternative that, regrettably, is still in extensive use. Gaia theory honors systems thinking on a planetary scale. James Lovelock and Lynn Margulis established the foundations of the paradigm decades ago, working assiduously and collaborating since those founding days to show its applicability across disciplines and even in everyday society.

The Gaian perspective emerged from the observation that physical and chemical conditions on Earth are inseparable from life's ubiquitous presence. Powerful influences crisscross living and nonliving domains binding them inextricably. With the birth of Gaian science some forty years ago, this intuitively grasped integration became the empirical subject matter of an ever-burgeoning body of researchers. At a theoretical level, the integration of living and nonliving domains was conceptualized as an amalgamation so profound as to form a biogeochemical entity that behaves as a self-regulating system. How the Earth system is best conceived, and what metaphors should be deployed to describe it, are matters of ongoing discussion and debate in the literature. James Lovelock has often drawn on cybernetics to represent this system; Lynn Margulis has called it a symbiotic planet and a global ecosystem; Tyler Volk has invoked the concept of holarchy. Regardless of what metaphors are chosen, and what power is ascribed (or not) to the Earth system's regulative abilities, Gaian thinkers converge on the idea that, as a whole, the Earth has emergent properties that make it a drastically different type of planet than a lifeless one (Lovelock 1979; Margulis 1998; Volk 1998).

Before the emergence of Gaian inquiry, conventional wisdom maintained that due to the wonderful serendipity of our planet being just the right distance from the sun, the appropriate chemical and physical conditions have existed for the emergence and continued presence of life on Earth. Based on a comparison of the three sister-planets (Venus, Mars, and Earth), this conception of a region in space favorable for life has been called the *habitable zone*—or, more playfully, "the Goldilocks view" in honor of Goldilocks' exclamations upon tasting the three bowls of porridge: Too hot! Too cold! Ah, *just right*!

What Gaian thinkers submit may one day be regarded as less extravagant than the Goldilocks view of life's persistence on Earth. Instead of conditions being assessed as "just right" on account of the good fortune of our planet's positioning and size, viable conditions are regarded as actively maintained by the biosphere.[2] To put it starkly, the biosphere is not simply *in* a habitable zone but also *makes* a habitable zone. Large-scale physical and chemical environments of atmosphere, hydrosphere, and upper lithosphere, along with the climates that these domains contribute to forging, have been—for 3.8 billion unbroken years of life's existence—viable contexts for an ever-changing, increasingly complex, and most often abundant biota. Gaia theory proposes that life's endurance during the unimaginable time span of over three and a half eons is unlikely to be just a matter of luck: alternatively, early in life's history living and nonliving matter became entangled as *a single entity* within which organisms themselves may have been shaping conditions to their adaptive advantage.

Many concepts have been used to describe this single entity: Gaia, biosphere, geophysiology, and Earth system, as well as (more controversially) living organism and superorganism. Originally the primal personification of the Earth in classic Greek mythology, *Gaia* has its counterparts in many prehistoric and historic cultures around the world: the Middle East, Rome, Europe, India, Mexico, the High Andes, and elsewhere. In its mythological guises, Gaia represents humanity's visceral grasp of origins, interdependency, and nurturing. The neologism *biosphere* was coined by geologist Eduard Suess in 1875 and elaborated by Russian geochemist Vladimir Vernadsky in his pioneering work, *The Biosphere* (originally published in 1929 but not available in English until 1979). Vernadsky elaborated a scientific argument for life as a geological force, and his ideas are now seen as anticipating Gaian science. *Geophysiology* was offered by Lovelock to highlight the interconnectedness of all the Earth's ecosystems on the analogy of the interrelations of organs and systems within the physiology of an organism. *Earth system* encompasses the planet's interacting domains of biota, atmosphere, lithosphere, and hydrosphere as a unity. *Earth system science* (inspired in part by Lovelock's thought) is the interdisciplinary inquiry into the complex workings of the Earth system, synthesizing such seemingly disparate disciplines as biochemistry, geology, climatology, microbiology, and ecology (see Wilkinson 2006).

Whatever name or conception best summarizes it, the Gaian perspective posits that "organisms and their material environment evolve as a single coupled system from which emerges the sustained self-regulation of climate and chemistry at a habitable state for whatever is the current biota" (Lovelock 2003: 769). While in ordinary language the concept of regulation connotes agency, in the context of Gaian science it is used analogically with the nonconscious, complex ways an organism's body regulates its own temperature and chemical parameters: not at set points but within acceptable ranges. According to Gaia theory, perturbations that would tend to shift conditions away from their relatively stable viable ranges are counteracted especially by means of negative feedback; such counteracting responses are termed the system's homeostatic tendencies. In the early days of Gaian thinking, most especially, *homeostasis* was identified—openly and implicitly—as the biosphere's signal feature. Over time, however, homeostasis has come to be seen as too static a paradigm to deliver the essence of a dynamic planet that has exhibited extremely varied physicochemical states and biota types over geological time. Homeostasis gave way conceptually to *homeorrhesis*, an idea cognate to the evolutionary model of punctuated equilibrium proposed by Niles Eldredge and Stephen Jay Gould: long periods of stable parameters (e.g., of temperature, atmospheric composition, and elemental cycling) are punctuated by planetary shifts, instigated by strong internal or external forcings, into new stable states (Eldredge and Gould 1985; Margulis and Lovelock 1989; Lovelock 2006).

Perhaps no event illustrates more crucially the biosphere's ability to respond in an apparently nonrandom manner to an external forcing than the Earth's maintenance of a viable surface temperature despite the sun's 25 percent increase in luminosity from the Archean to the present. (While this change is quantitatively substantial, it has obviously unfolded very slowly.) Prominent among the mechanisms of tuning temperature—in a way that has preempted the Earth from linearly tracking this heat increase—has been the gradual removal from the atmosphere of the greenhouse gas $CO_2$. How $CO_2$ is removed illustrates the exquisite choreography of the Earth's blended living and nonliving forces to yield a consequence favorable to life overall. Carbon dioxide is removed by rainfall that chemically reacts on land with calcium-silicate rock to form the soluble compound calcium bicarbonate, eventually flowing seaward. The chemical reaction is known as rock weathering—or, in Gaian terms, *biologically enhanced* rock weathering because the reaction is amplified, by several orders of magnitude, by soil (a biological phenomenon), plants, and other organisms (Schwartzman and Volk 1989; Williams 1996). But this is only part of the story of $CO_2$ reduction. After the carbon molecules of the once free-floating gas reach the seas, they are snatched up by organisms known as coccolithophores and by other marine creatures for use in constructing their exoskeletons. When these organisms die, their exoskeletons sink to the ocean floors. Through plate tectonics and volcanism some of that carbon eventually returns to the atmosphere as $CO_2$, but the net result over time has been the reduction of this key greenhouse gas, thereby countering—as Gaian scientists conjecture—the sun's increasing output (Westbroek 1991; Harding 2006).

The Earth story just described, involving the complex interplay of solar energy, rocks, soil, chemistry, plants, water in many forms, microorganisms, marine life, and gravity (to mention a few of the obvious factors), illustrates the seminal role life plays in shaping its environment.

Indeed Gaians propose that life can only prevail over long spells of time in the universe if it becomes chemically so powerful and physically so abundant as to contribute significantly to molding its planetary home. "In that sense, life is probably a property of planets rather than individual organisms" (Morowitz in Volk 1998: 107).

In the first two decades of the Gaia hypothesis, Gaian ideas became mired in scientific controversy and, to Lovelock's chagrin, were often greeted with silence and stonewalling. A piece of the chilly reception had to do with the name *Gaia*—and its train of association with such nebulous (or presumably disreputable) expressions as myth, metaphor, gender, spirituality, and New Age culture brought into the arena of straight facts and grounded theories. Another piece of the scientific establishment's initial recoil from Gaia involved its resurrection of an animistic view of the Earth. After 400 years of being virtually shelved by dominant mechanistic and reductionist perspectives, not only is *anima mundi* unabashedly expressed in Gaian literature, it has been turned into a research program within an interdisciplinary field charged to investigate it (see Barlow 1991). While neither the nontechnical naming after the Greek Earth goddess nor the extra-scientific intention to "animate Earth" (to cite Stephan Harding's recent title) have been abandoned, scientific representations of Gaia have changed and diversified since the early period of the 1970s. Changes ensued in response to critiques of the Gaia hypothesis, and also as a consequence of the natural unfolding of a scientific framework—in which numerous investigators have contributed to its elaboration and refinement.

The early Gaia hypothesis boldly proposed that the biota controls the global environment in order to keep planetary conditions habitable, stable, and even optimal for all life. This definition of Gaia came to be known as "strong Gaia" (and sometimes "optimizing Gaia"), and while it is often still recited in nonscientific arenas, it is now downplayed in the scientific literature for both conceptual and empirical reasons. The conceptual reason involved the teleological overtones of the idea that the biota can strive toward sustaining livable conditions. The critique of the first Gaia concept as teleological was offered by neo-Darwinians (Doolittle 1981; Dawkins 1982; Kirchner 1991), and it inspired greater care in conceptualizing Gaia so as to avoid the scientifically unsupportable implication that life, as a unified whole, can have a goal. (The neo-Darwinian critique also inspired the creation of the Daisyworld model by Andrew Watson and James Lovelock to be discussed shortly.)

The empirical reason for the rejection of strong Gaia involved the deepening recognition that catastrophe and instability have been such integral and reoccurring aspects of Earth's history that notions of the biota being in control, creating optimal states, or maintaining homeostatic conditions seem unsustainable (see Huggett 2006). Geologists, in particular, challenged the proposal that the biota—a "paper-thin" layer on the planet's surface—could possibly govern geological processes and cycles that act on far slower time scales and vaster spatial scales than biological systems (see Holland 1984). Goaded by astute biological and geological critiques, the Gaia hypothesis evolved into Gaia theory, while Lovelock's intention to unify Earth and life sciences inspired the emergence of Earth system science—a field that is friendly toward but not coextensive with Gaian thinking (e.g., Jacobson et al. 2000).

While strong Gaia has thus been on the wane for three decades, its antipode, "weak Gaia" (also known as "influential Gaia"), was always regarded as too self-evident to merit central status in the definition of Gaia. Weak Gaia simply states that life physically and chemically influences the global environment—a fact with which few can disagree (e.g., the oceans' microorganisms, alone, make 40 percent of the atmosphere's oxygen). James Kirchner (2002) pithily summarized the widely shared verdict on the two perspectives: strong (or optimizing) Gaia is new but not true while weak (or influential) Gaia is true but not new. This leaves the mid terrain for articulating an empirically robust and theoretically tenable understating of Gaia. Some have called this middle ground "co-evolutionary Gaia"—the view that, by constantly impinging on one another, geological and organismal domains form a coevolving unity that indeed has always been habitable (Schneider 1986). But are nonliving and living domains merely coevolving and otherwise coincidental influences, or are they coevolving as an integrated system that regulates planetary conditions to some degree or other? Co-evolutionary Gaia leaves the question unanswered but open.

As Jon Turney (2003) noted about the four decades of its transformations, Gaia theory has become more complex, richly associative, and open to modification. Gaian thinking evolved from the provocative hypothesis that life controls or optimizes planetary conditions for its own benefit to a more nuanced theoretical framework that submits life (within the co-evolving nexus of biotic and inorganic world) is a key player in shaping the planet. Working out the details of the intense interaction and feedbacks between the living and inorganic worlds, especially on large-scale and global levels, comprises the Gaian research program.

Perhaps the ultimate challenge of this program is to demonstrate that life's impact is so substantial as to be (or have been) the catalytic ingredient of keeping Earth livable in the face of inexorable, often stupendous cosmic, geophysical, and geochemical forces. To that end Gaian scientists examine to what extent, by what mechanisms, and by what patterns of (inter)action the biota may load the dice, so to speak, for its own persistence beyond the play of chance.

How might the biota contribute to its own persistence without purpose, intention, or as Richard Dawkins once quipped, public-minded collaboration for the good of all life? The creation of the computer model "Daisyworld" in the 1980s served to illustrate how organisms can tune global conditions to their own advantage simply by doing what organisms do best—growing abundantly (Watson and Lovelock 1983). In this model a hypothetical planet (like Earth), orbiting a star that is increasing in luminosity (like our sun), is seeded with daisies that come in black and white varieties. The black daisies absorb sunlight and thus do best in the early times of a cooler sun, while the white daisies reflect sunlight and thus prosper as the sun gets hotter. The average surface temperature of a Daisyworld without its daisies would directly correlate over time with the linear increase of the sun's output (assuming an unchanging atmosphere). In a Daisyworld with thriving daisies, however, the average surface temperature is stabilized over an extended period, within a daisy-friendly range, by the thermostat-mimicking play of black and white daisies growing; black ones predominating initially, followed by a black and white planetary tapestry, and concluding with mostly white-daisy cover. (The sun's overbearing heat eventually trumps all varieties.) The creation of Daisyworld *in silico* was a landmark moment in Gaian science. Its power did not lie in modeling the Earth but in representing conceptually and mathematically that a living mechanism on a planet—provided its global effects reinforce the benefits of its local effects—can literally tune a planetary variable such as temperature in an automatic, nondeliberate, and morally neutral (requiring neither collaboration nor competition) manner. Its simplicity notwithstanding, Daisyworld has remained a memorable biospheric model for its perspicacity in making a point.

Organisms' exquisite ability to adapt to environmental exigencies has been well established in the 150 years since the publication of Darwin's *On the Origin of Species*. The Gaian perspective complements this knowledge by investigating life's less explored capacity to tame the very exigencies that impinge on it. The biota can have global impact as a consequence of its abundant products and processes of metabolism, nutrition, respiration, and behavior. Its chemical and physical effects add up to a collection of forcings that tip the Earth into a state very different from what a lifeless one would be. A hypothetical Earth without life—but endowed with the same size, distance from the sun, and initial conditions—would be very different from the biosphere we know and biospheres past. So, while the evolution of life is largely driven by natural selection, Gaian scientists also insist on the significance of life itself modulating the selective forces that act upon it.

In an influential paper seeking to wed Darwinism with a Gaian understanding, Tim Lenton (1998) proposed that organisms altering their environment in ways that (happen to) benefit them could have greater likelihood of being favored by natural selection than those organisms creating effects that backfire on them (see also Lenton 2004; Lenton and Williams, chapter 5 of this volume). Organismal traits that benefit their carriers by increasing their short-term reproductive fitness certainly tend to be selected for. To this classic Darwinian view, Gaian thinking adds that if (many of) those same traits *also* perchance result in environmental effects (or by-products) that eventually provide positive feedback to their carriers, the latter may be doubly favored: for such traits will confer both short-term reproductive fitness and mid- to long-term reproductive fitness via environment-enhancing consequences.

The Gaian perspective has never diverged from the Darwinian tenet that life adapts to its conditions via, in large part, the mechanism of natural selection that favors those organisms better suited to their particular conditions. Gaian scientists have noted, however, that when natural selection is one-sidedly emphasized, as it is by some neo-Darwinian thinkers, the latent message is a representation of living organisms as more passive than they actually are: they are portrayed as bystanders within an environment that, on one extreme, rewards them with reproductive success, while on the other, wipes them out if they are misfits. Some critics of Gaia, for example, James Kirchner (2002), insist that the environment merely appears well-tailored to the needs of life on account of straightforward Darwinian adaptation—only those living organisms persist that were selected for their good fit to their conditions. Gaian scientists counter that physical and chemical variables are so inextricably entangled with the biological world—being either a product of the biological world or hugely modified by it—that it may make more sense to regard the environment as life's extended phenotype, than to conceptualize the environment as a straightforward independent variable that molds life.

The integrated framing of Earth as a biogeochemical entity has generated new forms of inquiry since the early days of controversy. Components of the biosphere can now be investigated for their potential roles within the whole; and the maintenance of those components within certain ranges can be queried for the systemic functions thereby served. Gaia theory famously drew attention, for example, to the long-term stability of oxygen at around 21 percent. Inquiring into the potential function of oxygen within the biosphere, Gaians pointed out how the respiration of animals, on the one hand, and the fire regimes of forests, on the other, are both well served at this proportion; scientists also posited mechanisms or feedbacks maintaining it in a 21 percent range for perhaps 200 million years (Lovelock 2003). Emphasis on elemental cycles and interconnections within the biosphere led Gaian scientists to further suspect the existence of a mechanism by which sulphur and iodine, drained into the seas by rain and rivers, are returned to land; this eventuated the discovery that the biogenic gases methyl iodide and dimethyl sulfide cycle those elements back to land. The connection between dimethyl sulfide and cloud formation later added another chapter to the ways that organisms—marine creatures, in this case—influence temperature and create climate (Lovelock 1991).

In brief, much of the value of Gaian epistemology lies in offering a framework within which new questions, new hypotheses, and new knowledge can emerge. At the same time, and crucially for the present day, the value of Gaian thinking lies in the ways scientific ideas, ethical realizations, and environmental implications intersect within it: Gaia renews the ancient understanding of the Earth as a living subject rather than an inanimate object. As David Abram offered, Gaia compels us "to recognize, ever more vividly, our interdependence with the countless organisms that surround us, and ultimately encourages us to speak of the encompassing Earth in the manner of our oral ancestors, as an animate living presence" (1996a: 302). This extra-scientific resonance of Gaia evinces in the broader culture and in spiritual inquiry—a resonance that involves tropes of intuition, sensing, love, religion, and compassion inside the planet's living presence (Abram 1990, 1996b; Primavesi 2000; Harding 2006).

The environmental dimensions of Gaia theory revolve around two fundamental concepts: consequences of human-driven perturbations of the biosphere, and implications of habitat destruction and fragmentation of the Earth's ecosystems. While small-scale disturbances can be absorbed by the biosphere, large-scale perturbations sooner or later trigger far-reaching and uncontrollable consequences. Consider the matter of great contemporary anxiety—$CO_2$-loading of the atmosphere. The anthropogenic (or volcanic, for that matter) injection of relatively small amounts of $CO_2$ can be countered by the biosphere via their absorption by the oceans and the stimulation of the growth of photosynthetic organisms: these responses are indeed conceptualized by Gaians as negative feedback mechanisms of Earth's global metabolism countering additional atmospheric $CO_2$ (Williams 1996; Lenton 2002). But when $CO_2$ amounts exceed the biospheric capacity to respond, then the forcing can make the Earth system's current equilibrium break down, shifting it into unknown territory. As many scientists have warned, human beings and countless other organisms are perched on the knife-edge of such a global shift. Moreover the carbon cycle is only the most obvious and most publicized of the element cycles that humans are disturbing; we are in fact profoundly disturbing all the cycles of the Earth's fundamental elements, including sulfur, nitrogen, and phosphorus (see Williams 1996; Volk 2008). In some cases we are seeing the effects of adverse synergies: for example, the recent increase of dead zones in coastal waters reflects the disturbance of both nitrogen and carbon cycles—as agricultural runoff is now spilling into waters warmed by excess $CO_2$ in the air (Juncosa 2008).

As for anthropogenic habitat destruction and fragmentation, this process began hundreds of years ago but has been escalated recklessly in the last few centuries and decades. In a Gaian context of the Earth as a global ecosystem, or a geophysiology, all ecosystems are interconnected on a planetary scale—analogously to the ways that all organisms are connected within their specific ecological communities. (The global interconnection of ecosystems mediates biogeochemical cycles, the creation of climatic regimes, and the propagation of biodiversity via gene flow and population migrations.) The demolition of natural habitats has reached a level where it no longer constitutes a set of destructive local or regional events, but reverberates into global repercussions—as indeed humanity is experiencing with the effects of deforestation and desertification, for example, reaching beyond their specific locales. Gaian scientists—especially Lovelock and Harding—have emphasized that the Earth cannot afford any more habitat destruction: if, following current trends, the planet is turned into an agricultural, aqua-cultural, and farm factory to feed increasing human consumption and population, then the interconnected wild ecosystems of the Earth will no longer fulfill their functions of creating familiar climate, cycling elements and nutrients, removing wastes, and birthing new life forms. From a Gaian perspective, we are perched on the knife-edge of converting the planet from a geophysiology—or a mantle of contiguous interwoven natural systems—into a sterile orb bearing life that merely serves or is compatible with narrow human interests.

No place exists in the Gaian paradigm for the inflated anthropocentric credo—be its origins religious or humanistic—that the Earth exists as an object of human dominion. To rip into the planet's rhythms, cycles, and interconnections, as the civilization we have created is doing, signals human folly not mastery. For one, the Earth system is ultimately unpredictable and more powerful than humanity's actions. Gaia theory proposes that organisms inflicting damage on their surroundings will eventually reap harsh consequences when feedback comes back to haunt them. We are currently experiencing such feedback in the form of climate change, ozone depletion, endocrine disruption, and desertification. Moreover there is no telling what other surprises await us, all the more as we are now disrupting the biology, physics, and chemistry of the oceans that cover three-quarters of the Earth's surface: they create and cycle huge components of the air we breathe, the climate we enjoy, not to mention the food we eat. As many scientists and analysts have noted, tempering so recklessly with the biosphere entrains the highest risks.

Further, by shredding the planet's rhythms, cycles, and interconnections, we forfeit a quality of human life that can be of the highest caliber in a world abundant in biodiversity and healthy ecosystems. Gaia teaches us that we live connected with all biotic and abiotic elements *inside* a planet that is more like a "physiology" than it is like a "spaceship" that carries a random crew of life-forms. Whatever we inflict on the biosphere does not only eventually have physical and survival consequences for human beings, it has immediate experiential repercussions. We submit that the increased entropy civilization is producing—through ecosystem destruction and impoverishment, habitat fragmentation, unending development, agro-industrial monocultures, and rampant extinction of species and subspecies—returns to us in the form of epidemics of violence, alienation, depression, disease, and nihilism across households, cultures, tribes, nations, and religions (Roszak et al. 1995; Fisher 2002; McKibben 2007).

"Human activities," Tim Lenton and his colleagues noted in a recent climate-change publication (2008: 1786), "may have the potential to push components of the Earth system past critical states into qualitatively different modes of operation, implying large-scale impacts on human and ecological systems." Such qualitative shifts can occur as a consequence of what are called *tipping points*, whereby relatively small changes in input have long-term, large-scale, and often irreversible output (ibid.). Improved climate models, recent climatic paleo-data, and on-the-ground observations and measurements are driving home the realization that such tipping points can make climate change manifest more like a switch than a dial (Linden 2006; Flannery 2006; Lovelock 2006). The anthropogenic amplification of the greenhouse effect underway is rapid and large enough that it may unleash positive feedback—via loss of albedo of light-reflecting surfaces (ice and snow), release of methane from the tundra (and possibly even sea floors), and other consequences: positive feedback, in turn, can trigger runaway heating. Such an eventuality will not only cause widespread human suffering, it will transform the Earth into a biological wasteland. Arriving at a time when the natural world is already severely wounded by human activities, rapid climate change is exacerbating biodepletion: it threatens to wipe out one million species or more and is jeopardizing entire classes of ecosystems, namely the Amazonian rainforest, coral reefs, boreal forests, polar landscapes, and marine microorganisms and krill at the base of the ocean food chain (Thomas et al. 2004; Lovejoy and Hannah 2005; Flannery 2006; Harding 2006).

While the specters of climate change now draw considerable attention from scientists, policy makers, politicians, and the general public, the equally if not more momentous event of the biodiversity crisis—which includes the current human-driven mass extinction—has yet to pass a critical threshold into collective awareness (Crist 2007). The impoverishment of ecosystems and the depletion of wild species have occurred for centuries (or longer), but these losses have escalated since the Industrial Revolution with consumption increase, population growth, and technological sophistication reaching dizzying levels. The Earth is estimated to be losing thousands or tens of thousands species yearly, and the 2005 Millennium Ecosystem Assessment found nearly two-thirds of the services provided by nature to humankind in decline worldwide (Watson et al. 2005). While the biodiversity crisis has yet to be assessed for its potential of destabilizing the Earth system—of overstepping a tipping point beyond which lies a different planet—such an event horizon should not be required to make the depletion of Earth's biological wealth a calamity of unthinkable proportions. Even though the mass extinction of species and the wholesale decline of ecosystems have yet to trump contemporary fixations on the economy, politics, peak oil, terrorism, and entertainment, biodepletion will undoubtedly be judged, in retrospect and not soon enough, as the most momentous, far-reaching event of our time.

We still live in the Holocene and should resist the sirens of realism that call for branding our human-dominated era by a new name.[3] We do not need the form of realism that surrenders to the seemingly unstoppable expansionism of human civilization in the biosphere, that resigns itself to more ecological losses, and that calls for coping in piecemeal fashion with consequences that come our way. Instead, we need an enlightened form of realism in order to undertake the tasks that can make the decisive difference: "at this point in our environmental freefall," as Paul Hawken (2007: 172) aptly surmised, "we need to *preserve* what remains and dedicate ourselves to *restoring* what we have lost" (emphasis added). While the tasks of preservation and restoration of Gaia's natural systems can be assisted by on and off the ground technologies, clearly, they cannot be effected by technological fixes. These tasks are rooted in a vision of conservation at landscape and seascape levels, involving the protection of natural areas and species, reconnecting fragmented habitats, reintroducing natives and removing invasives, growing and harvesting food ecologically and ethically, and allowing the richness of the biosphere to blossom again into a semblance of its erstwhile diversity and abundance. Such a conservation vision calls for concerted work at global, regional, and local levels. It requires what Lovelock (2006) has so frankly called *sustainable retreat*: we must scale down our consumption, shrink our ecological footprint, and generously share the biosphere with all living beings.

The attraction and power of Gaian inquiry have always extended beyond natural science to other academic disciplines and, of course, into the broader culture. Its interdisciplinary nature is evident in the welding of geological and life sciences, as reflected, for example, within the Gaia-influenced arena of Earth system science. The interdisciplinary nature of Gaia inquiry is also evident in the ongoing dialogues that Gaia has inspired between the natural sciences, social sciences, and the humanities, as reflected in major conferences as well as numerous edited works (e.g., Thompson 1987; Barlow 1991; Schneider and Boston 1991; Bunyard 1996; Schneider et al. 2004). A fascinating but also dismaying consequence of this intense interdisciplinarity is that "Gaia" is articulated in a bewildering diversity of ways, depending on the epistemological, political, ecological, or cultural contexts and purposes of its use. To mention a pointed example, the shorthand description of Gaia through the metaphor of "living planet" was first invoked by Lovelock himself (1979). Yet science is not equipped to address the question of whether the Earth is alive, since the question itself cannot be scientifically formulated. Even so expressions of the intuition of Earth-as-living abound in Gaia-inspired art, philosophy, spirituality, and even popularized science; such expressions are as much a part of the legacy of Gaia as, for example, strictly technical endeavors to describe Gaia as an emergent effect of organisms' waste by-products or to represent organisms' regulatory effects through computer modeling.

The present volume reflects Gaia's longstanding disciplinary richness and diversity of understanding. Some two dozen contributors—natural scientists, social scientists, philosophers, theorists, technologists, and educators among them—helped to shape it. We have partitioned the book into three sections. Chapters in part I focus on the science of Gaia: the fluxes of essential elements through the biosphere; the potentially critical role of life in retaining abundant water on the planet since the Archean; the interface between Earth-system thinking and levels of Darwinian selection; and Gaian feedback mechanics connecting canopy and soil organisms as a key ecological circuitry in the self-maintenance of forest systems. Contributions in part II examine global environmental quandaries: the urgent matter of biodiversity destruction, especially given the importance of biodiversity for the resilience of ecosystem functions and of the Earth system as a whole; the dangers of the rapid climate change underway, and the energy and policy shifts required to stabilize climate within familiar ranges; the imminent freshwater crisis poised to imperil millions (if not billions) of people, as well as freshwater species and natural systems; the need for large-scale, restorative conservation strategies—from assisted migrations in a world of shifting climate regimes and fragmented habitats, to rewilding landscapes for the protection species, ecosystems, and evolutionary processes. Chapters in part III explore the influence of Gaian thinking on sociocultural visions and discourses— environmental ethics, mind and experience, politics, technological systems, and education. Broadening Aldo Leopold's celebrated "land ethic" into an "Earth ethic" that can encompass—in thought and policy— the spatial and temporal scales of our global crises; remapping mind as a property of the Earth in which all beings participate, and considering the implications of such an understanding for human experience within the Earth's elemental moods and beauties, as well as within the Earth's troubled times—now and ahead; dreaming a new (and hopefully rising) political culture in which Gaian principles of symbiosis and embeddedness displace the psychosis of the growth imperative; querying how emerging information technologies—able to document whole Earth processes—once available to a growing grassroots environmental and justice movement, can become a potent political tool and educational medium for restoring the Earth; and critically dissecting trajectories and uses of systems theory for understanding the biosphere.

After reading an advance copy of Darwin's *On the Origin of Species*, Thomas Henry Huxley, the widely proclaimed "bulldog" for the nascent theory of evolution by natural selection, exclaimed: "How exceedingly stupid not to have thought of that!" (see Huxley 1900). Like many of the best ideas, evolution by natural selection seemed obvious once someone had formulated it. A first reading of basic Gaia literature often provokes the same emotional response: Isn't that obvious? Yet it is not obvious to everyone, and sometimes its presentation has required a near-combativeness in its defense among its varied advocates. We hope that this volume will provide readers a compelling understanding of Gaia as a way of knowing: Earth, home to countless and evolving species, diverse ecosystems, and complex biogeochemical processes, all interconnected and awaiting not only discovery but, even more crucially, the awakening of our gratitude and awe.

## Notes

1. See Wilson (1998).
2. Following Tyler Volk's convention, we use "Gaia" and "biosphere" interchangeably to signify the integrated whole of air, oceans, soil, and life that has emergent effects on the planet.
3. We are referring to the circulating ill-thought proposal to rename our era the Anthropocene.

## References

Abram, D. 1990. The perceptual implications of Gaia. In A. H Badiner, ed., *Dharma Gaia: A Harvest of Essays in Buddhism and Ecology*. New York: Parallax, pp. 75–92.

Abram, D. 1996a. *The Spell of the Sensuous: Perception and Language in a More-Than-Human World*. New York: Vintage Books.

Abram, D. 1996b. The mechanical and the organic: Epistemological consequences of the Gaia hypothesis. In P. Bunyard, ed., *Gaia in Action: Science of the Living Earth*. Edinburgh: Floris Books, pp. 234–47.

Barlow, C., ed. 1991. *From Gaia to Selfish Genes: Selected Writings in the Life Sciences*. Cambridge: MIT Press.

Bunyard, P., ed. 1996. *Gaia in Action: Science of the Living Earth*. Edinburgh: Floris Books.

Crist, E. 2007. Beyond the climate crisis: A critique of climate change discourse. *Telos* 141: 29–55.

Dawkins, R. 1982. *The Extended Phenotype*. Oxford: Freeman.

Doolittle, W. F. 1981. Is nature really motherly? *The CoEvolutionary Quarterly* 29 (spring): 58–63.

Eldredge, N., and S. J. Gould. [1972] 1985. Punctuated equilibria: An alternative to phyletic gradualism. In T. J. M. Schopf, ed., *Models in Paleobiology*. San Francisco: Freeman Cooper, pp. 82–115.

Fisher, A. 2002. *Radical Ecopsychology: Psychology in the Service of Life*. Albany: State University of New York Press.

Flannery, T. 2006. *The Weather Makers: How Man Is Changing the Climate and What It Means for Life on Earth*. New York: Grove Press.

Harding, S. 2006. *Animate Earth: Science, Intuition and Gaia*. White River Junction, VT: Chelsea Green.

Hawken, P. 2007. *Blessed Unrest*. New York: Viking.

Holland, H. 1984. *The Chemical Evolution of the Atmosphere and Oceans*. Princeton: Princeton University Press.

Huggett, R. J. 2006. *The Natural History of the Earth: Debating Long-term Change in the Geosphere and Biosphere*. London: Routledge.

Huxley, L. ed. 1900. *The Life and Letters of Thomas Henry Huxley*, vol. 1. London: Macmillan.

Jacobson, M., R. Charslon, H. Rodhe, and G. Orians. 2000. *Earth System Science: From Biogeochemical Cycles to Global Change*. San Diego: Academic Press.

Juncosa, B. 2008. Suffocating seas. *Scientific American*, October.

Kirchner, J. 1991. The Gaia hypotheses: Are they testable? Are they useful? In S. H. Schneider and P. Boston, eds., *Scientists on Gaia*. Cambridge: MIT Press, pp. 38–46.

Kirchner, J. 2002. The Gaia hypothesis: Fact, theory, and wishful thinking," *Climatic Change* 52: 391–408.

Lenton, T. 1998. Gaia and natural selection. *Nature* 394: 439–47.

Lenton, T. 2002. Testing Gaia: The effect of life on Earth's habitability and regulation. *Climatic Change* 52: 409–22.

Lenton, T. 2004. Clarifying Gaia: Regulation with or without natural selection. In S. Schneider, J. Miller, E. Crist, and P. Boston, eds., *Scientist Debate Gaia: The Next Century*. Cambridge: MIT Press, pp. 15–25.

Lenton, T., H. Held, E. Kriegler, J. W. Hall, W. Lucht, S Rahmstort, and H. J. Schellnhuber. 2008. Tipping elements in the Earth's climate system. *Proceedings of the National Academy of Sciences* 105 (6): 1786–93.

Linden, E. 2006. *The Winds of Change: Climate, Weather, and the Destruction of Civilizations*. New York: Simon and Schuster.

Lovejoy, T., and L. Hannah, eds. 2005. *Climate Change and Biodiversity*. New Haven: Yale University Press.

Lovelock, J. 1979. *Gaia: A New Look at Life on Earth*. Oxford: Oxford University Press.

Lovelock, J. 1991. *Healing Gaia: Practical Medicine for the Planet*. New York: Harmony Books.

Lovelock, J. 2003. The Living Earth. *Nature* 426: 769–70.

Lovelock, J. 2006. *The Revenge of Gaia*. London: Allen Lane.

Margulis, L. 1998. *Symbiotic Planet: A New look at Evolution*. New York: Basic Books.

Margulis, L., and J. Lovelock. 1989. Gaia and geognosy. In M. Rambler, L. Margulis, and R. Fester, eds., *Global Ecology: Towards a Science of the Biosphere*. San Diego: Academic Press, pp. 1–30.

McKibben, B. 2007. *Deep Economy: The Wealth of Communities and the Durable Future*. New York: Times Books.

Primavesi, A. 2000. *Sacred Gaia: Holistic Theology and Earth System Science*. London: Routledge.

Roszak, T., M. Gomes, and A. Kanner, eds. 1995. *Ecopsychology: Restoring the Earth, Healing the Mind*. San Francisco: Sierra Club Books.

Schneider, S. H. 1986. A goddess of the Earth? The debate on the Gaia hypothesis—An editorial. *Climatic Change* 8: 1–4.

Schneider, S., and P. Boston, eds. 1991. *Scientists on Gaia*. Cambridge: MIT Press.

Schneider, S., J. Miller, E. Crist, and P. Boston, eds. 2004. *Scientists Debate Gaia: The Next Century*. Cambridge: MIT Press.

Schwartzman, D., and T. Volk. 1989. Biotic enhancement of weathering and the habitability of Earth. *Nature* 340: 457–60.

Thomas, C., A. Cameron, R. E. Green, M. Bakkenes, L. J. Beaumont, Y. C. Collingham, and B. F. N. Erasmus. 2004. Extinction risk from climate change. *Nature* 427: 145–48.

Thompson, W. I., ed. 1987. *Gaia, A Way of Knowing: Political Implications of the New Biology*. Great Barrington, MA: Lindisfarne Press.

Turney, J. 2003. *Lovelock and Gaia: Signs of Life*. New York: Columbia University Press.

Vernadsky, V. 1998 [1979]. *The Biosphere*. New York: Copernicus.

Volk, T. 1998. *Gaia's Body: Toward a Physiology of Earth*. New York: Copernicus.

Volk, T. 2008. *$CO_2$ Rising: The World's Greatest Environmental Challenge*. Cambridge: MIT Press.

Wallace, A. R. 1876. *The Geographical Distribution of Animals*. New York: Harper.

Watson, A., and J. Lovelock. 1983. Biological homeostasis of the global environment: The parable of Daisyworld. *Tellus* 35B: 284–89.

Watson, R. T. et al. 2005. Living beyond our means. www.milleniumassessment.org/documents/document.429.aspx.pdf.

Westbroek, P. 1991. *Life as a Geological Force: Dynamics of the Earth*. New York: Norton.

Wilkinson, D. 2006. *Fundamental Processes in Ecology: An Earth Systems Approach*. Oxford: Oxford University Press.

Williams, G. R. 1996. *The Molecular Biology of Gaia*. New York: Columbia University Press.

Wilson, E. O. 1998. *Consilience: The Unity of Knowledge*. New York: Vintage Books.

Wilson, E. O. 2002. *The Future of Life*. New York: Little Brown.

# 2

## Our Sustainable Retreat

James Lovelock

It has been 42 years since the idea of a "living Earth" came to my mind at the Jet Propulsion Laboratory in California. Shortly after this, Nobel Prize winning novelist William Golding proposed that the hypothesis be called Gaia after the ancient Greek Earth goddess. There was nothing mystical in this proposal from a classical scholar since the name of the same goddess is the root of geo, geography, geology, geophysics, and so on. The concept of a live, self-regulating Earth was in the early 1970s welcomed by climatologists, by a few geologists, and by the eminent biologist Lynn Margulis, who joined with me in developing the science of Gaia. The first predictions of the hypothesis concerned the natural cycles of sulphur and iodine as were confirmed by direct measurements and established quantitatively by the ocean chemist Peter Liss.

Why therefore, despite successful predictions, mathematical models, and strong evidence, do many scientists still regard the concept of Gaia as New Age mysticism and not part of science? The answer lies mainly I think in the evolution of science during the two past centuries. The reductionist approach was a stunning success. It led to the triumphs in molecular biology and to the deconvolution of the code of life; in physics, from subatomic to cosmological levels, there were successes of comparable magnitude, all of this while science was integrating socially within the universities. The very natural ambitions of strong-minded professors encouraged and strengthened the separation of science into those tribal territories called "disciplines." In such a world there was no place for the holistic science of Gaia. At most, there were interdisciplinary gatherings that were oddly similar to international conferences of politicians—far more was said than done.

Somehow the systems sciences, physiology, and the theoretical side of engineering have managed to exist, despite their top-down not bottom-up approaches. Not surprisingly, the concept of Gaia found expression as Earth systems science.

This halting pedestrian evolution of Gaian or Earth system thinking would not much matter if we as a species had secure tenure on the Earth, but climate changes that we have set in motion appear to be moving our planet rapidly to one of its hot states, perhaps similar to the one that existed 55 million years ago (see Lovelock 2006). If this happens, humans could be joining the growing list of extinction candidates, or at best surviving as a few breeding pairs on oases, large islands, and the Arctic basin. It is painful to wonder if we would have avoided this fate had Darwin developed a Gaian view as part of his concept of evolution. When Darwin came upon the concept of evolution by natural selection, he was almost wholly unaware that much of the environment, especially the atmosphere, was a direct product of living organisms. Had he been aware, I think he would have realized that organisms and their environment form a coupled system and that what evolved was this system, the one that we call Gaia. Organisms and their environment do not evolve separately. If Darwin had known this, Gaia might have been part of his concept of evolution; we would have known sooner the consequences of changing forests to farmland and of adding greenhouse gases to the air.

If my pessimistic view seems stark, consider the clearly visible disappearance of floating ice from the Arctic and Antarctic oceans. The change of heat flux from this event now underway and accelerating will increase the Earth's heat flux by a quantity, more than one watt per square meter, which is comparable with that from the infrared absorption of all the fossil fuel $CO_2$ we have so far added. Other positive feedbacks from the Earth system in its spontaneous move to a hotter stable state are already adding heat at such a pace that the sum of them all may soon exceed anything that we have so far caused. It would seem that we have pulled the trigger that set in motion ineluctable climate change. Our only comfort is that hot states in which life survives do seem to have existed in the long history of the Earth, and there has always been recovery to a cooler, more fertile Earth.

Is there nothing that we can do to bring back the lush and comfortable Earth of a few hundred years ago? Probably not in times measured on a human scale. There are three courses of action we could undertake as part of a planned program for survival as a civilized animal on a changed and hotter planet. First would be to prepare to adapt to anticipated changes such as rising sea levels, intense storms, and unprecedented heat and drought. In doing so, we should pay special attention to those places likely to escape the worst consequences of climate change, such as the Arctic basin, large island nations, and high altitude places on the continents where rainfall is plentiful. At these refuges we would need new cities to house the flood of climate refugees, ample supplies of energy, raw materials, water, and food. It seems that we will need to swallow unpopular options such as the use of nuclear energy and food synthesized wholly or partially from raw chemicals. The second line of defense could be geoengineering to reduce the input of radiation from the sun. Geophysicists Bala Govindasamy and Ken Caldeira have proposed using sun shades in space. At different times climatologists Mikhail I. Budyko, Robert E. Dickinson, and Paul J. Crutzen have suggested artificial stratospheric aerosols of sulphuric acid, while physicist John Latham proposed a method for generating marine stratus clouds by spraying seawater. Direct geoengineering of this kind might buy us the time needed to carry through our planned evacuation and/or develop a more permanent way of restoring the status quo. Unfortunately, in themselves these measures resemble dialysis as a treatment for end-stage kidney failure—something useful but no cure. The third approach is to think of the Earth as a live self-regulating system and devise ways to alter the sign of the feedback of climate-regulating processes from positive to negative.

In a recent letter to *Nature*, Chris Rapley (former director of the British Antarctic Survey and current head of London's Science Museum) and I raised the possibility that feedback from the ocean ecosystems, 70 percent of the Earth's surface, might be made negative by mixing cool nutrient-rich subsurface water with the stable but barren floating layer of the ocean. This would feed algal growth, making the surface a more efficient sink for $CO_2$, while the algal growth would release DMS (a precursor of clouds). This feedback might be achieved by a relatively simple system of pipes and be driven automatically by wave energy. Small-scale attempts to do this have been described and they appear to work. We were well aware that there could be practical reasons why this simple idea might not work, such as that the waters of the deeper ocean are richer in $CO_2$ than the surface and to bring them up would add to the release of $CO_2$ to the atmosphere. We raised the possibility to show the value of thinking of the Earth as a living system whose gigantic stores of energy might be available for use in its and our interest. We hoped it might stimulate other proposals of this kind and that among them was one or more that could be effective. We also wanted to show that the Gaian approach of stimulating the Earth to cure itself was more than mere rhetoric.

Should we fail in our attempts to restrain the Earth's move to a hot but stable state we would have to devote all of our energies to sustaining a civilized way of life on those few remaining oases where human life could continue. Immense civil engineering projects would need to be devised to offset flooding, including the technology to synthesize food and ensure that an abundant and reliable supply of energy did not involve burning carbon fuel. France has solved this last problem and draws 80 percent of its electrical energy from nuclear fission and 20 percent from water power.

As well as technological needs there will be immense social problems. Civilization is the most fragile component of society, and in the past profound disorder is sometimes followed by primitive tribal societies run by gang leaders.

For those pessimistic environmentalists who regard humanity as a disease, as a pathology of the Earth, I offer the thought that although we may be as bad as that, we do learn from our mistakes. More than this, in the 3.5 or more billion years of evolution, Gaia has evolved a species with the ability to think and communicate its thoughts. This human species has allowed the Earth to see itself from space in all its beauty and has begun to understand its place in the universe and itself. Yes, we are part of Gaia, and therefore that top-down view was worth her waiting a quarter the age of the universe.

### References

Budyko, M. I. 1969. The effect of solar radiation variations on the climate of the Earth. *Tellus* 21: 611–19.

Crutzen, P. J. 2006. Albedo enhancement by stratospheric sulphur injection: A contribution to resolve a policy dilemma. *Climatic Change* 77 (3/4): 211–19.

Dickinson, R. E. 1996. Climate engineering: a review of aerosol approaches to changing the global energy balance. *Climatic Change* 33 (3): 279–90.

Govindasamy, B., and K. Caldeira. 2000. Geoengineering Earth's radiation balance to mitigate $CO_2$-induced climate change. *Geophysical Research Letters* 27 (14): 2141–44.

Latham, J. 1990. Control of global warming? *Nature* 347: 339–40.

Liss, P. S., and P. G. Slater. 1974. Flux of gases across the air-sea interface. *Nature* 247: 181–84.

Lovelock, J. 2006. *The Revenge of Gaia*. London: Allen Lane.

Lovelock, J. E., and C. G. Rapley. 2007. Ocean pipes could help the Earth to cure itself. *Nature* 449: 402.

# II

## The Science of Gaia

# 3

# How the Biosphere Works

Tyler Volk

The Earth's surface is a special system worthy of a name. As I will elaborate in this chapter, all life and the three environmental matrices of atmosphere, soil, and oceans form a closely integrated network that can be called the *biosphere*. Its upper boundary is clearly the top of the atmosphere. Its lower boundary is admittedly fuzzier. Groundwater reaches kilometers down into pores of rock, and bacteria have been found kilometers down as well. But for practical purposes, both in terms of providing a rationale for our concepts and for technical modeling of the impacts of organisms on the chemistry of the global environment on relatively short timescales, we can exclude from the definition of the biosphere the minerals in rocks underneath the soil because the elements in those rocks have been out of active circulation for millions or hundreds of millions of years.

## Defining the Biosphere within the Gaia Perspective

Some Gaia theorists, like James Lovelock (2006) and Tim Lenton and David Wilkinson (2003), use the word Gaia to be closely equivalent to this chapter's definition of the biosphere. But within Gaia they usually include the surface rocks that have been affected by organisms in Earth's geologic past, such as carbonate rocks (limestone) that were laid down from the accumulated shells of creatures many millions of years ago. A Gaia that is larger than the biosphere as defined above does help us grasp the fact that the effects of life stretch beyond any present slices of time. But the point remains that those carbonate rock minerals have been absent from active circulation for vast ages; as far as the organisms living today are concerned, it is almost as if the carbonate rock minerals did not exist. As will be shown, what is in or out of circulation is key to characterizing the biosphere as a unique system worthy of a name.

The air circulates globally in about a year. Such rapid mixing is evident from the fact that although most of the human-generated fossil-fuel injections of $CO_2$ into the atmosphere take place from nations in the Northern Hemisphere, the $CO_2$ at the South Pole has been rising at levels and rates that are almost identical to those at sites in the Northern Hemisphere—for example, at Mauna Loa, Hawaii, as is clear from nearly fifty years of data (figure 3.1).

From studies of deep ocean currents[1] the ocean is known to turn over and mix in about one thousand years. Soil, the third environmental matrix in the biosphere, is stirred by creatures and the chemical circuits of decomposition. Most of the matter in the soil is cycled over at time-scales of tens to hundreds of years (for the most part). Organisms themselves "turn over" on the intervals that bound their lives: from days to hundreds of years—though we should give a nod of honor to the much longer-lived Methuselah trees, such as Bristlecone pines.

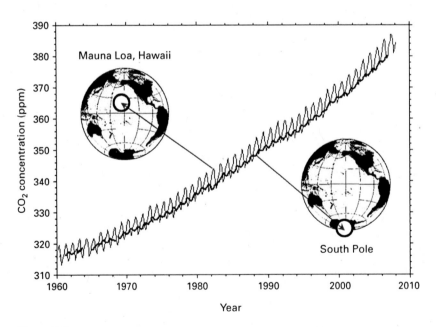

**Figure 3.1**
Atmospheric $CO_2$ data from the South Pole Observatory (90°S, dark line), compared with data from Mauna Loa, Hawaii (19°N, oscillating light line). Data from the Carbon Dioxide Information Analysis Center.

So what should we conclude from this survey of the timescales? Because the air, ocean, and soil interconnect, the entire system—including the organisms within these three largest environmental matrices—circulates on about the same timescale as that of the ocean, namely, about one thousand years. This is short with respect to the timescales of evolution, during which species come and go across millions of years, so the components inside the biosphere are interlinked like a single biochemical stew, synchronized in what are virtually evolutionary instants by their chemical connections to each other.

Sometimes, and I must emphasize the need to be wary about this point, the word "biosphere" is taken to mean all of life. For example, a discussion on how the atmosphere, hydrosphere, and biosphere interact would make no sense in my terminology, in which the atmosphere and oceans are internal parts of the greater biosphere. There is a perfectly good, unambiguous word for the sum of all life: the biota. In a second alternative meaning, which I suggest is also infelicitous, sometimes "biosphere" is used to refer to the zone where life is found, from the ocean sediments to the tops of mountains. But this is highly abstract and without physical meaning. The entire atmosphere is mixed both horizontally and vertically and so the chemical impacts of life are not confined to the air only up to the tops of mountains.

Sometimes both of these renegade meanings of "biosphere" are used in nearby sentences without even pointing out the incongruity to the reader. One geologist has recommended the term "ecosphere" (Huggett 1999), but that term has not taken off in the competition for word dominance, and so I will stick with "biosphere" as the integrated system of air, oceans, soil, and life. It has often been used in this way, and there are good reasons for wanting to think more about this united, well-mixed, and amazingly complex thin shell of a system—within which we and all other creatures live sandwiched between hard rock and black space.

**Fluxes of Bio-essential Elements inside the Biosphere**

With a definition in place, a good place to start inquiring into how the biosphere works is to look at the magnitudes of fluxes of matter. Carbon is a great choice for that because carbon is the core of the organic molecules of life, whether terrestrial or marine. Carbon is in a key atmospheric greenhouse gas ($CO_2$). It is in a major ion in the ocean (bicarbonate, $HCO_3^-$, as well as in other marine forms). It is crucial to the structure of soil, as humus, which provides nutrients, ion exchange, and moisture retention. The carbon cycle has been well studied, especially because of rising concentrations of $CO_2$ from the global combustion of fossil fuels, releasing new carbon atoms now circulating in the biosphere (figure 3.1). The cycle of carbon has received a lot of attention in field studies and global inventories.

The interconnectedness of all the biosphere's parts via the cycling of elements can be illustrated by considering the fate of one carbon atom that we exhale. We end up putting most of the carbon in the food we eat into the atmosphere as $CO_2$ metabolic waste gas. An airborne carbon atom from one of those waste molecules may end up in a couple years in a bicarbonate ion in ocean water, next in the body of a green phytoplankton, then expelled as organic waste from a crab-like tiny zooplankton that eats the algae, then consumed by bacteria and expelled as inorganic waste and thus passed into bicarbonate again; from there, it could be shunted back into the air as $CO_2$ in the process of air–sea gas exchange (perhaps all within a half dozen years), and then it might be placed by photosynthesis into the cellulose structure in the leaf of an oak tree in China.

Within this vast circuitry—which gets as convoluted as the paint strands in a Jackson Pollack—the tiny sizes of the molecules almost defy imagination. The waste molecules of $CO_2$ that we exhale mix globally throughout the atmosphere in about a year, across all those lands that we have ever traveled and those we have not yet seen. I have calculated that *every* green leaf that grows anywhere on the planet (e.g., in about a year from now, to allow for the complete mixing of the atmosphere) will contain a few dozen atoms of carbon from the 500 million trillion new $CO_2$ molecules that we released from each and every one of our exhalations.

The annual total release of $CO_2$ from all humans is relatively small, compared with the $CO_2$ that comes out of the soil each year, generated by soil organisms that feed on the organic carbon in the soil, which in turn comes mostly from dead plants, their leaves, branches, and roots. The respired $CO_2$ from those soil organisms enters the soil's air and then percolates up into the atmosphere. It comes from worms and millipedes, fungi and beetle larvae, but mostly from soil bacteria— a respired flux that totals about 60 billion tons of carbon in the form of $CO_2$ each year.

That number is only about half of the flux of carbon that enters terrestrial green plants each year (120 billion tons) because about half the

$CO_2$ that the plants take in for photosynthesis is respired back into the air as the plants burn their own newly formed sugar molecules as a source of energy to drive the subsequent chemical reactions they need to form their proteins, lipids, and nucleic acids for maintenance and growth. Another huge number comes from the exchange of carbon in the form of $CO_2$ between ocean and air (100 billion tons per year, the air-sea gas exchange referred to earlier). Figure 3.2 shows these numbers as fluxes within circuits in the biosphere. In this simplified big picture, I have ignored several levels of detail, such as the 80 billion tons of carbon photosynthesized within the ocean's surface by phytoplankton, the 40

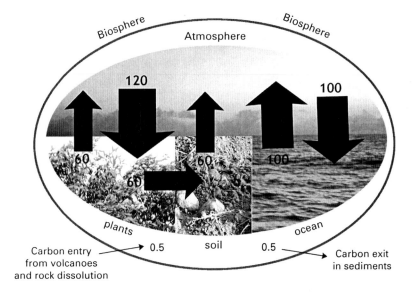

**Figure 3.2**
Massive fluxes of carbon in the cycle of nature: Air–ocean gas exchange of $CO_2$ (100 billion tons of carbon each way per year), photosynthesis by terrestrial plants that turns $CO_2$ into organic molecules of life (120 billion tons of carbon per year), respiration by plants to build more complex molecules inside their bodies and release of $CO_2$ (60 billion tons of carbon per year), transfer of carbon primarily by plants but also by animals as detritus into the soil (60 billion tons of carbon per year), respiration by soil organisms, mostly bacteria, releasing $CO_2$ into the soil and then up into the air (60 billion tons of carbon per year; for simplicity this includes respiration by land animals from insects to mammals of about 5 billion tons of carbon per year). The entry and exit fluxes to and from the biosphere are much smaller, as shown. Not shown is the cycle of photosynthesis and respiration in the ocean (see the text).

billion tons they respire, the 40 billion tons consumed by zooplankton and the other consumers in the ocean's food webs, and the several billion tons consumed by animals on land. There are also the other circuits of carbon in the oceans associated with water currents and regeneration by bacteria in deep water, in the details of how carbon circulates in ocean sediments and in different kinds of terrestrial soils, and in a relatively small flux that rivers carry to the ocean.

The simplified big picture allows me to get to a point that holds even if more detail is considered. In figure 3.2, I have added the fluxes of carbon that enter and leave the biosphere. These border fluxes are small compared to the major fluxes of carbon shown within the biosphere. Specifically, geologists estimate that half a billion tons of carbon enter the biosphere as $CO_2$ from volcanoes and as carbonate ions ($CO_3^{2-}$) released when rocks such as limestone dissolve and pass their chemicals into groundwater. About the same amount (which must be true on long enough timescales) leaves the biosphere during what geologists call carbon burial. Almost all burial takes place in ocean sediments, as carbon exits the biosphere in the form of carbonate shells from organisms, as well as in a smaller amount in unrecycled organic matter.

The amount of carbon the global hordes of photosynthesizers require each year to live and grow is much larger than the half billion tons of new carbon that enters the biosphere. Summing the net terrestrial plant photosynthesis of 60 billion tons and the net marine photosynthesis of 40 billion tons yields 100 billion tons a year required for the current biosphere's annual growth of green living things, and all other creatures are fully dependent for their livelihood on that primary amount that infiltrates the various food webs. That amount of 100 billion tons per year is a factor of 200 times larger than the fresh rate of carbon supply into the biosphere. The flow of carbon into global photosynthesis is therefore dependent on the cycles of carbon *within* the biosphere. We have already seen the source of this supply: it comes from the respiration of heterotrophs such as zooplankton, whales, fungi, centipedes, eagles, people, and many other creatures, including the all important marine and soil bacteria.

In considering the fluxes in the global carbon cycle, I have emphasized the unity of the biosphere—that it is truly a system somewhat isolated from the rocks below by its large internal system of fluxes compared to the small fluxes back and forth across its lower boundary. I do not want to make more of a tempting organic analogy than just what I will say here, but the biosphere's large internal fluxes and small border fluxes might be compared to the human body, with its large flows of blood around the internal organs and relatively small fluxes of food and water that go in and out each day. In addition to deriving a message about the unity of the biosphere, we can see a second point from the big picture: the workings of the biosphere are intricately tied to the fact that the wastes from certain groups of creatures become nutrients to other groups of creatures.

## Waste Networks of Biochemical Guilds

I have previously suggested the name "biochemical guilds" for groups of organisms that perform what are virtually equivalent metabolic transformations of certain elements in the biogeochemical cycles of the biosphere (Volk 2003a). For example, in the discussion about the magnitudes of fluxes, I lumped all photosynthesizers to get a global number. I can do this because they all take carbon from $CO_2$ and turn it into carbon chemically bound in organic molecules. Most heterotrophs, on the other hand, are members of the biochemical guild of respirers. Members of this guild take the carbon from organic matter and turn it into $CO_2$ waste, deriving material for their bodies and, crucially, energy for their metabolisms. Admittedly, denoting lines between groups can get complicated. For example, photosynthesizers perform some respiration as well. Different categories can be generated, depending on who is doing the analysis and for what purpose. The reality goes beyond any analysis, say of an ecosystem, biome, or globe. There really are common metabolisms out there. Furthermore in the cycles of the elements that are key to living things the biochemical guilds can link up to each other because wastes from one are nutrients to another.

The year 1958 marked the first real-time data for atmospheric $CO_2$ at Mauna Loa. That year also premiered a television program called *The Naked City* about police work in New York City. Each episode had a concluding epilogue which always began, "There are eight million stories in the naked city, this has been one of them...." If we were to look at each carbon atom, for example, in a place called The Naked Biosphere, then our narrator, in conclusion to any one story of those atoms would have to say, "There are two million billion trillion quadrillion stories in the global carbon cycle, this has been one of them...." (That is not a random big number, see note 1 in the appendix for the calculation.) But because the atoms follow certain statistics, we, as audience, might be more interested in how the life of an average atom of carbon compares, say, to one of nitrogen or phosphorus. Those different elements have substantially unique stories worth individual episodes. Furthermore, because there are only a couple dozen bio-essential elements, following an episode of big events in the biogeochemical cycle of carbon, the narrator might say: "There are twenty stories in The Naked Biosphere, this has been one of the them...."

So what about a second story? What gives with nitrogen? Here's a summary of several main steps. Nitrogen-fixing bacteria (and lightning, to a smaller extent) convert nitrogen gas ($N_2$) into ammonium ($NH_4^+$). Bacteria called nitrifiers change ammonium ($NH_4^+$) into nitrate ($NO_3^-$). Ammonium assimilation is performed by plants and algae that can alter ammonium ($NH_4^+$) into nitrogen-containing organic compounds, such as proteins (call it $N_{org}$). In ammonification, decomposers in the soil and oceans break down those organic compounds in wastes ($N_{org}$, e.g., proteins in decaying leaves of plants) into ammonium ($NH_4^+$). Bacteria called denitrifiers take in nitrate ($NO_3^-$) and excrete nitrogen gas ($N_2$). Finally, within the categories of nitrifiers and denitrification are groups of bacteria that create intermediate forms of nitrogen, such as nitrite ($NO_2^-$) and nitrous oxide ($N_2O$).

In the stories of the bio-essential elements in The Naked Biosphere, sometimes the elements join together, while at other times they travel separately. When carbon enters the leaf of a green plant as $CO_2$, that carbon atom might well get hooked up with an atom of nitrogen as a bonded neighbor in an amino acid inside what will become a protein molecule. The nitrogen atom came up into the plant through the roots, dissolved as an ion of ammonium or nitrate in the water from the soil. Then, after the decaying leaf (now on the soil litter) passes through the guts of an earthworm, the carbon atom could exit the earthworm as $CO_2$ and the nitrogen atom might go into the ammonium-salt waste. The paths part of the once-joined atoms.

There is a second main point to the stories of the bio-essential elements. In one crucial way, broadly speaking, the stories of carbon and nitrogen are similar. As already alluded to, the wastes from creatures in one biochemical guild can be nutrients to those in another guild. $CO_2$ is waste we expel, but it is airborne food to green plants. The nitrogen gas $N_2$ is waste from the denitrifying bacteria but a nutrient to the nitrogen-fixing bacteria. If you look at the parts of the nitrogen cycle outlined above, you can find several other pairings (or more complicated networks) in which a chemical form of an element is waste from one but nutrient to another: ammonium as waste to ammoniaficators, but nutrient to ammonium assimilators and nitrifiers; nitrate as waste from nitrifiers, but nutrient to denitrifiers.

Similar stories are found in the cycles of carbon with methanogens, which produce methane as waste, and methanotrophs, which feed on methane. Comparable stories are also in the cycle of sulfur, and more. It all seems so amazing, as if there is some superdesign in the workings of the biosphere knitting everything together.

The designer, of course, is none other than the blind watchmaker of evolution. A waste in the environment that was ejected from an organism was always, at minimum, a potential source of raw material for another type of organism that had (or could evolve) the metabolism to use that waste, either as a source of matter or energy (when coupled with other substances). Many details of when and how evolution played a role in forging the biogeochemical cycles are still under scientific scrutiny (e.g., there is no clear consensus about when oxygen-generating photosynthesis began). But without doubt the resulting biosphere is truly phenomenal. The waste–nutrient networking gives us pause to think anew about the mundane, elementary school story of the $CO_2$ photosynthesis and respiration cycle, since the message therein is so much more expansive—a systemic pattern.

### The Term "Gaia" Can Personalize a Relationship with the Biosphere

We all have a personal relationship with the biosphere whether or not we like it. With our breaths, our food intake, and our waste ejection, we participate in food webs and in the great life-supporting, global biogeochemical cycles that link us to the upper reaches of the atmosphere, to the deepest cold reaches of the ocean, to the dark, pungent places in the soil, as well as to every creature with which we cohabit all the corners of the biosphere. Our links reach back in time, too. Every one of us is a product of a 100 percent successful series of reproductive acts that go all the way back without break to the beginnings of life and the earliest cells. And all this continuous evolutionary unfolding took place within a biosphere that was (as it had to be by definition) hospitable to life, even if the conditions for the first two billion years, at least, were deadly inhospitable with respect to *today's* life, because of lack of oxygen and other "problems."

I have tried to provide some scientific facts about our togetherness with all other species and environmental matrices in the biosphere. How do Gaia and Gaia theory, according to the renowned British scientist James Lovelock, fit into the picture I have drawn?

Terms that trigger our bonding instincts aid the creation of ties with large entities that otherwise would be perceived as too abstract (Pinker 2002). Unions can be called brotherhoods, a nation might be hearkened as the motherland, and corporations are sometimes blatantly termed families. It seems clear that by labeling the biogeochemical entity that we share with other creatures with the name of an ancient Greek Earth goddess, Gaia, one evokes a greater sense of belonging than would be possible with technical biogeochemical terminology.

Without doubt Lovelock has helped foster feelings of togetherness with the bacteria and with the water of the oceans, to mention just a couple members of the biosphere. And his writings and technical papers have helped further a scientific focus on feedback loops within the biosphere. Knowledge about such loops, as a general principle of global biogeochemistry, were firmly in discussions of the global carbon cycle, for instance, for years preceding Lovelock's Gaia hypothesis in the 1970s. But his own approach has helped focus attention on properties at the biosphere scale.

Let's look in more detail at what Lovelock is currently saying about Gaia, using the definition of Gaia theory from his recent book (2006: 162), *The Revenge of Gaia*: "A view of the Earth that sees it as a self-regulating system made up from the totality of organisms, the surface rocks, the ocean, and the atmosphere tightly coupled as an evolving system. The theory sees this system as having a goal—the regulation of surface conditions so as always to be as favorable as possible for contemporary life. It is based on observations and theoretical models; it is fruitful and has made ten successful predictions."

I have no major squabbles with the first sentence. One could debate the meaning of "self-regulation," and I have found it to be a term that Gaia theorists like to use without careful definition. But one could take it to mean approximately what complexity theorists mean when they say "self-organization," which could be applied, for instance, to the formation of a hurricane. I have argued that any perceived stability in the biogeochemical system of the biosphere is simply the result of the way that any complex, dynamical chemical system would settle into zones of limited behaviors (see note 2 in the appendix).

I am less happy about calling Gaia "an evolving system." Richard Dawkins (1982) pointed out that Gaia is a population of one, and therefore by definition cannot evolve. Evolution requires a population of variants that can be selected based on their reproductive contributions to the next generations. So when I introduce Gaia in a classroom at New York University to nonscience students on a course about the biosphere, and compare the biosphere to an organism, I quickly follow up with a denial of similarity because organisms evolve but Gaia does not. I admit that I sometimes like to foster in the students a sense of togetherness and concern by employing the personalized term Gaia. However, I also want it known that Gaia does not evolve. Of course, in a loose sense, such language is acceptable. Astronomers speak of the evolution of galaxies, as they change from blobs to spirals.

My problems change from annoying nits I want to pick to issues more serious in Lovelock's second sentence. I reject as inherently problematic Lovelock's use of the word "goal." A goal is a term that is usually reserved for human-engineered cybernetic systems (computers, cars) that are designed to perform in certain ways and, more generally, for the representations that humans carry around in their minds about their future states. The concept of goal can be appropriate for living creatures, particularly those with nervous systems that change their behavior in the face of environmental conditions. Granted, abstract concepts in language can spread out like oil on the surface of a still lake and lead us to extended uses for a word such as "goal." But I wouldn't want to say, for example, that the water vapor in the sky has the goal of becoming a cloud, even though most of it will end up in clouds. I also wouldn't want to say that clouds have the goal of removing water from the sky. So what is the goal of Gaia? Is Gaia like a hungry fox chasing a rabbit or a cloud generating raindrops?

Lovelock maintains that the goal is "the regulation of surface conditions so as always to be as favorable as possible for contemporary life." Yet Stephen Schneider (1986) pointed out two decades ago the problem with positing a metric like "as favorable as possible." What does that mean? Paraphrasing Schneider, favorable for penguins or for tropical butterflies? Following my own analysis, environmental conditions in the biosphere could be a lot more favorable, if we consider the metric of global terrestrial photosynthesis (Volk 2003b). It is now about 60 billion tons of carbon from $CO_2$ fixed into organic tissue annually. But most plants would be more productive under higher $CO_2$ levels. That would provide more organic carbon for the food webs of animals, fungi, and bacteria. More rain might be helpful too, as would a more favorable supply of soil nutrients. Considering just these environmental constraints on current productivity and maintaining biochemical machinery of photosynthesis, the lands today are only about one-tenth as productive as they could be, were the conditions "more favorable." Speaking in the language of "as favorable as possible" makes it sound that we are under the care of a nurturing super-parent. Why hasn't Gaia delivered on that need for more rain? Or more nutrients?

A Gaia theorist might respond by saying "as favorable as possible" does not mean in any possible world. The theorist might say that it is not perfect, but only as good as it could be. But what does that mean? All in all, I do not see Lovelock's language as scientifically helpful. At the same time Lovelock's language has contributed to inculcating a personal relationship. His idea that we are part of a larger entity that has a goal of creating favorable conditions reminds me of many traditional religious viewpoints on the cosmos.

The biosphere has nurtured us in the sense that it consists of an integrated network of biochemical cycles. Crucial parts of those cycles are produced by guilds of organisms in which wastes from one become nutrients to another. These wastes are goals in the sense that organisms need to rid themselves of their wastes to detoxify—an important process of living metabolisms. The wastes, however, are not produced at cost to give to other creatures. (I just want to be clear, I am not hinting that Gaia theorists say this.) The wastes are simply by-products. The biosphere is a stupendous network of waste by-products that are also nutrients. For me, this view connects the daily ins and outs of my breaths to the hard-won knowledge about the global biogeochemical cycles in a way that is both rationally and emotionally fulfilling.

All, I hope, will soon know these basic principles about how the biosphere works. Engineers, politicians, agronomists, voters, gardeners, wilderness preservationists, and any global citizen who desires to seize responsibilities and joys of life in the biosphere will be led to contemplate and help collectively decide on courses of action in the specter of potentially huge climate change. Knowledge about the working of the biosphere is pragmatic knowledge for a shifting and uncertain future, and the root cause of these changes is the perturbation of the global carbon cycle.

The increase in the concentration of atmospheric $CO_2$ shown in figure 3.1 is only about half the amount of $CO_2$ that was released from the combustion of fossil fuels. The rest of the $CO_2$ went elsewhere. Indeed much more than half went elsewhere, but then some came back directly or came back as replacement fluxes from the oceans and land carbon subsystems of vegetation and soils. The dynamics of the whole cycle are shifting, resulting in such carbon-caused phenomena as ocean acidification and crop and forest fertilization, in addition to the raw physics of climate change. Issues such as carbon capture and storage for artificial sequestration, the possible carbon neutrality of bioenergy crops, reforestation as natural sequestration, and the multitude of promising energy systems that do not emit $CO_2$—all are intimately bound up with the carbon cycle, its present disruption, and potential solutions to the resulting climate change and other perturbations (see Volk 2008).

**Appendix**

1. Calculation of number of carbon atoms in the biosphere: about 40,000 billion metric tons of carbon in the biosphere (mostly in the ocean) = $40{,}000 \times 10^9 \times 10^6 \, \text{g/t} = 40 \times 10^{18} \, \text{gC}$. At $12 \, \text{gC/mole}$ and $6.02 \times 10^{23}$ atoms/mole (Avogadro's number), the biosphere then contains $2 \times 10^{42}$ atoms of carbon. Partitioning the exponent 42, that's $6 + 9 + 12 + 15$, thus about 2 million billion trillion quadrillion atoms of carbon.
2. To delve into more depth about some of the issues I raise here regarding how the networks of biochemical by-products work, see my book (Volk 2003a). I also recommend the recent book by Wilkinson (2006) that explores the inevitability of chemical cycles and other fundamental processes of the biosphere. Stephen Schneider, the editor of the journal *Climatic Change*, wrote a prescient editorial (1986) about the conceptual problems in formulating Gaia theory, and has recently sponsored the publication of views and debates about Gaia theory. See, for example, Kleidon (2002, 2004), Lenton and Wilkinson (2003), Kirchner (2002, 2003), Lenton (2002), Lovelock (2003), and Volk (2002, 2003b, 2003c, 2007).

**Note**

1. But primarily from the estimated age of marine radiocarbon (carbon-14), which diffuses into the ocean as part of the carbon cycle after its formation in the upper atmosphere.

## References

Dawkins, R. 1982. *The Extended Phenotype: The Gene as the Unit of Selection.* San Francisco: Freeman.

Huggett, R. J. 1999. Ecosphere, biosphere, or Gaia? What to call the global ecosystem. *Global Ecology and Biogeography* 8: 425–31.

Kirchner, J. W. 2002. The Gaia hypothesis: Fact, theory, and wishful thinking, *Climatic Change* 52: 391–408.

Kirchner, J. W. 2003. The Gaia hypothesis: Conjectures and refutations. *Climatic Change* 58: 21–45.

Kleidon, A. 2002. Testing the effect of life on Earth's functioning: How Gaian is the Earth system? *Climatic Change* 52: 383–89.

Kleidon, A. 2004. Beyond Gaia: Thermodynamics of life and Earth system functioning. *Climatic Change* 66: 271–319.

Lenton, T. M. 2002. Testing Gaia: The effect of life on Earth's habitability and regulation. *Climatic Change* 52: 409–22.

Lenton, T. M., and D. M. Wilkinson. 2003. Developing the Gaia theory: A response to the criticisms of Kirchner and Volk. *Climatic Change* 58: 1–12.

Lovelock, J. E. 2003. Gaia and emergence. *Climatic Change* 57: 1–3.

Lovelock, J. E. 2006. *The Revenge of Gaia.* New York: Basic Books.

Pinker, S. 2002. *The Blank Slate: The Modern Denial of Human Nature.* New York: Viking Penguin.

Schneider, S. H. 1986. A goddess of the Earth? The debate on the Gaia hypothesis—An editorial. *Climatic Change* 8: 1–4.

Volk, T. 2002. Toward a future for Gaia theory. *Climatic Change* 52: 423–30.

Volk, T. 2003a. *Gaia's Body: Toward a Physiology of Earth.* Cambridge: MIT Press.

Volk, T. 2003b. Seeing deeper into Gaia theory. *Climatic Change* 57: 5–7.

Volk, T. 2003c. Natural selection, Gaia, and inadvertent by-products: A reply to Lenton and Wilkinson's response. *Climatic Change* 58: 13–19.

Volk, T. 2007. The properties of organisms are not tunable parameters selected because they create maximum entropy production on the biosphere scale: A by-product framework in response to Kleidon. *Climatic Change* 85: 251–58.

Volk, T. 2008. *$CO_2$ Rising: The World's Greatest Environmental Challenge.* Cambridge: MIT Press.

Wilkinson, D. M. 2006. *Fundamental Processes in Ecology: An Earth Systems Approach.* Oxford: Oxford University Press.

# 4

# Water Gaia: 3.5 Thousand Million Years of Wetness on Planet Earth

Stephan Harding and Lynn Margulis

Without continuous flows of carbon, hydrogen, nitrogen, sulfur, phosphorus, and other essential elements, primarily as compounds in watery solution, no known life form continues to thrive. The purpose of life, much like other thermodynamic systems open to the flow of matter and energy, is to dissipate chemical and thermal gradients (differences across distances) as elegantly detailed by Schneider and Sagan (2006). The assurance of energy and matter flows in appropriate amounts, rates, and useable chemical form is a sine qua non of the living state. All living beings tend to overgrow their bounds and are invariably limited by appropriate availability of energy and matter. The many limitations to life's intrinsic capacity for growth and diversification is the process Charles Darwin (1809–1882) recognized as "natural selection."

**Our Thesis: Life Retained Planetary Water**

We champion the poorly developed Gaian view that life has vigorously helped maintain abundant water on the Earth's surface over the last three and a half thousand million years. We defend the idea that life's populations persist and continue to expand on Earth not because a "lucky accident" has situated our moist planet at an optimal distance from the sun; rather communities of living organisms have actively maintained wet local surroundings. The result has been the retention of moist habitability over geological time. We suggest that without life's involvement in complex geological, atmospheric, and metabolic processes, Earth would long ago have lost its water, becoming a dry and barren world much like Mars and Venus. Theoretical interpolation of a lifeless planet Earth between that of Mars and Venus shows that our planet now would be a dry, carbon dioxide-rich world with a temperature primarily determined by steady increase in solar luminosity (Lovelock 2000).

In recognition that independence from the biosphere is death and that life is a powerful geological force, V. I. Vernadsky (1863–1945) explained that all life is connected through Earth's fluid phase (Sagan 2007). This comprises the atmosphere (air, including that in soil, caves, and dissolved in water) and the hydrosphere (oceans, lakes, rivers, streams, springs, etc.). Early in the twentieth century, Harvard University scholar L. J. Henderson (1958) presented a persuasive but nearly forgotten argument that life would not exist on this planet without the water that sustains and supports it. He reviewed the salient features of life's "universal solvent system" in his chapter dedicated to the physics and chemistry of water. The thermal properties of water (its specific heat, latent heat, thermal conductivity, expansion before freezing) and its action upon other substances (as a solvent, and by virtue of its ionization and surface tension properties) are unique among solvents and are utterly required by the physiological and ecological systems of life on our planet. The eclipse of Henderson's virtually unknown work may be attributable to the tendency in evolutionary biology literature to overlook environmental chemistry in general and, in this case, the chemistry and biochemical involvements of water in particular. What is remarkable is the fact that Henderson's analysis is not at all obsolete: we find it germane to any Gaian analysis of the water anomaly on Earth relative to the other inner planets.

In the spirit of Ian McHarg's remarks we recommend that a modern detailed reappraisal of Henderson's concept of the "fitness of the environment" be undertaken (Margulis and Lovelock 2007). McHarg adds Henderson's concept of the environmental importance of water to Darwin's work on evolution in his search for understanding the creative survival of the living. For McHarg, there is a criterion by which living (and other) processes can be evaluated for their creativity (and destruction). He calls it "creative fitting in health," contrasting it with "reductive misfit revealed in pathology" (McHarg 2006). He points out that whereas Darwin emphasized that the organism "is fit for the environment," Henderson (whom McHarg admired as much as Darwin) maintained that "the actual environment, the actual world constitutes the fittest possible abode for life." McHarg unites Darwin's and Henderson's viewpoints when he concludes that "there is a requirement for any system—whether subcellular, cell, tissue, organism, individual, family, institution—to find the most fit of all environments and to adapt both the environment and the system itself" (2006). Survivors, on McHarg's analysis, adapt by actively and continuously changing their environment to accomplish fitting in a thermodynamically creative way. The sum of active and incessant local environmental alteration, in this case by the movement of water and matter with which life interacts, we recognize as "Water Gaia."

We expand the insights of our predecessors by elucidating the tight correlations between life and water. Life, aptly called *animated water* by Vernadsky and colleagues, is mandated by the presence and properties of water. Life ensures its own continuity by retaining and interacting with liquid water on our planet's surface.

## Water on Venus, Earth, and Mars

Scientists concur that all three inner planets Venus, Earth, and Mars, prior to the Archean eon over 3,500 million years ago, began with meteoric and probably subsurface water in abundance. Geomorphological observations of erosion by water, steady bombardment by water-rich comets, asteroids, and meteorites, along with other evidence attest to copious quantities of early water on Earth (Robert 2001). Further evidence for the presence of surface liquid water on the Hadean Earth comes from analyses of Hadean zircons (Wilde et al. 2001). Water must have out-gassed from ancient tectonic activity, as all these planets and their moons were bombarded by the water-rich bolides of the early solar system. The surface of Venus, closer to the Sun, and that of Mars, beyond Earth's orbit, reveal riverine, lacustrine, or marine features that suggest vast quantities of open water flowed on pristine active lithospheres of our early "sister planets." Recent analysis of phyllosilicates from Mars suggest that water-rich environments conducive to life were widespread during its earliest geological history (Mustard et al. 2008). Whereas much, perhaps even an ocean's-worth or more of water, was lost from both our neighbors, the early Earth retained its primordially wet conditions.

Our hypothesis that water retention is a Gaian phenomenon is testable. Venus probably lost its water because its proximity to the Sun meant that even early in the history of the solar system it would have received 40 percent more solar radiation than today's Earth. This high radiative flux would have evaporated huge amounts of water vapor into the atmosphere of Venus that set in train catastrophic positive feedback on heating due to the powerful greenhouse effect of water vapor; this is

known as the runaway greenhouse. Abundant water vapor in the stratosphere would have been photo-dissociated by ultraviolet radiation leading
to massive quantities of hydrogen loss to space (Kump et al. 2004).

Although Mars receives some 43 percent less solar radiation than
the Earth, it likely once had sufficient greenhouse gasses in its atmosphere to generate temperatures high enough to liquefy water on its
surface. Carbon stripped out into carbonate rocks would not have been
returned to the atmosphere because of the absence or early demise of
plate tectonics on the planet (Kump et al. 2004). Some of this water
would then have evaporated into the thin Martian atmosphere,
followed by photo-dissociation of water vapor and hydrogen loss to
space. The extent to which water ice exists in the Martian north pole,
the south pole, or trapped under large areas of the Martian surface is
the subject of vigorous current research.

By comparison to the 10 centimeters, or fewer, precipitable water
measured today on dry and barren Mars and Venus, the Earth is shock-
ingly wet. More than $10^4$ times the quantity of water expected on an
Earth without life is still here. From reconstruction of its past history
scientists conclude that throughout the geological eons our planet has
been watery. Today water is found mostly in its liquid phase within the
global oceans which cover some 70 percent of our planet's surface.
Quantitatively small but climatically crucial amounts of water also exist
in the gas phase as clouds and water vapor. In the solid phase as sea and
glacial ice, frost, hailstones, and snow, water augments the Earth's albedo
(i.e., greater reflectivity of solar energy to space).

The movements of water between these and other reservoirs constitutes the hydrological cycle—"the largest movement of any substance
on Earth" (Cahine 1992). The hydrological cycle has massive effects on
climate because of the ways water determines the exchange of heat and
moisture between the atmosphere and the planet's surface. Contemporary organisms actively configure the Earth's climate into a state suitable for water (and thus for the perpetuation of life) by influencing the
hydrological cycle through the process of evapotranspiration in trees
and plants. Evapotranspiration involves massive movements of water,
against gravity, from the entire root zone (rhizosphere) up a few to over
30 meters into the air. The flow of water up through tree trunks and
plant stems is powered by solar energy. Water is released as vapor
through the stomata—the active pores that open and close on the
undersides of leaves. Organisms also influence the hydrological cycle

in important ways by retaining water in soils and by emitting a variety of cloud-seeding chemicals over land and ocean (Hayden 1998; Bonan 2002). Furthermore bacteria such as *Pseudomonas syringae* that are commonly swept up into clouds in large numbers exert a massive influence on the hydrological cycle. Proteins on the outer surfaces of these bacteria facilitate the formation of ice crystals that eventually return significant quantities of water to the Earth's surface as rain and snow (Christner et al. 2008).

## Water and Gaia's Thirst

Earth's abundant water in comparison to Mars and Venus lead us to a Gaian analysis of this "water anomaly." Scientists often assume that environments are physicochemical givens to which organisms must adapt in order to survive. But unlike the prevalent assumption that life passively adapts to its environment, Gaia researchers propose that life may contribute to active regulation of biologically relevant aspects of Earth's surface within habitable limits (Lovelock 1972, 2000, 2005; Lovelock and Margulis 1974). This regulation is posited to emerge from tightly coordinated feedback subsystems that intrinsically and continuously embed the biota in its abiotic surroundings (Lovelock 2005; Lenton 1998). In a masterful analysis of the Earth's physicochemical history, NASA geoscientist Paul Lowman and astronaut Neil Armstrong show that during the Archean the major influences were the same as those prevailing on Mars and Venus. But from the Proterozoic eon (2,500 million years ago) until the present day, Gaia's unique signature is writ large: Earth became the Gaian planet. Paucity of water, failure to detect granite, vastly slower geochemical cycles of elements such as oxygen, carbon, and phosphorus, and much other evidence testifies to the fact that neither Venus nor Mars are Gaian (Lowman and Armstrong 2002). Lowman and Currier (2009) provide a short accessible summary explanation that connects Gaia theory with the uniqueness of water-dependent lateral plate tectonic movement on Earth.

Life's sensitivity to water quantity and saltiness seems to be the most elemental of all senses. Thirst and the knowledge of desiccation level is apparently universal. The universality of water detection and the response of living cells to this ubiquitous solvent, that some equate with life itself, lies apparently in the properties of the lipid-protein membrane, the bilayer semipermeable external boundary of all cells. When breached and membrane integrity is lost, the autopoietic entity known as the cell, whether a small bacterial cell or a large egg, dies. The ability for material and energy flow is irretrievably lost, as water leaks into the environment. This is what we call death. Self-maintenance and identity are replaced by an inert puddle of carbon-hydrogen-nitrogen compounds that immediately lose all signs of animation and become food for those who retain their membranes and, with them, the profound ability to sense water.

Life does indeed *adapt* itself and its environment as Henderson and McHarg insisted. Yet, when used in a way that implies a passivity of life and that ignores emergent synergies between our planet's physics, chemistry, and biology, the term adaptation can hinder our understanding of the Earth as a complex system.[1] We prefer statements of passive adaptation of organisms to their surroundings to be replaced by a conception of life's "active fitting." Gaia emerges directly from this active fitting, writ large, since all organisms impact each other and their surroundings through the exchange of heat, light, liquids, and gases, as well as a huge array of metals, salts, sugars, and myriad other chemical compounds (usually dissolved in water).

With regard to the hydrosphere, Gaia theory proposes a prospective research program: that organisms have actively retained water by thwarting its tendency to be lost. Without the involvement of life's complex and often metabolic innovations,[2] Earth would long ago also have lost its water to space by atmospheric photolysis and hydrogen escape. We propose that life does not regulate the amount of water on the planet through a specific feedback process, but rather that it greatly reduces the rate of water loss by metabolic hydrogen capture and by regulation of relevant variables such as planetary temperature.

Here we explore the major abiotic processes that drive the loss of water from our planet, including the photodissociation of water and methane by solar UV radiation at the top of the troposphere and the chemical reactions in seafloor basaltic rocks that strip out oxygen atoms from water molecules. We then go on to outline the various ways in which life prevents such processes from drying out the planet. We include a discussion of how, by contributing to the regulation of the planetary carbon cycle over geological time, organisms have kept the planetary temperature suitable for the existence liquid water despite an ever-brightening sun and ongoing outgassing of carbon dioxide from volcanic activity.

## Modes of Water Retention by Life

Any chemical or physical process that liberates hydrogen from water molecules may, in principle, lead to water loss from a planet. Hydrogen ($H_2$) gas has a mass so light that it reaches escape velocity from the Earth's gravitational field.

We summarize some chemical and biological processes that both liberate and capture free hydrogen over geological time in table 4.1. They exemplify our habitation of an Earth with abundant water and serve as a guide to further detailed investigation. Geochemical processes that result in the liberation of molecular hydrogen began at least in the Archean and have continued until the present. They occur in basalt, the major rock type of the world ocean bottom. Basalt contains ferrous oxide (FeO), which, in the presence of carbon dioxide, strips out oxygen atoms from seawater. The net effect is to remove oxygen and place it in solid form in carbonate rock, a process that liberates hydrogen gas (reaction 1, Lovelock 2005). Hydrogen liberation via loss to space may entirely desiccate an inner planet within two billion years (Lovelock 2005). Bacterial metabolic pathways also liberate hydrogen (e.g., anoxygenic photosynthesis, anoxic decomposition of dead organic matter [fermentation, reaction 2], anaerobic glycolysis, and many others release hydrogen on geologically instant time scales).

The Earth has evaded desiccation by many means that inspire further investigation. Since Archean times bacterial communities have released oxygen into the sediments, water, and air by oxygenic photosynthetic processes that split water (reaction 3), a reaction that to this day is limited to only three immensely talented inclusive taxa. In a purely abiological process, hydrogen gas (e.g., that released from reaction 1) combines with oxygen from photosynthesis, thereby regenerating water (reaction 4). Oxygenic photosynthesis (reaction 3) also captures and retains hydrogen extracted from water for carbon dioxide reduction, thereby renewing organic matter in the making of food, body parts, and energy storage molecules such as sugar and starch. New avenues of oxygen liberation were opened up during the Proterozoic eon (some 1,200 million years ago) by photosynthetic algal protoctists, as well as in the lower Phanerozoic eon (about 450 million years ago) by the first land plants. All these oxygenic photosynthetic processes continue today unabated. Even anti-Gaia scientists admit that chlorophyll *a* photosynthesis produced the oxygen-rich atmosphere that permanently altered

**Table 4.1**
Selection of key biological and abiological processes that influence the retention of water on planet Earth

| Reaction and domain in which it takes place | Reactants | Effect on Earth's water |
|---|---|---|
| (1) $2FeO + 3CO_2 + H_2O \rightarrow Fe_2(CO_3)_3 + H_2$<br>Abiological: geochemical | Ferrous oxide in sea floor basalt reacts with carbon dioxide and water | Desiccates the Earth by liberating free hydrogen |
| (2) $3CH_2O + H_2O \rightarrow CH_3COO^- + CO_2 + 2H_2 + H^+$<br>Biological: fermenting bacteria in anoxic environments | Organic matter and water | Desiccates the Earth by liberating free hydrogen |
| (3) $CO_2 + H_2O \rightarrow CH_2O + O_2$<br>Biological: oxygenic photosynthesis by bacteria, protoctists, and plants | Carbon dioxide and water reacted by photosynthesizers; organic matter and oxygen produced | Oxygen available for reaction with hydrogen, potentially reconstituting water |
| (4) $2H_2 + O_2 \rightarrow 2H_2O$<br>Abiological: atmospheric chemistry | Hydrogen and oxygen, producing water | Free oxygen from (3) reacts with free hydrogen, reconstituting water |
| (5) $S + H_2 \rightarrow H_2S$<br>Biological: bacterial reduction of elemental sulphur | Elemental sulphur and hydrogen | Sequesters hydrogen into hydrogen sulphide gas |
| (6) $2H_2S + O_2 \rightarrow 2S + 2H_2O$<br>Biological: aerobic chemautorophic bacteria | Hydrogen sulphide from reaction (5) with oxygen from reaction (3) | Reconstitutes water |
| (7) $CO_2 + 2H_2 \rightarrow CH_2O + H_2O$<br>Biological: anaerobic chemautorophic bacteria | Carbon dioxide and hydrogen | Organic matter produced, reconstituting water |
| (8) $CO_2 + 4H_2 \rightarrow CH_4 + 2H_2O$<br>Biological: anaerobic methanogenic bacteria | Carbon dioxide and hydrogen | Methane produced, reconstituting water |

Source: Data from Smil (2003) and Lovelock (2005).

Earth and its evolutionary course. Without these bacterial metabolic innovations, no animal would exist.

Another bacterial contribution to hydrogen capture comes from the activities of bacteria such as *Desulfovibrio* that live in ocean sediments in sulfur-rich habitats. *Desulfovibrio* and its many relatives liberate hydrogen sulfide gas (reaction 5) as they reduce elemental sulfur, thiosulfate, or the sulfate ion itself by "breathing." Water is reconstituted when hydrogen sulfide is oxidized by aerobic chemoautotrophic bacteria such as *Sulfolobus* or *Beggiatoa* that abide at oxygen-rich seawater/ sediment, caves, sulfur springs, and other interfaces (reaction 6).

An important metabolic pathway in certain bacteria hardly seems possible, in principle. These bacteria reconstitute water by reacting molecular hydrogen with carbon dioxide under conditions where oxygen gas is absent (reaction 7). Known as anaerobic chemoautotrophy, in this process hydrogen is used to reduce carbon dioxide to organic matter, and water is thereby reconstituted. Also in regions without any oxygen gas, methanogenic bacteria remove carbon dioxide and react it with free hydrogen to produce methane and water (reaction 8). Reactions 7 and 8 both require anoxic habitats, such as marine, lacustrine, and riparian sediments, or the intestines of insects and mammals.

### Water Loss via the Photodissociation of Methane

A physical process that is thought to have led to hydrogen escape during Earth's geological history is the photodissociation of water by ultraviolet radiation in the lower stratosphere. Yet relatively little hydrogen must have escaped via this route because of the "cold trap" in the tropopause (Catling et al. 2001). Since Archean times, water vapor molecules have frozen out in this region of very cold air and fallen back into the lower atmosphere before they could be photodissociated by stratospheric ultraviolet radiation. David Catling and his colleagues suggest that the photodissociation of methane provided the main exit route for hydrogen during the Archean eon, and hypothesize that abundant methane was the major greenhouse gas that counteracted the early lower solar luminosity. Methane's lower freezing point relative to water allowed it to transit into the stratosphere through the cold trap in gaseous form unaffected. There (much like the few water molecules that managed to reach the lower stratosphere above the cold trap) the methane was split by ultraviolet radiation, yielding molecular hydrogen that could escape to space and leaving carbon dioxide and oxygen in the atmosphere.

These reactions are simplified and summarized as (Catling et al. 2001)

$$CO_2 + 2H_2O \rightarrow CH_4 + 2O_2 \rightarrow CO_2 + O_2 + 4H \ (\uparrow \text{space}),$$

which is reaction 9. In this scenario the methane came from the bacterial decomposition of organic material in which hydrogen from water was originally fixed by oxygenic photosynthesis (reaction 3). One could thus argue that methanogenic bacteria could have been responsible for life-threatening water loss during the Archean (David Schwartzman, personal communication). However, life as a whole may have prevented this eventuality in at least two ways. First of all, during the Archean, carbon dioxide released mostly by decomposing bacteria would have permitted hydrogen to accumulate in the lower atmosphere via a newly proposed hydrodynamic mechanism that slows down the rate of hydrogen loss except when carbon dioxide levels are very low (Stevenson et al. 2008). Then, in the Proterozoic, biogenic oxygen would have captured the hydrogen. Thus it seems that there might have been a rather dangerous period during the Archean when biogenic methane production could have accelerated water loss, but that this danger was avoided early on thanks to biotic carbon dioxide release, and later on when biogenic oxygen became sufficiently abundant to reconstitute water via reactions 4 and 10. Clearly, a synergy between robust photochemistry and sound biology is required to further explore this issue.

Reaction 9 may have led to the so-called Great Oxidation Event that took place between 2,400 and 1,800 million years ago during the Proterozoic. This event involved a relatively rapid transition to an oxidizing atmosphere, and may have ultimately produced the high levels of oxygen gas (ca. 20 percent) in today's atmosphere. The rise of atmospheric oxygen gas during the Proterozoic has been amply documented in the geological record, especially by worldwide deposits of banded iron formations, or BIFs (Cloud 1989). Apparently a relatively small increase in the burial rate of organic carbon may have triggered a nonlinear switch to a high oxygen atmosphere at that time (Goldblatt et al. 2006). The stratospheric ozone layer that resulted has significantly influenced the effectiveness of the cold trap to this day (Nisbet 1991).

Whatever led to the surplus of free oxygen gas in the Proterozoic, it is agreed that hydrogen loss via the photo-dissociation of methane would have declined significantly when oxygen became sufficiently abundant to oxidize methane to carbon dioxide and water via the reaction

$$CH_4 + 2O_2 \rightarrow CO_2 + 2H_2O,$$

which is reaction 10. As the Archean atmosphere probably contained a thousand times more methane than today's value of 6 to 7 parts per million, the rate of hydrogen loss must have been approximately three hundred times greater than at present (Catling et al. 2001). The modern biosphere's effectiveness at preventing hydrogen loss, and hence planetary desiccation, is illustrated by the very low rate of hydrogen loss to space. The Earth currently loses a mere 50 tonnes of hydrogen per day from an atmosphere with a total mass of around $50 \times 10^{14}$ tonnes (Morton 2007: 182).

The Great Oxidation Event marked a shift from methane to carbon dioxide as the Earth's dominant greenhouse gas (Lovelock 2000). Other consequences for life and its effects on the planetary surface include the appearance of early eukaryotic cells and their obligate relation to oxygen respiration in symbiotic bacteria that became mitochondria (Margulis et al. 2006) and a Gaian redistribution of many chemical elements such as manganese, copper, phosphorus, lead, tin, and vanadium.

The metabolic versatility of bacteria permits oxidation of methane even in the absence of oxygen gas. Sulfate reducers, such as *Desulfovibrio* and some relatives, use oxygen in sulfate ions that are abundant in seawater to reconstitute water from methane:

$$CH_4 + SO_4^{2-} \rightarrow HCO_3^- + HS^- + H_2O,$$

which is reaction 11. Could these reactions (10 and 11) have produced water in sufficient quantity to increase the depth of the global ocean (S. Marashin, personal communication)?

## Water and Earth's Temperature

Why has Earth retained both life and abundant liquid water since the Archean despite at least two strong external factors that have conspired to enhance the similarities between Mars, Venus, and Earth? One external factor is the increase of luminosity of the Sun (with an energy output 25 percent greater than it was 3,500 million years ago), and the second is the continual eruption of carbon dioxide from volcanoes over the same period. These and other observations lead us to conclude that global temperatures have been actively regulated within the range suitable for liquid water by the Earth as a whole system. That the behavior, metabolism, and physiology of organisms are essential to this regulation is a central tenet of Gaia theory (Lovelock 2000; Margulis and Lovelock 2007).

Much remains to be learned, but we can now state with some confidence that organisms help to regulate the Earth's temperature by manipulating the ratios of greenhouse gases in the atmosphere and by altering the planetary albedo (reflectivity), primarily by emitting cloud-seeding chemicals. Other effects on temperature and hence water retention by organisms involve the albedo of living beings themselves, such as the extensive cover of dark coniferous trees in the far northern latitudes that help to warm the modern Earth (Bonan 2002). Organisms can also change the amount of surface water directly exposed to evaporation: elephant bodies carve out ponds and thus expose subsurface water to the surface; exudates of microbial mat organisms directly retard evaporation; caves made by water flowing through limestone, or the conversion of limestone to gypsum, protect water flow beneath the rocks.

We suggest that liquid water would have left the Earth's surface long ago if organisms had not regulated global temperatures by these and other means. Continued volcanic activity that puts methane, water vapor, carbon dioxide, and other greenhouse gases in the atmosphere in the face of an ever-brightening sun would long ago have led the Earth into a Venus-like runaway feedback on global heating. On the other hand, too little carbon dioxide would have caused the oceans to freeze over, with the consequent albedo increase plunging the planet into a permanent frozen state via positive feedback (Ward and Brownlee 2000). A major way in which life contributes to the regulation of global temperature is through its involvement in the long-term carbon cycle in which calcium carbonate from the weathering of basaltic and granitic (silicate) rocks is deposited in the oceans (table 4.2, reactions 12 and 13).

On the land, reaction 12 is enhanced by organisms: roots and hydrophilic microbial chemical exudates physically fracture the rock and thereby increase its reactive surface area; microbial and plant root respiration increase carbon dioxide levels in the soil, and bioturbation of the soil increases the flow of water onto particles of rock, taking water into places it would not otherwise be able to access. This process, first proposed by Lovelock and Whitfield (1982) and now referred to as "biologically assisted silicate rock weathering," amplifies the purely chemical weathering rate between 10 to 1,000 times depending on

Table 4.2
Reactions in the long-term carbon cycle

| Reaction | Effect on Earth's temperature |
| --- | --- |
| (12) $CaSiO_3 + 2CO_2 + 3H_2O \rightarrow$ $Ca^{2+} + 2HCO_3^- + H_4SiO_4$ | $CaSiO_3$ is wollastonite, a simple mineral representing the general chemical composition of all silicate rocks. Note that two carbon atoms are removed from the atmosphere for each calcium ion weathered out of the rock. |
| (13) $Ca^{2+} + 2HCO_3^- \rightarrow$ $CaCO_3 + H_2O + CO_2$ | Denotes the intracellular precipitation of calcium carbonate. Note that one carbon atom is released to the atmosphere for each calcium ion precipitated. The net effect of reactions 12 and 13 is thus to cool the Earth. |
| (14) $CaCO_3 + SiO_2 \rightarrow$ $CaSiO_3 + CO_2$ | Granite is regenerated, and carbon dioxide is liberated to the atmosphere via volcanoes, thereby warming the Earth. |

Source: Adapted from Kump et al. (2004).

location (Schwartzman and Volk 1989); it is greatest where high temperatures combine with abundant rainfall.

Thus carbon that once resided in the atmosphere finds itself in calcium bicarbonate flushed by rivers and groundwater into the oceans where it is precipitated intracellularly as calcium carbonate by coccolithophorids (haptophyte algae) and foraminifera in their scales and exoskeletons (reaction 13). When these organisms die the calcium carbonate accumulates in ocean sediments. Their fate is lithification into chalk and other limestones. Huge quantities of carbon have been sequestered in this way over geological time—the stock of carbon in the contemporary calcium carbonate reservoir is $4 \times 10^7$ GtC, almost four orders of magnitude greater than the carbon in present-day fossil fuel reserves (Kump et al. 2004). Chalk and limestone also contain significant quantities of silica (from the silicic acid in reaction 12) that may be deposited as radiolarite (chert rock that come from remains of radiolarian skeletons), or diatom tests (shells) and glass sponge spicules (Lovelock 2005).

Such dynamics imply negative feedback with respect to the carbon cycle (Lenton 1998) and hence surface temperatures suitable for liquid water: if surface temperature increases (because of volcanic inputs of carbon dioxide to the atmosphere, together with an ever-brightening sun) so does rainfall. In a wetter and warmer world biologically assisted silicate rock weathering transfers more carbon from the atmosphere to calcium carbonate in the ocean, which cools the Earth, potentially down to a stable but lifeless frozen state. However, in a cooler and hence drier world this fate is avoided because the terrestrial biosphere rapidly becomes less effective at weathering silicate rocks, and so carbon dioxide accumulates in the atmosphere from volcanoes, thereby raising the global temperature (Lovelock 2005). An emergent property of this feedback has been the regulation of planetary temperature within limits suitable for life (and hence liquid water) over geological time.

The carbon dioxide that returns to the atmosphere via volcanoes is regenerated when silica-rich carbonate sediments are subducted into the mantle as the raised portions of descending slabs (plates) of the seafloor (reaction 14, table 4.2). Here, at high temperature and under immense pressure, the sediments metamorphose and produce carbon dioxide and fresh granitic material that floats on top of the denser mantle to become new continental land mass available to be weathered (Kump et al. 2004). Without such recycling of Earth's crustal materials, no terrestrial biota would exist to enhance silicate weathering.

## Water and Plate Tectonics

The long-term carbon cycle thus cannot operate without volcanic activity, itself an integral component of the colossal processes of plate tectonics, with its mountain chains, subduction zones, and large granitic continents afloat on giant rafts of spreading seafloor basalt. These tectonic processes, which are essential for the maintenance of all organic life, cannot take place without huge quantities of liquid water.

Water infiltrates the laterally moving seafloor basalt, changing its chemical nature so that it is pliable enough to sink into the Earth's mantle when it collides with the edge of a continent at a subduction zone. Seafloor basalt becomes extensively hydrated at the mid-oceanic ridges. Here, magma chambers act as heat sources that drive local-scale convective systems that force hot seawater through fractures in the basalt. For it to be effective at hydration of seafloor basalt, the process requires an overlay of large amounts of water (Campbell and Taylor 1983). At subduction zones, water-rich slabs of seafloor basalt are carried deep into the mantle where the material melts to produce vast amounts of granitic magma that rises up to form the continents. This process adds to the granite generated by the metamorphism of silica-rich calcium carbonate sediments beneath subduction zones mentioned earlier. The volatility of limestone produces watery carbon dioxide–rich lubricant, which enhances the rates of plate tectonic activity. A vast amount of water has been required to generate the Earth's continents, 60 percent of which were almost certainly present since the beginning of the Proterozoic some 2.5 billion years ago (Taylor and McLennan 1995).

Without subduction, plate tectonics would stop because there would be no closure of the convective cycle that reaches down to the planet's outer core, in part driven by the decay of radioactive materials in the Earth's depths (Kump et al. 2004). Without plate tectonics, the return of carbon to the atmosphere would be severely curtailed or perhaps completely shut off. In tens of millions of years all the Earth's land masses would be removed by weathering, with no new granite to replace this loss. The long-term carbon cycle would cease, and the Earth would perhaps be plunged into a permanently frozen state (Ward and Brownlee 2000). We therefore propose an interesting and appropriately circular Gaian dynamic here: no life, no water→no water, no plate tectonics→ no plate tectonics, no life.

## Water and Culture

Western culture is expert at the abuse of our planet's watery heritage. Two examples will suffice to illustrate the scale of our misappropriation of water. First: *concrete*. Scientists focus, rightly, on the massive emissions of carbon dioxide liberated during the process of making this material, but we should also be aware that prodigious amounts of water are extracted from the water cycle when concrete is mixed, poured, and set. Each decade, we lock up about $3,400\,km^3$ of water in concrete—a volume approximately equivalent to that of Lake Huron.[3] How long it will take for these Huron-loads of water to return to the natural cycle is anyone's guess—clearly it depends on how timing of the weathering processes liberate the water from its prison of artificial rock. Second: *oil*. Natural hydrocarbon reservoirs (oil and gas wells) contain large amounts of water, which is often brought to the surface during extraction as "produced water." Much of this is pumped back down to extract more hydrocarbons, but some remains at the surface where it becomes a hazard to agriculture and other aspects of human and plant life due to its saltiness, its oil and grease content, its burden of chemical additives from extraction process, and sometimes its radioactivity from radio nuclide contamination. A recent estimate by Sergio Maraschin (personal communication) suggests that up to $74.4\,m^3$ of this oil water remains on Earth's surface annually. Sadly, every year, approximately $49\,km^3$ of clean surface water is captured and used to force oil out of the ground. In some places, notably the Middle East, so much surface water is pumped into oil wells that rivers such as the Euphrates in Syria are in danger of drying up. On the plus side, or so it would seem, our economic activities also liberate water. As much as $62.7\,km^3$ per year is released when hydrocarbons are burned in our engines and generators. Thus, in total, the hydrocarbon industry injects some $88.1\,km^3$ per year of water into the natural water cycle (S. Maraschin, personal communication). This sounds good, until one realizes that much of this tailpipe water carries a contaminant burden that affects human well-being in the global ecosystem.

Fortunately our culture is also capable of engaging in more benign relationships with water. From the facts of a watery Gaian Earth can be inferred knowledge and wisdom that extends beyond science (Harding 2006). Recognition of the complex relationship of water, life, and Earth history has recently become available in two oversized, gorgeous books: *Water* (also published identically as *Agua* in 2006) and *Water Voices from Around the World* (Marks 2007). The frontispiece of the first states: "We need to create a new culture that acknowledges and respects the value of water. The survival of future generations of humans and all other species on this planet depends on such a new culture." The second is dedicated to our ancestor, Water, and bears testimony of citizens from fifty countries around the world. Nobel laureates figure in both books and the color photographs from satellite to microscopic levels are remarkable. In *Water Voices* we learn about Lake Sarez in Tajikistan formed by the 1911 earthquake's landslide, and kept in place by the largest natural dam in the world. Tajikistan's reverence for fresh water is palpable. The song of this Central Asian country is joined by many human and nonhuman voices: a cayman from Cuba most of whose close relatives have been extinguished, a Red Eye tree frog from a Central American rain forest, wild salmon from Kamchatka, clown fish and corals, and the tail of a humpback. The spectacular photographs in *Voices* and those of Antonio Vizcaino in *Agua* (*Water*) need no admonishment to induce us to protect our home planet. We commend both these magnum opuses; they speak louder than our words in search of Water Gaia. They represent a step, along with others expressed in this volume, toward actions that respect the

Earth. We end with a suggestion: that we properly rename our third from the Sun inner planet after the humble, crucial chemical compound that sustains us: Water!

## Notes

We thank Richard Betts, Tim Lenton, James Lovelock, James MacAllister, Sergio Maraschin, Will Provine, David Schwartzman, and Bruce Scofield for useful discussion pertinent to the writing of this chapter. LM thanks The Tauber Fund, Abe Gomel and the University of Massachusetts Graduate School for support.

1. In fact common claims of adaptation, with its passive connotations, may impede investigation of the evolution of the Earth's environment through geological time. We recommend a re-examination of this ambiguous term. Usually biologists study specific correlations of behavior, morphology, or chemistry of a given organism to its immediate environment. But the assertion that any organism is well adapted to its habitat has little meaning, since the adaptation is not measurable nor even estimable in a communicable way. All organisms alive today are adapted by virtue of the implied continuation of their ancestors from the past to the present.

2. Examples are lipid monolayer biosynthesis, calcium ion extrusion that induces changes in carbonate, bicarbonate, and $CO_2$ equilibria, oxygenic photosynthesis, and reversible protein absorption and release of water.

3. Unpublished experiments with three different kinds of concrete and calculations showed these to be repeatable results. B. Wartski, North Carolina, 2008 (personal communication).

## References

Bonan, G. 2002. *Ecological Climatology*. Cambridge: Cambridge University Press.

Cahine, M. T. 1992. The hydrological cycle and its influence on climate. *Nature* 359: 373–80.

Campbell, T. H., and S. R. Taylor. 1983. No water no granites—no oceans no continents. *Geophysical Research Letters* 10 (11): 1061–64.

Catling, D. C., J. Z. Zahnle, and C. P. McKay. 2001. Biogenic methane, hydrogen escape, and the irreversible oxidation of early Earth. *Science* 293: 839–43.

Cloud, P. E. Jr. 1989. *Oasis in Space: Earth History from the Beginning*. New York: Norton.

Christner, B. C., C. E. Morris, C. M. Foreman, R. Cai, and D. C. Sands. 2008. Ubiquity of biological ice nucleators in snowfall. *Science* 319 (5867): 1214.

De la Macorra, X., ed. 2006. *Water (Agua)*. México City: Grupo Modelo, America Natural Publishers.

Day, W. 2007. Life as growth. In L. Margulis, and E. Punset, eds., *Mind, Life and Universe: Conversations with Great Scientists of our Time*. White River Junction, VT: Chelsea Green Publishing, pp. 239–43.

Goldblatt, C., T. M. Lenton, and A. J. Watson. 2006. Bistability of oxygen and the Great Oxidation. *Nature* 443:683–86.

Harding, S. 2006. *Animate Earth: Science, Intuition and Gaia*. White River Junction, VT: Chelsea Green Publishing.

Hayden, B. P. 1998. Ecosystem feedbacks on climate at the landscape scale. *Philosophical Transactions of the Royal Society (London)* B353: 5–18.

Henderson, L. J. [1958] 2007. Fitness of the environment. In I. McHarg's *Conversations with Students, Dwelling in Nature*. Boston: Beacon Press.

Kump, L. R., J. F. Kasting, and R. G. Crane. 2004. *The Earth System*, 2nd ed. Upper Saddle River, NJ: Pearson Prentice Hall.

Lenton, T. M. 1998. Gaia and natural selection. *Nature* 394: 439–47.

Lovelock, J. E. 1972. Gaia as seen through the atmosphere. *Atmospheric Environment* 6: 579–80.

Lovelock, J. E. 2000. *The Ages of Gaia*. New York: Oxford University Press.

Lovelock, J. E. 2005. *Gaia: Medicine for an Ailing Planet*, 2nd ed. London: Gaia Books.

Lovelock, J. E., and L. Margulis. 1974. Biological modulation of the Earth's atmosphere. *Icarus* 21: 471–89.

Lovelock, J. E., and M. Whitfield. 1982. Life span of the biosphere. *Nature* 296: 561–63.

Lowman, P. D. Jr., and N. A. Armstrong. 2002. *Exploring Earth, Exploring Space*. New York: Cambridge University Press.

Lowman, P. D. Jr., and N. Currier. 2009. Platetectonics and Gaia. In L. Margulis, C. Asikainen, and W. E. Krumbein, eds., *Chimera and Consciousness: Evolution of the Sensory Self*. White River Junction, VT: Chelsea Green Publishing.

Margulis, L., and M. F. Dolan. 2001. *Early Life*, 2nd ed. Sudbury, MA: Jones and Bartlett.

Margulis, L., and J. E. Lovelock. 2007. The atmosphere, Gaia's circulatory system. In L. Margulis and D. Sagan, eds., *Dazzle Gradually: Reflections on the Nature of Nature*. White River Junction, VT: Chelsea Green Publishing, pp. 157–70.

Margulis, L., M. Chapman, R. Guerrero, and J. Hall. 2006. The last eukaryotic common ancestor (LECA): Acquisition of cytoskeletal motility from aerotolerant spirochetes in the Proterozoic eon. *Proceedings of the National Academy of Sciences* 103: 13080–85.

Marks, W. E., ed. 2007. *Water Voices from Around the World*. Edgartown, MA: William E. Marks Publisher.

McHarg, I. 2006. *Conversations with Students: Dwelling in Nature*. L. Margulis, J. Coner, and B. Hawthorne, eds. Princeton, NJ: Princeton Architectural Press.

Morton, O. 2007. *Eating the Sun*. London: Fourth Estate.

Mustard, J. F., et al. 2008. Hydrated silicate minerals on Mars observed by the Mars Reconnaissance Orbiter CRISM instrument. *Nature*. 454: 305–309.

Nisbet, E. G. 1991. *Leaving Eden: To Protect and Manage the Earth*. Cambridge: Cambridge University Press.

Robert, F. 2001. The origin of water on Earth. *Science* 293: 1056–58.

Sagan, D. 2007. *Notes from the Holocene: A Brief History of the Future*. White River Junction, VT: Chelsea Green Publishing.

Schwartzman, D. W., and T. Volk. 1989. Biotic enhancement of weathering and habitability of Earth. *Nature* 340: 457–60.

Schneider, E. D., and D. Sagan. 2006. *Into the Cool: Energy flow, Thermodynamics and Life*. Chicago: University of Chicago Press.

Smil, V. 2002. *The Earth's Biosphere*. Cambridge: MIT Press.

Stevenson, D. J., et al. 2008. Slow hydrogen escape on the early Earth. *Astrobiology* 8(2): 463.

Taylor, S. R., and S. M. McLennan. [1995] 2005. Model of growth of continental crust through time. In J. V. Walther, ed., *Essentials of Geochemistry*. Sudbury MA: Jones and Bartlett.

Ward, P. D., and D. Brownlee. 2000. *Rare Earth*. Berlin: Copernicus/Springer.

Wilde, S. A., J. W. Valley, W. H. Peck, and C. M. Graham. 2001. Evidence from detrital zircons for the existence of continental crust and oceans on the Earth 4.4 Gyr ago. *Nature* 409 (6817): 175–78.

# 5

# Gaia and Evolution

Timothy M. Lenton and Hywel T. P. Williams

We present in this chapter a search for Gaia in computer-generated model worlds. The computer may seem like an odd place to be looking for a planetary-sized phenomenon when we could be examining the real world. However, with a sample size of only one Earth the inferences that can be drawn about the likelihood of certain features are necessarily limited. In particular, our existence as observers who can look back on and marvel at Earth history is only consistent with a history in which atmospheric oxygen built up to the levels necessary to support animals with large brains (Watson 2004) and the climate became (or remained) relatively cool (Schwartzman 1999). One can imagine myriad other scenarios for Earth-like planets in which life never reaches the stage of conscious observers, but by definition no observer is there to see such histories. In the next few decades we may be fortunate to learn about the atmospheric composition of planets orbiting other stars, and from that information we may learn something about the presence or absence of life on them (Lovelock 1965). This could increase the sample size of inhabited planets above one. In the meantime, by creating many virtual worlds in the computer, we can begin to examine whether features we see on Earth, such as abundant recycling and environmental regulation, are likely or unlikely phenomena once life has emerged on a planet.

## Simulating Gaia

The search for "Gaia in the machine" (Downing 2004) began with Daisyworld (Watson and Lovelock 1983; Lovelock 1983), and there have been many variants of it since (Wood et al. 2007). Perhaps because of Daisyworld's elegant simplicity and great adaptability, relatively few alternative model worlds have been developed. Notable exceptions are the Guild (Downing and Zvirinsky 1999) and Metamic (Downing 2002) models, although these share some key assumptions with Daisyworld. There are also stripped-down models based on Daisyworld that remove many of its key assumptions (Staley 2002; McDonald-Gibson et al. 2008). The danger with modeling is that it is always possible to make a model to illustrate a particular point. However, it is now possible to make stochastic, evolutionary models with very many degrees of freedom and a huge range of possible outcomes, many of which cannot be anticipated by the modeler. This is the approach we take here, of "testing Gaia theory with artificial life" (Lenton 1999). In this respect we follow the pioneering work of Keith Downing who first applied "artificial life" techniques to tackle Gaia questions (Downing and Zvirinsky 1999; Downing 2002, 2004).

Gaia theory posits that a Gaia system will tend to self-regulate in a habitable state, one in which life can survive. Here we follow our previous definitions of a Gaia system, using the terminology of regulation and self-regulation (Lenton 2004). A "Gaia system" is a type of planetary-scale, open thermodynamic system, with abundant life supported by a flux of free energy from a nearby star. *Regulation* describes the return of a variable to a stable state after a perturbation. *Self-regulation* describes a system automatically bringing itself back to a stable state, rather than an external agent imposing regulation or any conscious purpose (teleology) within the system bringing about regulation. For internal consistency of the definitions, and to distinguish a Gaia system from a planet in which, for example, a few extremophiles survive below the surface, we narrow the definition of "habitable state" to one which supports abundant life. Here we also consider nutrient recycling, which can be defined as occurring when the flux of a given nutrient through primary producers exceeds the input flux of that nutrient into a system (Volk 1998).

## Why Evolutionary Biologists Don't (or Didn't) Like Gaia

We begin by reviewing the evolutionary critiques of Gaia. At first glance one might have expected biologists to greet with enthusiasm the idea that life plays a major role in the regulation of the planet. After all, it gives extra prominence to their subject area. Of course, eminent microbiologist Lynn Margulis championed Gaia (Margulis and Lovelock 1974; Lovelock and Margulis 1974a; Lovelock and Margulis 1974b), and she was not (completely) alone in confronting the many neo-Darwinian evolutionary theorists who reacted with vehement opposition when Gaia was first proposed (Doolittle 1981; Dawkins 1983). While generally recognizing that the Earth does display some remarkable stabilizing properties, the critics contended that there was no mechanism by which "selfish" genes and organisms could come to regulate the planet. In other words, Gaia may work in practice but it will never work in theory!

The first protest was that the notion of "atmospheric homeostasis by and for the biosphere" (Lovelock and Margulis 1974a) implies teleology—some conscious foresight or planning on the part of unconscious organisms. This was convincingly answered by the Daisyworld model (Watson and Lovelock 1983; Lovelock 1983), which shows that self-regulation can occur without teleology in a feedback system of life coupled to its nonliving environment. This should surprise no one trained in systems theory. In any coupled system with a mixture of positive and negative effects forming multiple feedback loops there is a good chance that the system will settle down in a negative feedback regime. However, most evolutionary biologists (in contrast to their now distant colleagues in physiology) seem blissfully unaware of systems theory. As Jim Lovelock once put it (in conversation with T.M.L.); "I know professors of biology who have trouble with the concept of self-regulation, but they have no difficulty walking!"

The second protest was that while the self-regulating properties of organisms have been refined by natural selection, the Earth exists in a population of one, and therefore any self-regulation it displays cannot have been shaped by natural selection at the global scale. Given the obvious power of natural selection to engineer well-adapted individuals, this may imply that self-regulation in organisms will be more finely honed than any seen at the planetary scale. However, it does not deny the existence of regulation at the planetary scale. Even in organisms, natural selection cannot *create* self-regulation; it can only favor those individuals that happen to display self-regulation in the sense that they leave more descendants. Evolution by natural selection really comprises three parts: the tendency toward exponential growth (creating competition and selection pressure), heritable variation based on (near) faithful replication, and some source of innovations (new traits) that give differential survival rates. The "innovation" of self-regulation need not stem from a point mutation at some genetic locus; it may simply be the tendency of a sufficiently complex feedback system to settle in a negative feedback regime, a mechanism that is just as applicable to the planet as it is to organisms.

The third protest was that any system in which certain "altruistic" life forms expended energy contributing to making a better global environment would be vulnerable to "cheats"—organisms that enjoyed the benefits of a better global environment but did not contribute to it. Cheats would save energy and thus outcompete altruists, ultimately destroying the regulatory system. The limitation of this argument is that it is predicated on the ideas that organisms can "choose" whether or not to alter the environment and that it will cost energy to do so. In fact as anyone aware of thermodynamics should know, altering the environment is an unavoidable consequence of being alive and can be part of a process that transfers energy to the organism (Schrödinger 1944). Living organisms are highly ordered (low-entropy) structures, and to maintain an ordered state, they must take in matter and free energy (often combined together as "food"), transform the matter, and excrete waste products that are of a lower free energy (higher entropy) state, with some energy always degraded as heat. Plants and other photosynthesizers are an exception in that they can use the free energy in sunlight to turn lower energy compounds from the environment into higher energy compounds, but even they must then break down these low-entropy compounds to fuel their own metabolism.

An evolutionary biologist would rightly counter that in addition to the inevitable change of the environment that comes about with metabolism, most forms of Gaia (early hypothesis or later theory) postulate that life forms have traits that have been selected for their environment-altering properties. Daisyworld clearly invokes such traits—the blackness or whiteness of daisies in Daisyworld is (presumably) not an inevitable consequence of their being alive. These traits alter the local temperature of each daisy and its progeny, and they also alter the global temperature in the same direction. This means that traits that are good or bad at the level of individual selection are correspondingly good or bad for global temperature regulation (relative to the shared optimum of all the daisies). As many evolutionists have pointed out, this makes the model a special case, and one that is "designed" to give regulation. Many variants of the model have been created and they generally show regulation (Wood et al. 2007). The main exception is when the optimum growth temperature of the daisies is allowed to adapt unbounded, but that is an unrealistic scenario given the thermodynamic constraints on real biochemistry (Lenton and Lovelock 2000).

Interestingly, when the system is rewired so that black daisies produce white clouds that cool the global environment, regulation still occurs (Watson and Lovelock 1983). This is far from being an altruistic world—the white daisies are outcompeted and barely get a look in. However, evolutionary biologists have persisted in seeing Gaia as involving some form of altruism, and since William Hamilton's seminal work in the 1960s it has been clear that the conditions under which altruism can flourish are rather restrictive (Hamilton 1964, 1972). Furthermore, when Gaia hit the popular consciousness in the late 1970s, neo-Darwinists were struggling to rid their subject of arguments predicated on group selection. Gaia appeared to be perhaps the most extravagant example of altruism demanding higher level selection, consequently it was lumped in with the worst examples of arguments "for the good of the species," and summarily dismissed. In the last few years there has been a resurgence of interest in the evolution of cooperation (Nowak 2006) and multilevel selection (Sober and Wilson 1997; Goodnight 2005; Bijma et al. 2007). While by no means all of the evolutionary biology community approves of explanations involving selection above the level of the individual (or even above the gene), they are at least up for open discussion again. Thus, in some sense, Lovelock and other early proponents of Gaia were unlucky in their timing.

The final reason why evolutionary biologists and many other scientists don't like Gaia is the mythological name. Opinion varies considerably from scientist to scientist—for example, John Maynard Smith described it (in conversation with T.M.L.) as "an awful name for a theory," but Bill Hamilton had no objection and was happy to write a paper with "Gaia" in the title (Hamilton and Lenton 1998), so long as it was recognized that a Gaia system is a different class of system than an organism or a superorganism such as a termite mound. Unfortunately, Lovelock's likening of Gaia first to an organism and then to a super-organism agitated evolutionary theorists like Hamilton and Maynard Smith, because they wished to reserve the terms "organism" and "super-organism" for systems that can show adaptations due to natural selection. Kin selection can help explain the system level properties of the wasp nest and the bee colony, and it is for such systems that evolutionary biologists want to reserve the term "superorganism" (Hamilton 1964; Maynard Smith 1964).

## Current Status of the Debate

The crux of the present Gaia debate is: Do we need to invoke individual-level traits that have been selected for their environment-improving properties in order to account for observed Gaian properties of the global system? (And if we do, how can we avoid the problem of "cheats"?) Or, can we construct a reasonable theory of regulation based entirely on environment-altering properties that are simply by-products of metabolic traits selected for other reasons? The concept of by-products was independently introduced to the Gaia debate in the late 1990s by three different authors—Lenton (1998), Volk (1998), and Wilkinson (1999); and it was Volk (1998) who promoted the term "by-product." If regulation arises in a system built entirely on by-products, this makes Gaia theory much less vulnerable to criticisms from evolutionary theory. However, some authors are skeptical that regulation via by-products can occur, emphasizing recycling instead as the key Gaian property (Volk 1998).

To help understand the mechanisms at play in global regulation, one of us (T.M.L.) introduced the distinction between "feedbacks on growth" and "feedbacks on selection" (Lenton 1998). Feedback on growth occurs when a trait alters the environment in a way that affects the growth of its carriers and noncarriers in equal measure such that there is no change in the forces of selection. Feedback on selection occurs when a trait alters the environment in a way that affects the growth of its carriers and non-carriers differentially, and thus influences its own selection. In either case the responsible trait may initially arise as a selectively neutral by-product. In the case where the trait generates only feedback on growth it will remain a selectively neutral by-product, since it can offer no individual-level selective advantage. However, if the trait generates feedback on selection, it may become adaptive or maladaptive depending on its environmental effects, and its neutral by-product status is lost.

In ecology the related concepts of "extended phenotype" (Dawkins 1983), "niche construction" (Odling-Smee et al. 2003), and "ecosystem engineering" (Jones et al. 1994) involve the idea that genes or organisms can shape their abiotic environment in a way that alters their selective environment. We think feedback on selection may be most relevant in systems with environmental heterogeneity at such intermediate scales. However, we begin our modeling by removing the possibility of feedback on selection.

## Our Search for Gaia

Our approach has been to build a system where environmental alteration is entirely a no-cost by-product of metabolism, and to see what
conditions lead to the emergence of nutrient recycling and regulation.
We base our model around bacteria, since they are ubiquitous, have
a 3.8 billion year pedigree, run the global biogeochemical cycles,
and are highly adaptable. For reasons of tractability (of analysis as
well as computation) we look at a model microcosm, rather than the
global macrocosm, but the microcosm approach has other advantages.
We believe that similar principles must be at play in systems of all
scales so that we can learn something useful about the macrocosm
from the microcosm. Microcosm studies also offer the possibility of
empirical testing in the laboratory, something that would clearly be
impossible with a global model. We incorporate evolution by modeling individuals that each have a genetic code, and allowing selection
pressures on these individuals to emerge from the dynamics of the
microbial ecosystem.

The flask model (figure 5.1) simulates an evolving population of
microbes suspended in a flask of liquid, and hence the name (Williams
and Lenton 2007a, b). There is a prescribed supply flux of different
nutrients into the flask and corresponding removal fluxes proportional
to the concentration of each nutrient in the flask. There are also non-
nutrient "abiotic" environmental variables. The flask is seeded with a
clonal population of model microbes. Each microbe has a genetic code
that prescribes its uptake and release patterns for nutrients, its preference for the abiotic environment, and its metabolic by-product effect
on abiotic environmental variables. We place only one genetic constraint on metabolism: an organism is not allowed to consume what it
excretes. However, an important constraint on metabolism is built into
the model in the form of a peaked metabolic rate function that declines
smoothly to zero as the state of the environment moves away from
optimal growth conditions. Reproduction is asexual, and at each reproduction event there is the possibility of random mutation. We remove
the possibility that individuals can differentially benefit from altering
their environment by assuming that the liquid environment inside each
flask is well mixed such that any environmental change is experienced
equally by all individuals.

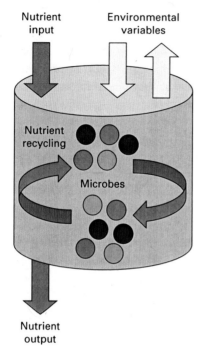

**Figure 5.1**
Schematic of the flask model.

## The Emergence and Disruption of Nutrient Recycling

The emergence of nutrient recycling loops is a robust result in our model system (Williams and Lenton 2007a). We measure recycling ratios (Volk 1998; Downing and Zvirinsky 1999) as the biological uptake flux of a given nutrient at each time step divided by the input flux of that nutrient to the flask. A value of 1 indicates efficient uptake of that nutrient. A value greater than 1 indicates that recycling is occurring. Initially we ran the model without any constraints on growth from the abiotic environment (and hence no environmental feedback) in order to concentrate on the emergence of nutrient recycling.

Selection acting on nutrient consumption traits causes the community to be dominated by specialist organisms adapted to take up a single nutrient (Williams and Lenton 2007a). This occurs because organism growth depends on the total quantity of nutrients consumed. Different nutrients are always taken up in fixed proportions set by the genetic code of each individual such that overall consumption is limited by scarcity of any required nutrient. Specializing on a single nutrient minimizes the risk of limitation. Nutrient release patterns do not directly affect the fecundity of individuals; therefore there is no selection of release traits. Organisms typically excrete a mixture of nutrients (excluding the one they consume).

In a typical run with four nutrients (figure 5.2), after initialization one nutrient is efficiently taken up and the population is limited by its supply. After around 8,000 time steps, adaptation of consumption traits allows a second nutrient to be efficiently utilized, some nutrient recycling of the two fully utilized nutrients begins, and the population increases. When further adaptation allows a third nutrient to be efficiently utilized (after around 18,000 time steps), the population size and recycling ratios rise again, but it is the efficient uptake of the fourth nutrient (after around 30,000 time steps) that really boosts the population size and recycling ratios. The population then reaches a carrying capacity that is set by the prescribed inputs of nutrients and the assumption of a fixed fraction of energy being lost as heat in metabolism.

When we introduce constraints on growth from the abiotic environment this leads to selection on the environmental preferences of the organisms (Williams and Lenton 2007a). Once again single-nutrient consumers dominate and efficient uptake and recycling of all nutrients becomes established. Typically, after some initial meandering, the environment settles in a fairly stable state and the community converges on shared preferences for that environment. There is no selection of abiotic effects on the environment because these cannot give differential benefit to individuals (because of the well-mixed shared environment). Consequently this part of the gene pool shows high diversity and genetic drift. Genetic drift can cause the environment to shift to a different state and the population's environmental preferences adapt in response.

As we tighten the environmental constraints on growth a new phenomenon emerges—the system becomes vulnerable to population crashes (figure 5.3). Although "cheating" (in the sense meant by evolutionary biologists) is not possible in a system built on by-products, "rebel" species can appear that disrupt the system by rapidly shifting the environment into an inhospitable state as they exploit previously unused nutrient stocks (Williams and Lenton 2007a). Vulnerability to such changes is worsened by genetic convergence of the population around a shared preferred environmental state. This convergence can cause a population

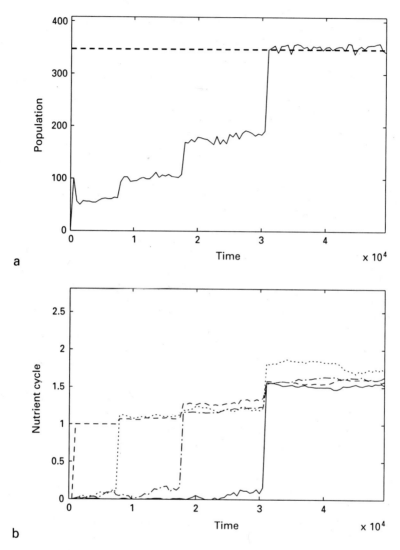

**Figure 5.2**
Establishment of nutrient recycling in a typical run of the flask model with no constraints from the abiotic environment: (a) Population size (solid line) and analytically derived carrying capacity (dashed line), (b) nutrient recycling ratios for the four nutrients in the system (each with a unique line type). Time is measured in $10^4$ time steps (the plots show 50,000 time steps). Image adapted from Williams and Lenton (2007a).

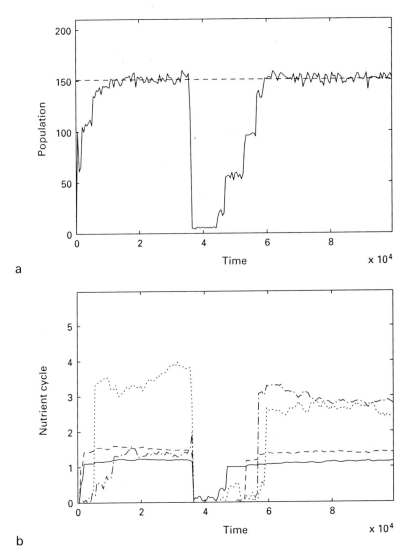

a

b

**Figure 5.3**
Establishment, collapse, and subsequent recovery of nutrient recycling in a run
of the flask model with relatively tight environmental constraints on growth:
(a) Population size (solid line) and analytically derived carrying capacity (dashed
line), (b) nutrient recycling ratios for the four nutrients in the system (each with
a unique line type). Time is measured in $10^4$ time steps, (the plots show 100,000
time steps). After around 35,000 time steps there is a crash in population size
and nutrient recycling caused by a "rebel" organism that utilizes an underused
nutrient while shifting the environment away from the state to which the
population is adapted. Image adapted from Williams and Lenton (2007a).

crash when the environment moves too far from the condition to which the population is adapted. Sometimes the population gradually adapts to the new environmental state and efficient nutrient consumption and recycling re-emerge, but in the worst case, extinction can occur. In this latter case members of the "rebel" species destroy the ecosystem and themselves with it—like an overvirulent parasite.

We ran large ensembles of runs for many versions of the model to quantify the robustness of the results (Williams and Lenton 2007a). The emergence of recycling is a robust result regardless of the tightness of constraints from the abiotic environment. The average recycling ratio typically asymptotes after around 50,000 time steps at a value that depends on the efficiency of conversion of nutrients into biomass during metabolism. For example, for a nutrient conversion efficiency of 60 percent, recycling ratio asymptotes at around 2, meaning that approximately the same amount of consumed material is from nutrient recycling as from external supply. Extinction rates increase with tightened constraints from the abiotic environment (figure 5.4). These rates count both

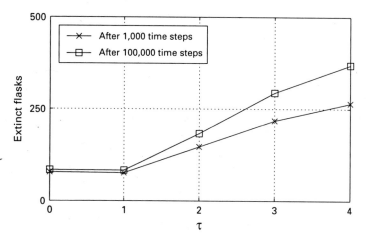

**Figure 5.4**
Number of extinct flasks in an ensemble of 500 runs as a function of the tightness of environmental constraints on growth (determined by the parameter $\tau$ in the model). The line with crossed markers shows initial extinctions due to communities being mismatched to their environment. The line with square markers shows total extinctions at the end of a long run. The difference between the lines indicates the internally generated extinctions due to "rebel" organisms during the model runs. The number of initial extinctions and internally generated extinctions both increase with tightening environmental constraints on growth.

extinctions at initialization because the seed community is unsuited to its environment and endogenously created extinction events caused by rebel species. However, even for very tight constraints from the abiotic environment, a significant fraction of systems survive.

In a single flask the abiotic environment can exhibit long intervals of quasi-stability with relatively small variation, punctuated by brief episodes of large shifts in state. However, this is not true regulation because the systems do not recover from perturbations. The dominant behavior in the single-flask system is robust nutrient cycling with stochastic variation in environmental state caused by genetic drift in the environment-altering traits of the population, sometimes with occasional discontinuities caused by population crashes when abiotic constraints on growth are tight.

## The Emergence of Environmental Regulation

To continue our search for environmental regulation in our virtual worlds, we constructed a spatial version of the flask model with multiple interconnected flask ecosystems (Williams and Lenton 2008). Typically we have ten flasks arranged in a ring with mixing between nearest neighbors. Mixing occurs at each time step by transferring a fixed volume of liquid between adjacent flasks. (Imagine simultaneously dipping and swapping cups of liquid from neighbor to neighbor.) This simple approximation to diffusive mixing transfers nutrients, abiotic factors, and organisms between the flasks. We retain perfect mixing within each flask but vary the degree of mixing between the flasks. Imperfect mixing between flasks introduces environmental heterogeneity into the global system: although the mixing process would eventually homogenize the global environment, it does not happen on a fast enough timescale to overcome the continual, differentiating metabolic activities of the local microbe populations.

We have explored this system with and without adaptation of the environmental preferences of the organisms. The same qualitative results emerge but the interpretation is much easier in the case where we switch off adaptation of environmental preferences and give all organisms the same fixed environmental preference. We measure the "environmental error" as the difference between the average environmental state and the shared preference of the organisms. Thus with fixed preferences, any reduction in environmental error must be due to the organisms collectively shifting the environment toward their preference. Local endogenous extinctions can still occur within individual flasks, as they did in the single-flask system. However, the rebel species responsible rarely succeed in spreading to destroy the global system. It is more usual to see instead the denuded flask being rapidly recolonized from its neighbors—an example of metapopulation dynamics (Hanski 1998).

More interestingly, for intermediate mixing rates we see a reduction in environmental error over time (figure 5.5), even when the system is perturbed by periodically changing the input fluxes of nutrients and the level of abiotic factors. The environmental error in an individual run is not reduced to zero, but instead the system undertakes behavior

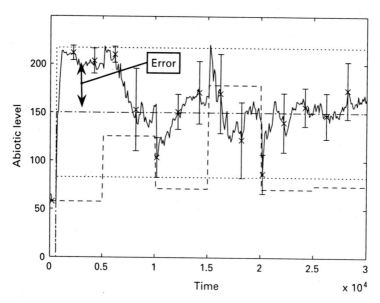

**Figure 5.5**
Case study of the emergence of environmental regulation in a spatial system of 10 flask ecosystems connected in a ring. The solid line with error bars denotes the actual state of the abiotic environment (mean value across all 10 flasks). The dashed line indicates the calculated abiotic environmental state in the absence of life. The dash-dot line indicates the universally shared microbial preference for the abiotic environment. The dotted lines indicate the bounds of the habitable range, found where metabolism exactly balances the maintenance costs of living. The system is subject to random perturbations to the external forcing of the abiotic environment every 5,000 time steps. The mismatch between the actual state of the environment and the preferred state generally reduces over time and the system counteracts perturbations.

approximating to bounded stochastic variation between the upper and lower limits of habitability. Here the bounds of habitability correspond to the environmental states at which metabolism brings in just enough energy to meet the essential maintenance costs of being alive. Across an ensemble of runs (figure 5.6), the mean error asymptotes to a value that represents the average distance from optimality of a system that is undertaking a random walk between the habitability bounds.

Measurements of the variation of several system metrics (population size, growth rate, nutrient availability) against environmental error suggest that we have a system with two regimes, one in which nutrients limit growth and the environment has no effect, and one in which the environment limits growth and nutrients are abundant. The environment-limited regime exists at and outside the bounds of habitability. When the

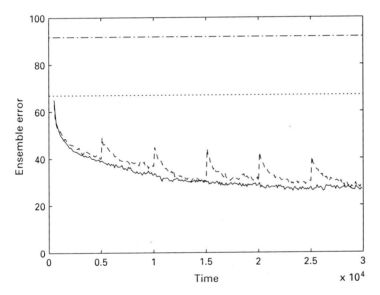

**Figure 5.6**
Reduction in environmental error over time averaged over ensembles of 500 unperturbed runs (solid line) and 500 perturbed runs (dashed line) (where figure 5.5 is an example of a perturbed run). The "error" is simply the magnitude of the difference between the preferred and the actual state of the abiotic environment. The dot-dash line indicates the expected size of the environmental error in the absence of life. The dotted line indicates the boundary of the habitable range, that on average the system would be in an uninhabitable state in the absence of life.

system strays outside these bounds then any ecosystem that collectively improves the environment (drags it back toward the habitable region) will experience positive feedback on growth, whereas any ecosystem that collectively degrades the environment (pushes it further away from the habitable region) will experience negative feedback on growth. However, this does not explain why we typically observe environment-improving local communities dominating the global system.

The observed dominance of environment-improving communities requires a mechanism by which they outcompete environment-degrading communities (figure 5.7). In the spatial flask model, this mechanism is selection at the level of the local ecosystem based on their differential rates of proliferation. Simply put, environment-improving communities become larger because they reduce the limiting environmental constraint on their growth, while environment-degrading communities become

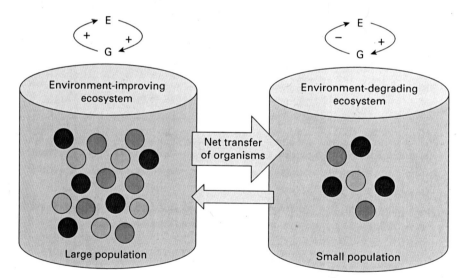

**Figure 5.7**
Schematic of the mechanism by which environment-improving ecosystems tend to dominate the global system. In the feedback loops at the top, *E* denotes the environmental state and *G* denotes growth. An ecosystem that collectively improves its environment generates positive feedback on growth increasing its population. An ecosystem that degrades its environment generates negative feedback on growth restricting its population size. Mixing between flasks occurs by an exchange of equal volumes of fluid and whatever microbes they contain. Thus the environment-improving ecosystem with higher population density tends to spread more rapidly than the environment-degrading ecosystem with lower population density.

smaller. Larger (and thus denser) communities spread more effectively than smaller ones because the fixed-volume transfer mechanism between flasks carries more individuals from a higher density population source than it does from a lower density population. This imbalance allows the members of environment-improving communities to eventually take over the global system, shifting the global environment toward the optimum for growth.

The hard-headed evolutionary biologist might suspect that the phenomena we observe in our model do not involve higher level selection operating on whole ecosystems, and that they may instead be explained by the implicit selection of a single "super-species" within an ecosystem that is alone responsible for improving the environment. Indeed we have seen such super-species in artificial ecosystem selection experiments with the flask model (Williams and Lenton 2007b). However, inspection shows that in the spatial system local communities are always highly diverse, with many species interacting in complex ways subject to individual level selection. There is high local diversity in values for the environment-altering traits, showing that these traits are selectively neutral at the individual level, but the global mean value of environment-altering traits clearly alters to counteract changes in external forcing. This demonstrates that selection pressure on these traits is active at some level, which our analysis shows to be the level of the ecosystem. We do not claim any long-term community-level adaptation (given the ongoing disruption from migration and mutation) but only the presence of selection acting over short timescales to promote the spread of communities that improve their environment over those that degrade their environment.

**Where Next?**

Our search for Gaia in the computer appears to have succeeded. The multiple-ecosystem flask model self-regulates (in the sense that it maintains a habitable environment and counteracts perturbations), despite being built on by-products. Ensembles of runs with the spatial model show a progressive improvement in the environment over time. These results are especially interesting given arguments by others questioning whether a global system built on by-products would tend to regulate (Volk 1998). The mechanism of regulation involves multiple levels of selection without needing selection at the level of the planet, and we believe that it represents a novel mechanism for generating environmental regulation. Although reproduction of an ecosystem is clearly less faithful in its replication than reproduction of an organism, there is still enough short-term heritability of collective environment-improving properties for communities with them to spread across the global system. Interestingly our mechanism involves shifting the localization of the environment to the ecosystem level so that our flask ecosystems become somewhat akin to the daisies of Daisyworld.

We have run the model with evolvable preferences and obtain qualitatively similar results. It is harder to disentangle what is causing the behavior when preferences evolve. This is because moving preferences toward the environment has an environmental error-reducing effect equivalent to that of moving the environment toward preferences. We have done extensive parameter sensitivity studies of the model system in both scenarios, and the results are robust. We have also reflected on how evolutionary biologists might critique the model. Clearly, we have built into the spatial model the necessary level of population structure for multilevel selection to operate. Environmental heterogeneity at some scale and corresponding selection of communities based upon it is critical to getting environmental regulation. However, such structure and environmental heterogeneity exist in the real world.

Having homogenized the local environment in the model, and banished environment-altering traits that cost their carriers, one next step would be to relax these assumptions. A more general model would allow the possibility of individual-level environmental alteration and selection of the traits responsible. Our contention is that there are cases in the real world where costly environment-altering traits are selected because they beneficially alter the immediate environment of the organism sufficiently to outweigh the cost and bring a net fitness increase. This scheme may be extended to improving the environment of one's offspring or other relatives, provided that the cost of such altruism is balanced by the benefit accruing to kin (Hamilton 1964). Niche construction (Odling-Smee et al. 2003) and its close relation, ecosystem engineering (Jones et al. 1994), describe cases where the environment-altering activities of an organism alter the selection pressures faced by the individual and its descendants. Examples where costly environmental alterations offer a selective benefit include the beaver's dam, ant colonies, termite mounds, and many more (Odling-Smee et al. 2003). There are also a host of phenomena in terrestrial ecosystems, such as sphagnum moss forming a peat bog that excludes trees, where it is unclear which mechanism to invoke to explain what is going on. Hamilton argued (personal communication to T.M.L.) that multilevel selection might play a role in shaping such systems. By allowing for a wider range of possibilities in the model, we could address what is the most likely explanation for specific scenarios in the real world.

**Wider Implications**

We have taken the constructive critics of Gaian regulation seriously and built a model with environmental alteration based only on by-products of metabolism. Recycling emerges as a robust Gaian property and in our spatial system environmental self-regulation also occurs. This suggests that what critics argue is the most "acceptable" version of Gaia theory may need revising. Even if "Gaia is life in a wasteworld of by-products" (Volk 2004), such a Gaia can still self-regulate, keeping the environment within habitable bounds and counteracting perturbations. A necessary condition for regulation is heterogeneity in the environment at some scale (in our model, between local flask ecosystems). This means we cannot apply the argument to truly well-mixed global variables such as the dominant gases in the atmosphere, but in principle, it could apply to any biologically-influenced non-nutrient variable that exhibits spatial gradients between different parts of the planet, such as aspects of the climate (e.g., temperature).

We also suggest that one should not be too dogmatic in focusing only on the version of the Gaia theory that is acceptable to critics—we strongly suspect that there are more than just by-products at play in shaping the real Gaia system. It is conceivable that an environment-altering trait may start life as a by-product but then be selected for its environmental effects, thus becoming adaptive. This was the argument put forward by Hamilton and one of us (T.M.L.) when thinking about the benefits of dimethyl sulphide emission and biogenic ice nucleation to aerial dispersal of spora (Hamilton and Lenton 1998).

In our model worlds, seemingly "cooperative" enterprises such as recycling and regulation robustly appear. In the spatial system, when the environment is limiting to growth, ecosystems or communities that "foul their nest" lose out to those that improve their local conditions. This has implications beyond our model, such as for life on the early Earth, for modern ecosystems, and for the human situation. There is no "tragedy of the commons" (Hardin 1968) in our model worlds. All individuals change their environment, but when the environment is limiting, environment-improving ecosystems or communities come to dominate.

It is pertinent to ask whether any parallels can be drawn between our model mechanism and the current predicament of human communities causing and experiencing climate change. Up until now, we humans have been a "rebel" species, altering the global environment as a by-product of more locally selected activities; for example, carbon dioxide emissions are a by-product of fossil fuel burning to produce energy. The "community-selection" mechanism cannot help solve the problem of rising carbon dioxide levels because carbon dioxide is a globally well-mixed variable. In this case the "tragedy of the commons" applies: communities that lower their carbon dioxide emissions will not see a differential climate benefit to those that increase theirs. However, if communities use localized mechanisms to lower their temperature, then there may be some scope for them to feel a differential benefit. Already there is inadvertent aerosol cooling due to fossil fuel (especially dirty coal) burning, which is masking greenhouse warming across some regions. If aerosol cooling were taken away, these regions would experience the greenhouse warming unmasked. In California, measures are being implemented to make building roofs and vehicles more reflective in order to counteract the combined global warming and urban heat island effects. More radically the prospect of climate engineering has been proposed, in particular, injecting aerosol into the stratosphere to cool the surface. If nations, for example, chose to deploy such strategies for their own benefit, this could introduce interesting dynamics in the global system. Whether such strategies are "selected" depends on their cost relative to the savings from avoided climate change damages. However, it is now clear that climate change damages could be expensive enough to restrict the growth of economies (Stern 2006), raising the possibility that a form of economic "community selection" could occur. The communities selected would be those that deploy strategies to keep themselves cool that cost less than the climate damages they manage to avoid. The community could be at a range of scales, from villages to continents, depending on the strategy used.

## Conclusion

Our search for Gaia in the computer has proved illuminating. In our virtual worlds, recycling and self-regulation of the environment within habitable bounds robustly emerge. If we draw an analogy between our model and microbial life on the early Earth, then we have some confidence that once life got started it would have soon solved the problem of nutrient recycling. Once there was a population of bacterial ecosystems, perhaps in ponds in different meteorite craters, or in different gyres of the ocean, or around different hydrothermal vents, environmental self-regulation could have emerged. Self-regulation would have worked best for somewhat heterogeneous environmental variables. We are still a long way from addressing the question (raised at the start of this chapter) of how probable a regulating and recycling Gaia system like the one we inhabit today might be, given the existence of life on a planet and 4 billion years of evolution. It is also important to remember the limitations of computer modeling and that artificial life *in silico* is fundamentally different from biological life. However, with these caveats in place, we nonetheless have increased confidence that a self-regulating "microbial Gaia" is a probable outcome once life gets started on a planet.

## References

Bijma, P., W. M. Muir, and J. A. M. Van Arendonk. 2007. Multilevel selection. 1: Quantitative genetics of inheritance and response to selection. *Genetics* 175: 277–88.

Dawkins, R. 1983. *The Extended Phenotype*. Oxford: Oxford University Press.

Doolittle, W. F. 1981. Is Nature really motherly? *CoEvolution Quarterly* (spring): 58–63.

Downing, K. L. 2002. The simulated emergence of distributed environmental control in evolving microcosms. *Artificial Life* 8 (2): 123–53.

Downing, K. L. 2004. Gaia in the machine: The artificial life approach. In S. H. Schneider, J. R. Miller, E. Crist, and P. J. Boston, eds., *Scientists Debate Gaia: The Next Century*. Cambridge: MIT Press, pp. 267–80.

Downing, K., and P. Zvirinsky. 1999. The simulated evolution of biochemical guilds: Reconciling Gaia theory and natural selection. *Artificial Life* 5 (4): 291–318.

Goodnight, C. J. 2005. Multilevel selection: The evolution of cooperation in non-kin groups. *Population Ecology* 47: 3–12.

Hamilton, W. D. 1964. The genetical evolution of social behaviour. *Journal of Theoretical Biology* 7: 17–52.

Hamilton, W. D. 1972. Altruism and related phenomena, mainly in social insects. *Annual Review of Ecology and Systematics* 3: 193–232.

Hamilton, W. D., and T. M. Lenton. 1998. Spora and Gaia: How microbes fly with their clouds. *Ethology Ecology and Evolution* 10: 1–16.

Hanski, I. 1998. Metapopulation dynamics. *Nature* 396: 41–49.

Hardin, G. 1968. The tragedy of the commons. *Science* 162: 1243–48.

Jones, C. G., J. H. Lawton, and M. Shachak. 1994. Organisms as ecosystem engineers. *Oikos* 69: 373–84.

Lenton, T. M. 1998. Gaia and natural selection. *Nature* 394: 439–47.

Lenton, T. M. 1999. Testing Gaia theory with artificial life. In D. Floreano, J.-D. Nicoud, and F. Mondada, eds., *Proceedings of the 5th European Conference on Artificial Life*. Berlin: Springer, pp. 9–10.

Lenton, T. M. 2004. Clarifying Gaia: Regulation with or without natural selection. In S. H. Schneider, J. R. Miller, E. Crist, and P. J. Boston, eds., *Scientists Debate Gaia: The Next Century*. Cambridge: MIT Press, pp. 15–25.

Lenton, T. M., and J. E. Lovelock. 2000. Daisyworld is Darwinian: Constraints on adaptation are important for planetary self-regulation. *Journal of Theoretical Biology* 206 (1): 109–14.

Lovelock, J. E. 1965. A physical basis for life detection experiments. *Nature* 207 (4997): 568–70.

Lovelock, J. E. 1983. Gaia as seen through the atmosphere. In P. Westbroek and E. W. de Jong, eds., *Biomineralization and Biological Metal Accumulation* Dordrecht: Reidel, pp. 15–25.

Lovelock, J. E., and L. M. Margulis. 1974a. Atmospheric homeostasis by and for the biosphere: the Gaia hypothesis. *Tellus* 26: 2–10.

Lovelock, J. E., and L. M. Margulis. 1974b. Homeostatic tendencies of the Earth's atmosphere. *Origins of Life* 5: 93–103.

Margulis, L., and J. E. Lovelock. 1974. Biological modulation of the Earth's atmosphere. *Icarus* 21: 471–89.

Maynard Smith, J. 1964. Group selection and kin selection. *Nature* 201: 1145–47.

McDonald-Gibson, J., J. G. Dyke, E. A. Di Paolo, and I. R. Harvey. 2008. Environmental regulation can arise under minimal assumptions. *Journal of Theoretical Biology* 251: 653–66.

Nowak, M. A. 2006. Five rules for the evolution of cooperation. *Science* 314: 1560–63.

Odling-Smee, F. J., K. N. Laland, and M. W. Feldman. 2003. *Niche Construction: The Neglected Process in Evolution*. Princeton: Princeton University Press.

Schrödinger, E. 1944. *What is Life?* Cambridge: Cambridge University Press.

Schwartzman, D. 1999. *Life, Temperature and the Earth: The Self-organizing Biosphere*. New York: Columbia University Press.

Sober, E., and D. S. Wilson. 1997. *Unto Others: The Evolution and Psychology of Unselfish Behaviour*. Cambridge: Harvard University Press.

Staley, M. 2002. Darwinian selection leads to Gaia. *Journal of Theoretical Biology* 218 (1): 35–46.

Stern, N. 2006. *The Economics of Climate Change: The Stern Review.* Cambridge: Cambridge University Press.

Volk, T. 1998. *Gaia's Body—Toward a Physiology of the Earth.* New York: Copernicus.

Volk, T. 2004. Gaia is life in a wasteworld of by-products. In S. H. Schneider, J. R. Miller, E. Crist, and P. J. Boston, eds., *Scientists Debate Gaia: The Next Century.* Cambridge: MIT Press, pp. 27–36.

Watson, A. J. 2004. Gaia and observer self-selection. In S. H. Schneider, J. R. Miller, E. Crist, and P. J. Boston, eds., *Scientists Debate Gaia: The Next Century.* Cambridge: MIT Press, pp. 201–208.

Watson, A. J., and J. E. Lovelock. 1983. Biological homeostasis of the global environment: The parable of Daisyworld. *Tellus* 35B: 284–89.

Wilkinson, D. M. 1999. Is Gaia really conventional ecology? *Oikos* 84: 533–36.

Williams, H. T. P., and T. M. Lenton. 2007a. The flask model: Emergence of nutrient-recycling microbial ecosystems and their disruption by environment-altering "rebel" organisms. *Oikos* 116 (7): 1087–1105.

Williams, H. T. P., and T. M. Lenton. 2007b. Artificial selection of simulated microbial ecosystems. *Proceedings of the National Academy of Sciences* 104 (21): 8918–23.

Williams, H. T. P., and T. M. Lenton. 2008. Environmental regulation in a network of simulated microbial ecosystems. *Proceedings of the National Academy of Sciences* 105: 10432–37.

Wood, A. J., G. J. Ackland, J. Dyke, H. T. P. Williams, and T. M. Lenton. 2007. Daisyworld: A review. *Reviews of Geophysics* 46: RG1001.

# 6

# Forest Systems and Gaia Theory

H. Bruce Rinker

Land, then, is not merely soil... [I]t is a sustained circuit, like a slowly augmented revolving fund of life.
Aldo Leopold, *Sand County Almanac*, 1949

Gaia theory relies on the notion of feedback (Margulis and Sagan 1997; Rapport et al. 1998). Clearly, for forest systems an association exists between nutrient concentrations of tree foliage and the nutrient content of forest soils (Innes 1993). The mineral nutrition of trees closely reflects the availability of nutrients in specific soils that is determined by the composition of bedrock. Soil quality has long been known to influence forest productivity (Innes 1993; van Breeman 1992; Binkley and Giardina 1998): "Soil supports the plants and animals that in turn create and maintain the myriad hidden processes that translate into soil productivity" (Maser 1993). At the same time many lines of evidence corroborate the effects of trees on various soil types (Binkley and Giardina 1998), including those in marginal or degraded lands. Thus forests and soils are coupled feedback systems wherein changes in one element affect changes in the other, which subsequently feed back on the original change (van Breeman 1992).

## Forest Systems and Gaia Theory

In this chapter we will see that canopy insects and soil fauna—tiny organisms such as springtails and mites—interact via inputs from the "upstairs" to the "downstairs." That story of interactions will be told through two wide-ranging studies: one in the temperate forests of western North Carolina, the second in the eastern rainforests of Puerto Rico. Until these studies the complex interactions between "upstairs" and "downstairs" processes had not been quantified sufficiently in temperate and tropical systems. The feedback mechanisms inherent in Gaia theory that operate between forest canopies and forest soils, though mostly unformulated at present, may prove to be important considerations for our conservation strategies on local and global levels. No matter the scale of our scrutiny, canopy herbivores and soil fauna may indeed be the warp and woof for the ecological circuit of forests.

## Ecological Links between Canopy and Ground

Canopy processes, such as decomposition and nutrient cycling, are coupled to those of the forest floor through inputs of leaf and twig litter, rainwater, and insect droppings (Schowalter and Sabin 1991; Schowalter et al. 1991; Lovett and Ruesink 1995). Herbivory by insects in forests may impact primary productivity and nutrient cycling (Mattson and Addy 1975; Kitchell et al. 1979; Schowalter et al. 1991) and even alter foliar chemistry (e.g., Feeny 1970; Mattson 1980; Swank et al. 1981; Schultz and Baldwin 1982; Ritchie et al. 1998; Stadler and Michalzik 2000). Phytophagous, or leaf-eating, insects in forest canopies drop materials into the soil community through two major pathways. First, herbivores introduce frassfall (insect excreta), greenfall (fragmented leaf tissue dropped during herbivory), and leaves abscised prematurely to the forest floor (Schowalter and Sabin 1991; Schowalter et al. 1991; Risley and Crossley 1993; Lovett and Ruesink 1995; Fonte and Schowalter 2004). Second, throughfall (rainwater modified by its passage through the forest canopy) is altered by the combination of dissolved frassfall and modified leachates, or dissolved organic material, from damaged leaves (e.g., Stadler and Michalzik 2000; Reynolds and Hunter 2001). These pathways sometimes combine to introduce increased amounts of carbon (C), nitrogen (N), and phosphorus (P) into the soil community (Reynolds et al. 2000; Reynolds and Hunter 2001). Increased activity by mites, springtails, and other soil fauna, comminuting these herbivore-derived inputs in the soil, then produce increased levels of leaf litter decomposition. Ecological links between canopy herbivores and soil fauna, however, such as herbivory and decomposition, have long remained unquantified (Schowalter et al. 1986; Risley and Crossley 1993; Reynolds and Hunter 2001; Rinker 2004a).

Organisms that comprise the decomposer food web include microflora (e.g., bacteria and fungi), microfauna (e.g., protozoa and nematodes), mesofauna (e.g., mites and springtails, also known as microarthropods), and macrofauna (e.g., earthworms and termites). Numerous researchers (e.g., Seastedt 1984; Moore 1988; Moore et al. 1988, 2003; Laakso and Setälä 1999; Wardle 2002) have summarized the integral roles of soil mesofauna in ecosystem processes such as decomposition and nutrient cycling. Unlike the "green food web" aboveground, where the primary drivers are the autotrophs, the "brown food web" on the forest floor is dominated by heterotrophic organisms. Either these multiscaled organisms break down complex carbohydrates in plant-derived detritus directly, mineralizing the associated nutrients by converting them from an organic to an inorganic state, or they govern microbial processes by feeding upon microbes and each other (Wardle 2002). The autotrophs determine the amounts of carbon entering the food web, but the heterotrophs are responsible for governing the availability of nutrients required for plant productivity: thus two subsystems locked into a kind of obligate mutualism. Because of the vast range of body sizes of soil fauna, they determine soil processes across broad spatial scales (Moore et al. 1988; Laakso and Setälä 1999; Wardle 2002). Soil microbes often perform the initial breakdown of organic material, acting upon substrates to concentrate nutrients, modify toxic and recalcitrant substrates, and make substrates more palatable (Moore et al. 1988). After bacteria and fungi colonize the litter, arthropod-mediated comminution helps to accelerate microbial activity and decomposition via facilitative succession (Moore et al. 1988).

Forest soil mesofauna, sometimes called soil plankton (Johnston 2000), consume fungi and bacteria as they comminute detritus, thereby affecting primary production, decomposition, nutrient cycling, microbial structure and activity, and food-web stability (Moore et al. 1988; Heneghan et al. 1999; Rinker 2004b). Ecologically, collembolans or springtails are classified as fungivores (Wardle 2002), but are also consumers of decaying vegetation and associated microbes in ways that defy exact placement in trophic groups. Most authors agree, however, that these opportunistic microarthropods, as "*r*-selected" specialists,[1] play an important role in rhizosphere dynamics (Coleman and Crossley 1996). As "*K*-selected" specialists,[2] mites (especially the oribatids) are mixed feeders in detrital food webs. Oribatid mites are usually fungivores or detritivores (Woolley 1960; Wallwork 1983). More numerous in temperate climates than in tropical ones, prostigmatid mites have several feeding habits: many species are predatory, but some are fungivorous or feed on microbes (Luxton 1981). The mesostigmatid mites are nearly all carnivorous (Hunter and Rosario 1998). Pseudoscorpions, of course, are miniature predators in forest litter layers; more numerous in tropical and sub-tropical climates than in temperate ones, these diminutive cryptozoans prey on worms, mites, and other small arthropods (Coleman and Crossley 1996). Thus springtails and oribatids fragment forest litter that then provides new surface areas for microbial colonization. Prostigma-tids, mesostigmatids, pseudoscorpions, and other upper level mesofauna affect the densities of the collembolans and oribatids, whose populations may in turn influence rates of decomposition.

The hypothesis for the projects in North Carolina and Puerto Rico was simple and straightforward: herbivore-derived inputs from the canopy influence the decomposition of plant materials on the forest floor. Further the researchers predicted that changes in frassfall, greenfall, and throughfall from herbivory change the decomposition rates and the abundance and diversity of soil mesofauna. Manipulating the pathways between the "green food web" and the "brown food web" (viz., addi-tions or exclusions of frassfall, greenfall, and throughfall) helped to quantify the ecological links between aboveground and belowground components of temperate and tropical forests and, thereby, demon-strated their relevance for the systems' overall health and conservation (Rinker 2004a). Such links between the "green food web" and the "brown food web" also hint at Gaian-type regulation of forest ecosys-tems. Canopy leaves provide sugars for the ever-hungry herbivores that then produce a plethora of materials to nourish the decomposers far below on the forest floor, allowing them in turn to influence the trees that produce the sugar-filled leaves.

## A Temperate Forest in North Carolina

Barbara C. "Kitti" Reynolds and Mark D. Hunter, researchers from the University of Georgia's Institute of Ecology, conducted a comprehensive ecological study over several years at the Coweeta Hydrologic Labora-tory operated by the US Forest Service (see Reynolds and Hunter 2001; Reynolds et al. 2003). The laboratory is located in the Nantahala Mountain Range of western North Carolina (N35°03′ W83°25′) within the Blue Ridge Physiographic Province (Swank and Crossley 1988). The researchers selected three sites with similar physical aspects and vegeta-tion, but ranging in elevation from 795 m (low elevation) through 1,000 m (mid-elevation) to 1,347 m (high elevation) (Reynolds and Crossley 1995). Common tree species on the low elevation, north-facing, cove terrain site included *Liriodendron tulipifera* (tulip poplar, Magnoliaceae), *Acer rubrum* (red maple, Aceraceae), *Quercus rubra* (northern red oak, Fagaceae), *Betula lenta* (sweet birch, Corylaceae), and *Carya* sp. (hickory, Juglandaceae). The mid-elevation mixed-oak forest faced northeast and was dominated by *A. rubrum*. Other important tree species were *Q. rubra*, *Q. prinus* (chestnut oak, Fagaceae), *B. lenta*, *A. pennsylvanicum* (striped maple, Aceraceae), and *Tsuga canadensis* (eastern hemlock, Pinaceae) with *Rhododendron maximum* (great rhododendron, Ericaceae) as the primary understory. The high-elevation site was a northern hardwood stand that also faced northeast. The most common trees were *Q. rubra*, *A. rubrum*, *A. pennsylvanicum*, and *B. lutea* (yellow birch, Corylaceae) with an understory of *R. maximum*, *R. calendulaceum* (flame azalea, Ericaceae), and *Clethra acuminata* (sweet pepper bush, Clethraceae). Soil types, annual precipitation, and average temperatures also varied with elevation.[3]

Litterbags and Tullgren extractors are standard equipment for quantitatively sampling leaf litter for microarthropods (Crossley and Hoglund 1962). Litterbags containing *Q. rubra* and *A. rubrum* litter were placed at the three elevations along a moisture/productivity gradient and sampled monthly for two years (see Crossley and Hoglund 1962). Microarthropods, nematodes, and litter mass loss responses to the productivity gradient were measured. The relative abundance of springtails and mites was compared across the gradient. Herbivore inputs simulating the effect of canopy herbivory on soil processes included frassfall additions, throughfall additions, greenfall exclusions, and total litter exclusions as experimental treatments. Treatment did not have a significant effect on decomposition rates, but mass loss was greatest at the middle and high elevations and greater on two-year-old litter than on one-year-old litter. Nematode densities were also greater on the older litter. Experimental additions of frassfall to plots on the low- and mid-elevation sites resulted in an increase in springtail abundance in litterbags from those plots. Plots with frassfall and throughfall additions also showed increased numbers in some types of nematodes (bacterial and fungal feeders) in some months. Numbers of some types of mites (i.e., oribatids and prostigmatids) were reduced in litter exclusion plots. Thus results from the North Carolina study suggest not only significant influences of elevation on litter decomposition, and the abundance and diversity of soil fauna, but positive correlations between canopy herbivory and responses in the population densities of forest floor biota.

## A Tropical Forest in Puerto Rico

Steven J. Fonte, at the time a researcher from Oregon State University, and I conducted this study in Puerto Rico's Caribbean National Forest near the El Verde Field Station at roughly 400-m elevation. While Steve focused on nutrient cycling, I examined the links between canopy herbivory and soil microarthropods such as mites and springtails (see Rinker 2004a). Because of constraints of time, budget, and staffing we studied a single location rather than multiple elevation sites. The El Verde station is located in the northwestern portion (N18°10′ W65°30′) of Luquillo Experimental Forest, a long-term ecological research site (LTER) within the national forest in eastern Puerto Rico. Prior to its designation as public land around 70 years ago, the site was part of a small farm for coffee, fruit, and charcoal production. Vegetation is shallow-rooted with the greatest root biomass in the upper 10 cm of soil. The subtropical wet forest is dominated by *Sloanea berteriana* (motillo, Elaeocarpaceae), *Dacryodes excelsa* (tabonuco, Burseraceae), *Prestoea montana* (sierra palm, Arecaceae), and *Casearia arborea* (rabo ratón, Flacourtiaceae). The ecosystem is classified as subtropical moist and subtropical wet forest in the Holdridge Life Zone System (Holdridge 1947; McDowell et al. 1996). Its soil type, annual precipitation, and average temperature varied dramatically from the North Carolina site.[4]

Litterbag samples, filled with recently senesced leaves of *D. excelsa*, were measured for mass loss due to decomposition at six sample dates through a 36-week treatment period at a single elevation site. They were also analyzed for their abundance and diversity of springtails, three suborders of mites (oribatids, prostigmatids, and mesostigmatids), pseudoscorpions, nematodes, and "other" soil mesofauna. As herbivore-derived inputs, additions of frassfall, greenfall, and throughfall (or, collectively, the experimental treatments) promoted the abundance and diversity of some soil microarthropods and other mesofauna. No significant treatment effects were observed, however, on litter decomposition. During the sample period numbers of most organisms increased except at the transition between dry and wet seasons; the numbers of mesostigmatids and "other" mesofauna, however, continued to rise. A positive response was observed among total mesofauna to frassfall additions. Frassfall also had a dramatic positive effect on the densities of microarthropods relative to those of the control groups. Pseudoscorpions increased in response to throughfall additions. Numbers of nematodes were negligible, so it was difficult to ascertain a treatment effect on these organisms. In terms of treatment effect on estimated numbers of meso-fauna per meter square of forest litter, frassfall had the greatest relative impact, followed by greenfall and then throughfall. When contrasted against the controls, oscillations in the densities of some mesofauna in the treatment litterbags indicated a predator–prey feedback system, espe-cially for frassfall and greenfall additions. Further the densities of "other" mesofauna increased over time, especially for frassfall additions; impor-tantly, these "other" fauna included the larvae of numerous kinds of macroinvertebrates, suggesting detrital succession in the litterbag micro-cosms. Hence, like the North Carolina study, herbivore-derived inputs also play a significant role in the spatial and temporal dynamics of soil mesofauna in Puerto Rico. Such effects have important consequences for decomposition and, ultimately, for the health of the entire forest ecosystem.

**Discussion and Future Direction**

Wardle (2002) reported a wide range of consumer responses to plant productivity in decomposer food webs due to (1) hidden factors that co-vary with treatments, (2) context-dependent top-down or bottom-up ecological forces, (3) competition between plants and decomposers for carbon resources and other nutrients, and (4) active feedback systems between decomposers (not just donor-driven) and plants (not just passive providers). Terrestrial ecosystems consist of a producer subsystem and a decomposer subsystem (Wardle 2002). These plant and decomposer subsystems are obligate mutualists in carrying out processes required for the long-term maintenance of forests—and thus the health of both is an important consideration for ecological restoration and land-management practices. The experimental design of the projects in North Carolina and Puerto Rico partitioned different temporal and spatial components involved in the transfer of crown, or canopy, materials to the floor in temperate and tropical systems. Although these studies did not track the entire ecological loop (i.e., the feedback from forest floor to forest crown), their results helped quantify the first turn in the loop—the degree to which herbivory in the treetops influences soil mesofauna over a period of time. In the two experiments the positive effects of experimen-tally manipulating canopy frassfall, greenfall, and throughfall on soil invertebrates in temperate and tropical forests were unequivocal but varied by study site. Table 6.1 provides site comparisons for treatment effects from the North Carolina and Puerto Rico studies.

**Table 6.1**
Site comparisons for Coweeta Hydrological Station in North Carolina and Luquillo Experimental Forest in Puerto Rico

| Aspect | Coweeta | Luquillo |
|---|---|---|
| Location | N35°03' W83°25' | N18°10' W65°30' |
| Forest type | Deciduous | Tropical |
| Dominant canopy species | *Liriodendron tulipifera* *Quercus rubra* *Acer rubrum* *Carya* spp. | *Sloanea berteriana* *Dacryodes excelsa* *Prestoea montana* *Casearia arborea* |
| Elevation for experiment | Multiple (795, 1,000, and 1,347 m) | Single (350 m) |
| Average rainfall | 194 cm (low elevation) to 245 cm per year (high elevation) | 350 cm per year |
| Average temperature | 17°C to 20.5°C (July) | 21°C (Jan) to 25°C (May) |
| Decomposition rate | Moderate; no treatment effect | High; no treatment effect |
| Density for mites and springtails | Varied by date and elevation | 111 to 787 per g DW litter; 22,000 to 98,000 per m² litter (controls) |
| Frassfall addition | Increased collembola and prostigs | Increased total mesofauna and increased "other" mesofauna |
| Frassfall exclusion | NA | NA |
| Greenfall addition | NA | No treatment effects |
| Greenfall exclusion | Increased collembola at low elevation | NA |
| Throughfall addition | Increased collembola in August | Increased pseudoscorpions |
| Throughfall exclusion | NA | NA |
| Total litter exclusion | Decreased oribatids, more at mid- than at low elevation | NA |

Source: National Science Foundation grant DEB-9815133.

Experimental additions of canopy inputs to the research plots in North Carolina and Puerto Rico did not affect the rates of decomposition in forest litter. Of course, the sample period for the tropical site represented only a portion of a year, overlapping both the wet and dry seasons. These findings are consistent with reports from other researchers in temperate and tropical regions (e.g., Seastedt 1984). Canopy inputs did seem, however, to shape the abundance and diversity of mesofauna at both locations. Such findings may indeed be another confirmation from the field that ecological redundancy (a functional aspect of many detrital food webs for both aquatic and terrestrial systems in which multiple species seem to fill the same niche and, by definition, are individually superfluous; Laakso and Setälä 1999; Oldeman 2001) holds true on multiple taxonomic levels for soil mesofauna, particularly for the rainforests of Puerto Rico: decay proceeds steadily and at a fairly constant level, apart from canopy inputs and species composition. In other words, collective system physiology—not community morphology—may be primary to its ecological processes: it's what they do, not what they are, that may matter ultimately.

The densities of microarthropods per meter square of forest litter in both the control and the treatment groups in Puerto Rico exceeded previously reported estimates. Seastedt (1984) found densities of microarthropods in the tropics (and other areas with low amounts of soil organic material) to be less than 50,000 per meter square. Pfeiffer (1996) put forward an estimated 17,000 organisms per meter square for springtails and mites combined for Puerto Rico. He attributed this relatively low number to two factors: the scant levels of resources afforded by a relatively sparse litter layer and the abundant array of arthropod predators concentrated in a habitat with limited refuges for prey. Our experiment in Puerto Rico, however, reported higher densities of mesofauna (including microarthropods) per meter square of rainforest litter that generally increased over time, whether among the controls or the treatment groups, with some oscillations that may signal the emergence of a detrital successional community (table 6.1). A number of possibilities exist for the apparent discrepancy between the results of this study and that of other researchers: an enhanced leaf litter due to intervening hurricanes and tropical storms, the proximity of a nutrient-rich stream in the study site, abundant and heretofore undocumented refuges for prey species, and the existence of biological "hotspots" for soil mesofauna previously unmeasured. All of these possibilities deserve further evaluation.

When the densities among the treatment groups in Puerto Rico were compared to that of the controls, an intriguing pattern emerged. At the onset of the experiment, nearly all groups exhibited a decline in density except for the prostigmatids, whose numbers seemed enhanced by frass-fall and throughfall. Then the densities of the microarthropods started to oscillate, as compared to that of the controls, with vague initial resemblance to a standard Lotka–Volterra feedback model (see Odum 1971): collembolan and oribatid populations had a positive effect on the population sizes of predatory prostigmatids, mesostigmatids, and pseudoscorpions, which in turn had an inhibiting effect on their prey (figure 6.1).

For example, the positive correlation between throughfall and numbers of pseudoscorpions could have reflected, at least in part, the increased density of oribatids and other potential prey species that responded similarly to treatment inputs. Did treatment trigger (or even accelerate) predator–prey interplay as the litterbag communities became established? Toward the end of the tropical experiment, however, these oscillations in the litterbags—if in fact they continued—were obscured by the entry of "other" arthropods, including macrofauna such as immature or larval spiders, beetles, termites, isopods, and hemipterans, and thereby enhanced the complexity of the litterbag microcosms by providing for an advanced detrital succession (figure 6.2).

Soil moisture is usually considered the most important ecological factor for those species, such as springtails, that cannot withstand low humidity (Badejo and van Straalen 1993). As populations of these moisture-sensitive species declined in Puerto Rico through the dry season, so did many of the organisms that fed upon them. Exceptions included mesostigmatids and "other" arthropods that, accordingly, could have switched to other prey items. The data indicate that complexity in the detrital food web developed in the litterbag microcosms, eventually attracting macrofauna such as spiders, beetles, termites, isopods, and hemipterans to the system (or at least their larval stages). These larger organisms, residing in higher trophic levels, seemed less dependent on humidity and more dependent on other ecological co-variables in the forest litter for their survival. In the comminution of forest litter, then, bacteria are the colonizers, microarthropods the pioneers, and larger invertebrates a kind of staid occupier that enters the system only after it has been primed by its smaller cousins—corroborated here by the emerging picture in Puerto Rico of the spatial-temporal links between forest canopy and forest floor.

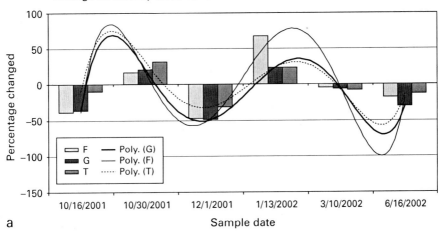

a

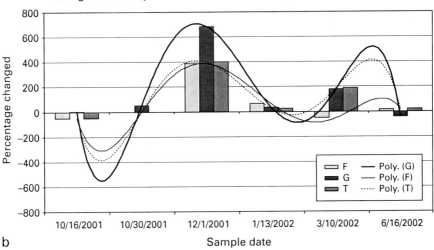

b

**Figure 6.1**

Comparison between oscillations in litterbag populations of collembolans (prey) and mesostigmatid (predator) mites from Puerto Rico, suggesting a Lotka–Volterra feedback model, especially in early detrital successional stages. Trend lines for frassfall (*F*), greenfall (*G*), and throughfall (*T*) manipulations are provided for purposes of illustration.

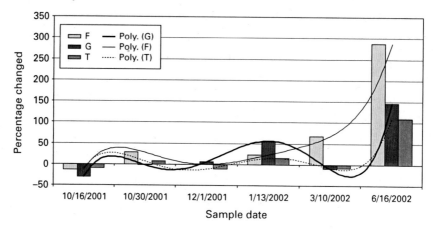

**Figure 6.2**
Changes in the litterbag populations of "other" mesofauna throughout the 36-week study in Puerto Rico, indicating an advanced stage of detrital succession as macrofauna enter the system to replace mesofauna; "other" mesofauna included the larvae of macrofauna. Trend lines for frassfall (F), greenfall (G), and throughfall (T) manipulations are provided for purposes of illustration.

Canopy inputs for Puerto Rico had other, somewhat ambiguous effects on the abundance and diversity of soil fauna. Frassfall tended to affect positive changes in abundance and diversity in comparison to other canopy inputs. This finding paralleled similar studies in the same tropical forest (e.g., Fonte 2003) and in North Carolina (e.g., Reynolds et al. 2003). Probably awash in bacteria and fungi and more readily decomposed than greenfall, insect excreta may present "islands of fertility" (Wardle 2002), a spatial variability of organisms and processes in the aboveground/belowground ecology of the forest that links herbivory to decomposition. In the tropical study, pseudoscorpions responded positively to throughfall additions. This may have reflected a hidden treatment effect on their major prey species (e.g., oribatids and larval forms of macroinvertebrates) more than any direct influence on the predators themselves. On the other hand, totals for collembolans and mites were not significantly affected by treatment additions. For this specific project the impact of canopy herbivory sometimes exceeded that of litter; that is, enhanced frassfall, greenfall, and throughfall showed positive effects on the abundance and diversity of soil mesofauna relative to the control groups. Somewhat ambiguous results point, however, to the need for a longer term study of the top-to-bottom ecological links in tropical forests. Further, given that Puerto Rico is a tropical island (with its inherent resource limitations, climate, and endemic species), alternative sites in locations such as the Amazon and Congo Basins for comparative studies are strongly recommended. Such research might allow the more general, or universal, links between herbivory and soil decomposition to emerge.

The results of these experiments in North Carolina and Puerto Rico have added to the growing evidence that forest canopies and soils indeed represent linked feedback loops through the processes of herbivory and decomposition. Any study of forest ecosystems encompassing just a few weeks or years cannot hope to address all the variation that occurs in the system due to changes in microclimates, populations of herbivores and their predators, interactions between microbes and soil nutrients, and a host of other hidden, or unknown, factors and processes. Such a comprehensive perspective necessitates long-term ecological study of these systems. On the other hand, the projects described herein demonstrated that variations in the canopy can have significant influence on the abundance and diversity of some soil invertebrates in temperate and tropical forests, though not necessarily on the rates of decomposition. Manipulations of frassfall, greenfall, and throughfall, as herbivore-derived inputs, can play a quantifiable role in the spatial and temporal dynamics of microarthropod populations. These are expected to have important consequences for soil processes, reverberating back to the canopy and, ultimately, affecting the health of the tropical forest ecosystem as a whole. Food webs often serve as a basis for the development of ecosystem models (especially for patterns of nutrient and energy flow) and can be used as efficacious tools for management and decision making (Johnston 2000). As feedback loops, they may also illustrate aspects of Gaia theory.

Numerous questions emerged from these studies that identify paths for future study, providing a map for researchers as we continue our analysis of the ecological links between the "green food webs" and the "brown food webs" in temperate and tropical systems. Microarthropods, particularly springtails and oribatid mites, probably contribute much to the aboveground food webs in temperate and tropical forests. What vertebrates and macroinvertebrates feed upon these microarthropods directly? How are their populations affected when vertebrates and macroinvertebrates consume their major predators? What are some of the effects of decomposer community structure on primary productivity in temperate and tropical forests? That is, how are soil mesofauna linked back quantitatively to plants and their canopy herbivores? Is ecological redundancy in edaphic communities (see Oldeman 2001) a reflection of scale? In other words, is redundancy a reality or just an artifact of spatial/temporal scale inherent in the study? Further, is ecological redundancy a hint of Gaian self-regulation of ecological services on a local scale? These questions about ecological redundancy are not meant to be spurious arguments to ameliorate the regretful consequences of human-caused extinction but an encouragement to learn whether such redundancy is a reality as it seems in other living systems (e.g., codons and protein synthesis). How can we best compare the strength and vigor of feedback loops? Are top-down influences more or less potent than bottom-up ones in temperate and tropical forests? How are these influenced by extrinsic factors such as temperature and moisture? And, finally, how can one best measure the overall health of any biological system, whether on an organismal or an ecological level? Both natural systems are open and thus vulnerable. Are species richness and trophic dynamics sufficient indicators of health for scales above the level of organism? What is the best way to quantify and then compare these ecological links (e.g., via their vigor, resiliency, and degree of trophic interconnectedness)?

## Conclusion: "Upstairs" and "Downstairs" Processes in Forest Systems

Few studies have directly compared ecological processes between tropical and temperate systems (but see Coley and Aide 1991; Lowman and Wright 1994). One major difficulty of across-site comparisons has been the lack of standardized protocols for field measurements. At a National Science Foundation workshop (1995) on database management, participants stressed the value of standardized protocols for accurate comparisons among sites. Another recognized difficulty is the strong influence that climate and vegetation exert on soil properties (Hobbie 1992). Tropical rainforests typically have low-nutrient soils while temperate forests generally have high-nutrient soils. Rates of nutrient allocation and demand, along with local climate and soil microbes, make across-site comparisons difficult. Plant and animal assemblages in tropical forests also exhibit extreme patterns when compared to those in temperate forests (MacArthur 1969), compounding any comparative ecological studies for these biomes. For example, because the tropics have more intense competition, predation, parasitism, and disease than temperate areas, a lower ceiling exists on the abundance of any given species, thus allowing more species to fit into equatorial forests.

Furthermore few studies have linked ecological processes between forest canopies and ground-level soils via feedback loops. Intuitively and anecdotally, scientists have noted numerous differences in the micro-environments for various vegetation strata yet few studies exist that quantify possible links among these layers. Especially intriguing is the apparent connection between canopy herbivory and soil decomposition in temperate and tropical forests. Insect herbivores exert enormous, well-documented pressure on photosynthetic tissue in treetops during out-break and nonoutbreak settings. They also produce numerous canopy inputs, including excrement and other droppings from their relentless foraging. Soil decomposers are vital for nutrient cycling. Canopy inputs seem essential for the ground-level nutrient cycle. If so, then conservationists must know something about this linkage for effective long-term forest management.

Canopy ecologists are shifting their emphasis from a descriptive aut-ecology of individuals to a more complex ecological approach including the development of conceptual models (Lowman and Wittman 1996; Reynolds and Hunter 2001). The complexity of temperate and tropical systems demands such an interdisciplinary avenue of study. Canopy researchers have predicted for some time that the emphasis of their work would adjust inevitably to address relationships between plants and animals (Schowalter et al. 1986) and between canopy and forest floor (Lowman and Wittman 1996). Ecological studies that link vegetation strata, address intrinsic differences in biota and seasonality, and compare temperate and tropical associations are not only timely studies; they are also indispensable ones for a pragmatic grasp of the energy circuit oper-ating in the world's forests. Such multifaceted linkages are embedded in Gaia theory from their local to global levels of association. Even among tiny canopy herbivores and soil decomposers munching away in temper-ate and tropical forests, evidence of Gaia is measureable and ever-evolving. More germane than reductive approaches to the complexities of forest ecology, Gaia theory provides an appropriate conceptual model for coupling the temporal-spatial aspects of forests in terms of intricate feedback loops among the subsystems operative within them.

## Notes

These studies were supported, in part, by the National Science Foundation (DEB-9815133), the Robert and Patricia Switzer Foundation, the New Hampshire Charitable Foundation, the Marie Selby Botanical Gardens (Sarasota, FL), Coweeta Hydrologic Laboratory (Otto, NC), El Verde Field Station (Luquillo,

Puerto Rico), and several private grants. Much appreciation is extended to Rita Aughey, Sunny Birdsong, Eileen Crist, Amanda Durbak, Steven J. Fonte, Steven J. Harper, Mark D. Hunter, Beth Kaplin, Donna J. Krabill, Saul Lowitt, Thomas E. Lovejoy, Margaret D. "Canopy Meg" Lowman, Virginia Miller, Lynn Margulis, Martin Ogle, Alonso Ramírez, Barbara C. "Kitti" Reynolds, Timothy D. Schowalter, Rachel Thiet, and the Pinellas County (FL) Department of Environmental Management.

1. An r-selected specialist—from the r term in the logistic equation—refers to those species whose populations are controlled primarily by density-independent factors. In general, such species have many offspring that are small, mature rapidly, and receive little or no parental care.

2. A K-selected specialist—from the K term in the logistic equation—refers to those species whose populations are controlled primarily by density-dependent factors. In general, such species have few offspring that are large, mature slowly, and often receive intensive parental care; consequently K-selected specialists tend to be more vulnerable to extinction than r-selected specialists.

3. At the low-elevation site soils were typic and humic hapludults, at the mid-elevation site typic hystrochrepts, and at the high-elevation site typic haplubrepts. Mean annual precipitation increased from 193.9 cm for the low-elevation site to 245.08 cm for the high-elevation site. Temperature decreased by 1°C to 2°C during the growing season between the low- and mid-elevation locations, and another 2°C between the mid- and high-elevation sites (20.5°, 19°, and 17°, respectively, as the mean hourly reading in July).

4. Soils are generally acidic clays with nutrient content typical of tropical montane forests (McDowell 1998). They are dominated by a zarzal-cristal complex with deep oxisols of volcanic origin and are mostly well-drained clays and silty clay loams. Hydrologic exports of N, P, and dissolved organic C are modest, and weathering rates are high in many of the montane catchments (McDowell et al. 1996; McDowell 1998).The area receives approximately 350 cm annual precipitation (mostly orographic rains or precipitation from tropical storms and hurricanes) that varies seasonally with 20 to 25 cm per month from January through April (dry season) and 35 cm per month during the remainder of the year. Precipitation exceeds evapotranspiration in all months (Fonte 2003). The average monthly temperature varies from 21°C in January to 25°C in August and September (Waide and Reagan 1996).

## References

Badejo, M. A., and N. M. van Straalen. 1993. Seasonal abundance of springtails in two contrasting environments. *Biotropica* 25: 222–28.

Binkley, D., and C. Giardina. 1998. Why do tree species affect soils? The warp and woof of tree-soil interactions. *Biogeochemistry* 42: 89–106.

Coleman, D. C., and D. A. Crossley. 1996. *Fundamentals of Soil Ecology*. San Diego: Academic Press.

Coley, P. D., and T. M. Aide. 1991. Comparisons of herbivory and plant defenses in temperate and tropical broad-leaved forests. In P. W. Price, T. M. Lewinsohn, G. W. Fernandes, and W. W. Benson, eds., *Plant-Animal Interactions: Evolutionary Ecology in Tropical and Temperate Regions.* New York: Wiley, pp. 25–49.

Crossley, D. A., and M. P. Hoglund. 1962. A litter-bag method for the study of microarthropods inhabiting leaf litter. *Ecology* 43 (3): 571–73.

Fenny, P. 1970. Seasonal changes in oak leaf tannins and nutrients as a cause of spring feeding by winter moth caterpillars. *Ecology* 51 (4): 565–81.

Fonte, S. J. 2003. The influence of herbivore generated inputs on nutrient cycling and soil processes in a lower montane tropical rain forest of Puerto Rico. Master's thesis. Oregon State University, Corvallis.

Fonte, S. J., and T. D. Schowalter. 2004. Decomposition of greenfall vs. senescent foliage in a tropical forest ecosystem in Puerto Rico. *Biotropica* 36 (4): 474–82.

Heneghan, L., D. C. Coleman, X. Zou, D. A. Crossley, and B. L. Haines. 1999. Soil microarthropod contributions to decomposition dynamics: Tropical-temperate comparisons of a single substrate. *Ecology* 80 (6): 1873–82.

Hobbie, S. E. 1992. Effects of plant species on nutrient cycling. *Trends in Ecology and Evolution* 7: 336–39.

Holdridge, L. R. 1947. Determination of world plant formations from simple climatic data. *Science* 105: 367–68.

Hunter, P. E., and R. M. T. Rosario. 1998. Associations of mesostigmata with other arthropods. *Annual Review of Entomology* 33: 393–417.

Innes, J. L. 1993. *Forest Health: Its Assessment and Status.* Wallingford, UK: CAB International.

Johnston, J. M. 2000. The contribution of microarthropods to aboveground food webs: A review and model of belowground transfer in a coniferous forest. *American Midland Naturalist* 143: 226–38.

Kitchell, J. F., R. V. O'Neill, D. Webb, G. A. Gallep, S. M. Bartell, J. F. Koonce, and B. S. Ausmus. 1979. Consumer regulation of nutrient cycling. *Bioscience* 29: 28–34.

Laakso, J., and H. Setälä. 1999. Sensitivity of primary production to changes in the architectures of belowground food webs. *Oikos* 87: 57–64.

Leopold, A. 1949. *A Sand County Almanac.* Oxford: Oxford University Press.

Lovett, G. M., and A. E. Ruesink. 1995. Carbon and nitrogen mineralization from decomposing Gypsy moth frass. *Oecologia* 104: 133–38.

Lowman, M. D., and P. K. Wittman, 1996. Forest canopies: Methods, hypotheses, and future directions. *Annual Review of Ecology and Systematics* 27: 55–81.

Lowman, M. D., and S. J. Wright. 1994. A comparison of herbivory in the rain forest canopies of Panama and Australia. *Selbyana* 15 (2): A14.

Luxton, M. 1981. Studies on the prostigmatic mites of a Danish beech wood soil. *Pedobiologia* 22: 277–303.

MacArthur, R. H. 1969. Patterns of communities in the tropics. *Biological Journal of the Linnaean Society* 1 (April): 19–30.

Margulis, L., and D. Sagan. 1997. *Slanted Truths: Essays on Gaia, Symbiosis, and Evolution.* New York: Springer.

Maser, C. 1993. Unexpected harmonies: Self-organization in liberal modernity and ecology. *Trumpeter* 10 (2). http://trumpeter.athabascau.ca/index.php/trumpet/article/view/394/631.

Mattson, W. J. 1980. Herbivory in relation to plant nitrogen content. *Annual Review of Ecology and Systematics* 11: 119–61.

Mattson, W. J., and N. D. Addy. 1975. Phytophagous insects as regulators of forest primary production. *Science* 190: 515–22.

McDowell, W. H. 1998. Internal nutrient fluxes in a Puerto Rican rain forest. *Journal of Tropical Ecology* 14: 521–36.

McDowell, W. H., C. P. McSwiney, and W. B. Bowden. 1996. Effects of hurricane disturbance on groundwater chemistry and riparian function in a tropical rain forest. *Biotropica* 28 (4a): 577–84.

Moore, J. C. 1988. The influence of microarthropods on symbiotic and non-symbiotic mutualism in detrital-based below-ground food webs. *Agriculture, Ecosystems, and Environment* 24: 147–59.

Moore, J. C., K. McCann, H. Setälä, and P. C. de Ruiter. 2003. Top-down is bottom-up: does predation in the rhizosphere regulate aboveground dynamics? *Ecology* 84 (4): 846–57.

Moore, J. C., D. W. Walter, and H. W. Hunt. 1988. Arthropod regulation of micro- and mesobiota in below-ground food webs. *Annual Review of Entomology* 33: 419–39.

National Science Foundation. 1995. *Proceedings of the NSF Canopy Database Workshop.* Evergreen State College, Olympia, WA.

Odum, E. P. 1971. *Fundamentals of Ecology.* Philadelphia: Saunders.

Oldeman, R. A. A. 2001. Canopies in canopies in canopies. *Selbyana* 22 (2): 235–38.

Pfeiffer, W. J. 1996. Litter invertebrates. In D. P. Reagan and R. B. Waide, eds., *The Food Web of a Tropical Rain Forest.* Chicago: University of Chicago Press, pp. 137–81.

Rapport, D., R. Costanza, P. R. Epstein, C. Gaudet, and R. Levins 1998. *Ecosystem Health.* Oxford: Blackwell Science.

Reynolds, B. C., and D. A. Crossley. 1995. Use of a canopy walkway for collecting arthropods and assessing leaf area removed. *Selbyana* 16: 21–23.

Reynolds, B. C., D. A. Crossley, and M. D. Hunter. 2003. Response of soil invertebrates to forest canopy inputs along a productivity gradient. *Pedobiologia* 47: 127–39.

Reynolds, B. C., and M. D. Hunter. 2001. Responses of soil respiration, soil nutrients, and litter decomposition to inputs from canopy herbivores. *Soil Biology and Biochemistry* 33 (12/13): 1641–52.

Reynolds, B. C., M. D. Hunter, and D. A. Crossley. 2000. Effects of canopy herbivory on nutrient cycling in a northern hardwood forest in western North Carolina. *Selbyana* 21 (1/2): 74–78.

Rinker, H. B. 2004a. The effects of canopy herbivory on soil microarthropods in a tropical rainforest. PhD dissertation. Antioch University Graduate School, Keene, NH.

Rinker, H. B. 2004b. Soil microarthopods: belowground fauna that sustain forest systems. In M. D. Lowman and H. B. Rinker, eds., *Forest Canopies*, 2nd ed. San Diego: Elsevier Press, pp. 242–50.

Risley, R. S., and D. A. Crossley. 1993. Contribution of herbivore-caused green-fall to litterfall nitrogen flux in several southern Appalachian forested watersheds. *American Midland Naturalist* 129: 67–74.

Ritchie, M. E., D. Tilman, and J. M. H. Knops. 1998. Herbivore effects on plant and nitrogen dynamics in oak savanna. *Ecology* 79: 165–77.

Schowalter, T. D., W. W. Hargrove, and D. A. Crossley. 1986. Herbivory in forested ecosystems. *Annual Review of Entomology* 31: 177–96.

Schowalter, T. D., and T. E. Sabin. 1991. Litter microarthropod responses to canopy herbivory, season and decomposition in litterbags in a regenerating conifer ecosystem in western Oregon. *Biology and Fertility of Soils* 11: 93–96.

Schowalter, T. D., T. E. Sabin, S. G. Stafford, and J. M. Sexton. 1991. Phytophage effects on primary production, nutrient turnover, and litter decomposition of young Douglas-fir in western Oregon. *Forest Ecology and Management* 42: 229–43.

Schultz, J. C., and I. T. Baldwin. 1982. Oak leaf quality declines in response to defoliation by gypsy moth larvae. *Science* 217: 149–51.

Seastedt, T. R. 1984. The role of microarthropods in decomposition and mineralization processes. *Annual Review of Entomology* 29: 25–46.

Stadler, B., and B. Michalzik. 2000. Effects of phytophagous insects on microorganisms and throughfall chemistry in forested ecosystems: Herbivores as switches for the nutrient dynamics in the canopy. *Basic and Applied Ecology* 1: 109–16.

Swank, W. T., and D. A. Crossley. 1988. *Ecological Studies. Volume 66: Forest Hydrology and Ecology at Coweeta.* New York: Springer.

Swank, W. T., J. B. Waide, D. A Crossley Jr., and R. L. Todd. 1981. Insect defoliation enhances nitrate export from forest ecosystems. *Oecologia* 51: 297–99.

van Breemen, N. 1992. Soils: Biotic constructions in a Gaian sense? In A. Teller, P. Mathy, and J. N. R. Jeffers, eds., *Responses of Forest Ecosystems to Environmental Changes* (European Symposium on Terrestrial Ecosystems: Forests and Woodland). New York: Elsevier Applied Science, pp. 189–207.

Waide, J. B., and D. P. Reagan. 1996. The rain forest setting. In D. P. Reagan, ed., *The Food Web of a Tropical Rain Forest*. Chicago: University of Chicago Press, pp. 1–16.

Wallwork, J. A. 1983. Oribatids in forest ecosystems. *Annual Review of Entomology* 28: 109–30.

Wardle, D. A. 2002. *Communities and Ecosystems: Linking the Aboveground and Belowground Components*. Princeton: Princeton University Press.

Woolley, T. A. 1960. Some interesting aspects of oribatid ecology (Acarina). *Annals of the Entomological Society of America* 53: 251–53.

# III

## Imperiled Biosphere

# 7

# Gaia and Biodiversity

Stephan Harding

Biodiversity is the diversity of life at various levels of organization, ranging from genes, species, and ecosystems to biomes and landscapes. As far as we can tell, the Earth just before the appearance of modern humans was the most biodiverse it had ever been during the 3.8 billion years of life's tenure; indeed before we began to upset things the Earth hosted a total of 10 to 100 million species (Wilson 1992). The fossil record reveals that there have been five mass extinctions in the last 500 million years or so, all due to natural causes such as meteorite impacts and flood basalt events, or possibly because of drastic internal reorganizations within biotic communities. The most recent mass extinction is happening now and is entirely due to the economic activities of modern industrial societies.

## Losing Life's Richness

We are hemorrhaging species at a rate up to 10,000 times the natural rate of extinction (Wilson 2002); more prosaically, every day we are losing about 80 species, mostly in the great tropical forests, because of our endless desires for timber, soya, palm oil, and beef. Coral reefs and the marine realm, in general, are not exempt from our destructive attentions. The list of atrocities our culture has perpetrated on the living world makes for chilling reading. Hundreds of thousands of species will be driven to extinction in the next 50 years or so (Primack 2006). According to the IUCN's Red List of Threatened Species, by 2000 about 11 percent of all bird species, 18 percent of mammals, 7 percent of fish, and 8 percent of all the world's plants were threatened with extinction. By 2008 the Red List estimated that some 14 percent of bird species, 25 percent of mammals, and 32 percent of amphibians were in danger of extinction. According to the Living Planet Index, between 1970 and 2000 populations of forest species declined by 15 percent, those of freshwater species by a staggering 54 percent, and those of marine species by 35 percent.

Does the current mass extinction really matter? What does biodiversity do for Gaia and for us? To anyone who is deeply in touch with the natural world it is absurd to ask these questions—clearly, the current mass extinction is a crime of vast proportions. Our intuitions and deep experiences of belonging to the more than human world tell us that biodiversity gives us three key benefits: integrity, stability, and beauty. But what does science have to say about the importance of biodiversity? To explore this question, we need a systems approach in order to assess whether or not biodiversity contributes to the well-being of Gaia (figure 7.1).

First, human influences act either directly on biodiversity or indirectly by changing Gaian processes such as climate and biogeochemical cycles. Human-induced changes to biodiversity then affect aspects of ecosystem health, such as how well an ecosystem resists and recovers from disturbances, how well it recycles its nutrients and how reliably, and how much biomass it produces over a given period of time. These various aspects of ecosystem health could feed back to influence biodiversity, as changes in nutrient cycling or productivity have an impact on the species in the ecosystem. Ecosystem health could also have repercussions for Gaian

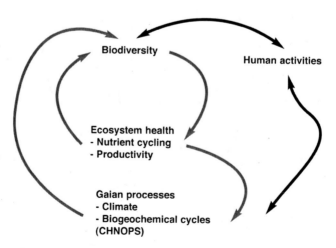

**Figure 7.1**
The importance of biodiversity for the health of Gaia.

processes, such as the abundance of greenhouse gasses in the atmosphere and the albedo of the planet, both of which influence climate. Every species has a preferred climate in which it feels most comfortable, so Gaian processes feed back to influence biodiversity. Last, altering biodiversity could expose human activities to feedbacks from two directions: directly from changes to biodiversity, and indirectly if ecosystem health and Gaian processes have been affected.

Let's look at each of these relationships. First, how are human activities influencing biodiversity? The answer has been summarized in the famous acronym HIPPO, which tells us that our lethal impacts on biodiversity are habitat destruction and fragmentation, invasive species, pollution, population, and overharvesting.

## Habitat Destruction and Fragmentation

Humans have been destroying habitats for a long time—we need only think of the deforestation of regions such as the Mediterranean, North Africa, and even China in ancient times to confirm this assertion. But widespread habitat destruction became a well-orchestrated global phenomenon only during the nineteenth century, with the onset of the Industrial Revolution. Before the beginning of widespread destructive human impact during the nineteenth century, Gaia was clothed with a continuous cover of wild habitats that melded gently into each other according to how climates varied over her surface. If we had been standing in Britain after the last ice age was well and truly over some 10,000 years ago, we could have walked all the way from the south coast of England to the north of Scotland without ever leaving the great mosaic of wild forest and natural meadows that covered most of the country. We would have experienced the same continuum on every continent. Crossing the channel to France, we could have walked all the way across Eurasia to the great rainforests of Burma, Thailand, and Vietnam without ever encountering a major disturbance to nature's vast wild domain. The abundance of flying, leaping, and swimming creatures in this pristine state astonished the first European settlers all over the world, who quickly set about logging, hunting, fishing, and clearing for agriculture with a demonic destructiveness that beggars the imagination (Pontin 2007).

Today, there is no habitat on Earth that has not been seriously degraded by humans. All the great biomes face increasing threats, including the mangrove swamps, the wetlands, the tropical dry forests, the tundra, and the boreal forests—the future for all of them looks bleak. When humans attack the great wild, they generally leave a few fragments of the original habitat here and there, perhaps out of laziness, or because of a pang of conscience, or, most likely, because no money could be made out of them. To begin with, these fragments are the last refuges for the wild organisms that once spread freely over the untamed Earth, but they soon turn into death camps for many of them as the effects of fragmentation begin to bite. Each fragment is an island, often surrounded by inhospitable habitats such as agricultural land, buildings, and roads that for many creatures create insurmountable barriers to foraging, dispersal, and colonization—even a small road in a nature reserve can be a daunting obstacle to tiny insects. The refugees may not be able to find the food they need in their fragments, or a good mate, or even a good place to sleep. Edge effects creep into the fragments, particularly the smaller ones, making things too dry or too hot or too cold. Pests and diseases can strike down the refugees more easily in the fragments, and even if there are enough breeding individuals to keep a population going, eventually lack of colonization from outside can lead to seriously damaging inbreeding depression.

You never know who the big players are in the wild world—seemingly insignificant, the dung beetles of the Amazon are critically important for the health of the whole forest (Klein 1989). Near Manaus, in the Amazon region of Brazil, a small dung beetle searches for food on the dry leafy floor of a small forest fragment left behind when the surrounding forest was cleared for pasture in 1982. In the old days, when the forest was entire, a whole host of dung beetle species, large and small, killed off parasites, buried seeds, and ensured that precious nutrients were quickly recycled as they fed their underground larvae on buried dung. But in the forest fragment there is little dung around, for most of the monkeys and birds that provided it in abundance before the forest was fragmented died or left a long time ago. Now there are fewer kinds of dung beetle, and those that remain are smaller and not very numerous.

The dung beetle extinctions happened in many ways. Hot, dry winds searing in from the pasture outside the fragment wiped out several species by killing off their larvae. For many species there just weren't enough good quality mates to go around and the inhospitable pasture prevented beetles from colonizing the fragment to boost numbers and bring in new blood. The consequences for the fragment's remaining denizens have not been good. There are more diseases among the few birds and mammals that remain, nutrients are washed away by heavy rains before roots can capture them, and the seeds of many plants have not been able to germinate. Seemingly insignificant, the dung beetles of the Amazon are major players in their ecological community—they are one of the keystone species of the forest.

## Invasive Species

Invasive species can cause extinctions even in areas where there has been little habitat fragmentation, and they wipe out more species than pollution, population pressures, and overharvesting put together. They come from all over the world—the goats, pigs, cats, rabbits, and many others—brought to places they could never have reached without the help of humans. According to the USDA Forest Service, about 4,000 exotic plant species and 2,300 exotic animal species have been brought to the United States alone, threatening 42 percent of species on the endangered species list and causing billions of dollars of damage every year in sectors such as forestry, agriculture, and fisheries. Introduced species often do well in their new locales in the absence of natural predators and diseases. Most don't do much damage, but a small minority take hold and do massive harm. Some are predators that exploit defenseless native prey species. A famous example is the brown tree snake, *Boiga irregularis*, a native of the Solomon Islands, New Guinea, northern and eastern Australia, and eastern Indonesia (Wilson 2002). Introduced to some of the Pacific islands, it has wiped out many endemic bird species. On Guam alone it is responsible for driving twelve to fourteen endemic bird species beyond the point of no return. Other introduced species are powerful competitors like the American gray squirrel, *Sciurus carolinensis*, that has pushed out the native red squirrel, *Sciurus vulgaris*, in most parts of Britain (Reynolds 1985).

## Pollution

Rachel Carson's book *Silent Spring* was instrumental in starting the green movement by bringing the dangers of pesticides to our attention in 1962. Since then pollution of many kinds have become alarmingly widespread. We are only too aware of gender-bending chemicals in water, and are well informed about atmospheric pollution such as acid rain from power stations and cancer-causing soot particles. One of the most insidious pollutants today is carbon dioxide gas. This is not commonly thought of as a contaminant because it is an essential nutrient for photosynthetic beings that they harvest directly from the atmosphere. But carbon dioxide is also greenhouse gas, and too much of it causes the climatic mayhem that is escalating the extinction crisis (Lovejoy and Hannah 2005; Lovelock 2006).

## Population

The human population has grown explosively, especially since the Industrial Revolution. The current world population stands at about 6.7 billion, and is projected to level off at around 10 billion by 2150 (Wilson 2002)—provided that we curb carbon emissions and put in place policies that are socially just and equitable. Otherwise, climate change could trigger a massive reduction in global population. People need land, water, food, and shelter, and often satisfy these needs by destroying wild nature. But it is not just a matter of sheer numbers, for the amount of resources consumed by each person is what really makes a difference to our impact on the planet. Paul Ehrlich devised his famous $I = P \times A \times T$ equation (pronounced IPAT) to make this point (O'Neill et al. 2004). $I$ stands for impact, $P$ for population, $A$ for affluence, and $T$ for technology. Human impact is a product of the last three terms, so that it is possible to have a high population, so long as people do not overconsume. In the current economic climate all the terms on the right-hand side of the equation are increasing alarmingly. Today the world's middle class numbers about 20 percent of the population but consumes about 80 percent of the available resources. An oft-quoted fact uncovered by the New Economics Foundation in the United Kingdom: if everyone in the world were to consume as much as the average Briton, roughly three extra planets would be required to provide the raw materials. Alternatively, a solar-based energy infrastructure could stabilize population and raise quality of life for all. For the moment the huge pressures of the human population in a swiftly industrializing global society is the underlying drive behind all the other causes of extinction, including overharvesting.

## Overharvesting

About one-third of endangered vertebrates are threatened in this way— by unsustainable, direct killing. Often overharvesting is carried out by poor rural people left with no other means of surviving after they have been forced off their lands by global economic forces. The rich countries of the North are also responsible for overharvesting and are especially responsible for driving several key fisheries to the point of extinction— the Grand Banks and the North Sea cod fisheries are sad examples. Many of the world's great whales, the right, the bowhead, and the blue had been pushed to the edge extinction by the early twentieth century. Detailed mathematical models designed to calculate "maximum sustainable yield" for some of these species were spectacular failures that led to catastrophic declines (Gulland 1971). Illegal whaling has been blamed for this, but the difficulties of observing and quantifying whale behavior in the wild were also responsible. Many whale species have been protected to some extent since 1946. A few, like the Minke whale, are recovering, but many smaller cetaceans (e.g., dolphins) are killed every year when they become entangled in the nets of the fleets that are decimating the world's fisheries.

## The Impact of Biodiversity Losses

Is it conceivable that the huge losses in biodiversity could feed back to influence the human enterprise in particular localities? To answer this question, we need to explore two aspects. Do organisms living in a specific place link up into an ecological "superorganism"—with valuable emergent properties such as climate regulation, better water retention, nutrient cycling, and resistance to diseases—or are they no more than collections of selfish individuals, each out to exploit as many of the available resources as possible, even to the detriment of the ecological community that enfolds them? If the former is true, then we will need to protect entire ecological communities in order to preserve the ecosystem services they provide. If the latter is the case, then we need only bother to look after the key players, or to introduce those of our own choosing.

These questions occupied the minds of the founders of ecology in the first half of the twentieth century. The American ecologist Frederick Clements, one the most influential ecologists of his day, studied how plants colonize bare ground. He noticed that there was a series of stages, beginning with an inherently unstable plant community and ending up in a stable climax community in balance with its environment. In Devon, from where I write, bare ground is first colonized by annual herbaceous plants, then by brambles and shrubs, and eventually by oak forest, which grows here because the mix of soil, temperature, rainfall, and wind are just right. For Clements, the development of vegetation resembled the growth process of an individual living being, and each plant was like an individual cell in our own bodies. He thought of the climax community as a *complex organism* in which the member species work together to create an emergent self-regulating network in which the whole is greater than the sum of the parts (Worster 1994).

Within the scientific community, a struggle ensued between the organismic views of Clements, and the objectivist approach of the Oxford botanist Sir Arthur Tansley and the American ecologist Henry Gleason. Tansley declared that plant communities couldn't be superorganisms because they are nothing more than random assemblages of species with no emergent properties. Tansley found Clements's views difficult to accept because they challenged our legitimacy as humans to remake nature. Tansley wanted to remove the word "community" from the ecologist's vocabulary because he believed, in the words of Donald Worster (1994), that "there can be no psychic bond between animals and plants in a locality. They can have no true social order." Tansley represented a breed of ecologists who wanted to develop a completely mechanistic understanding of nature, in which, according to Worster, nature is seen as "a well-regulated assembly line, as nothing more than a reflection of the modern corporate state." For Tansley, agricultural fields were no better or worse than wild plant communities. To paraphrase Worster (1994), the reduction of nature to easily quantified components removed any emotional impediments to its unrestrained exploitation. Ecology, he argues, took on the economic language of cost–benefit analysis, while economics learned nothing from ecology.

Which approach best describes biotic communities—organism or mechanism? Out in the flatlands of Minnesota, at a place called Cedar Creek, a long-term experiment is in progress that could have a bearing on this question. A strange chequerboard of meter square plots filled with prairie plants dots the landscape, tended by David Tilman, one of the world's leading ecologists. He has spent years investigating the relationship between the biodiversity in his plots and the ability of the small ecological communities they contain to produce more biomass by capturing sunlight and to survive stress. Tilman and his colleagues have set up hundreds of plots, each with a different number of species chosen from the native flora of the immediate locality. Halfway through one of these experiments, Minnesota experienced a severe drought, and to Tilman's amazement the plots that survived best were those with the highest biodiversity (Tilman and Downing 1994). This was evidence in favor of Clements and the organismic view, for the most diverse plots seemed to have developed a powerful emergent protective network, as their various members melded their individual survival skills into a greater whole linked by tight bonds of the plant kind. But there were critics. They pointed out that because Tilman had fertilized his plots with different amounts of nitrogen, the differences in drought resistance were due to this and not to the effects of species diversity (Huston 1997).

To eliminate this possibility Tilman established a more extensive experiment using 489 plots of two sizes with different amounts of plant biodiversity seeded in identical soil and chosen from a maximum of four "functional groups": broad-leaved perennial herbs, nitrogen-fixing legumes, warm season grasses, and cool season grasses (Tilman et al. 1997). This time the more diverse plots produced more biomass, fixed more nitrogen, were better at resisting weed invasions, and were less prone to fungal infections. The best plots were those that hosted a variety of species from each of the four functional groups. Once again, here is evidence that diverse biotic communities resemble organisms with powerful emergent properties. But the news was not all good because Tilman found that the benefits of having extra species in the community peaked at around five to ten species. Beyond that, extra species didn't seem to improve ecological performance—what mattered most was having at least one member of each functional group. One interpretation of these results is that most species in wild ecosystems are dispensable, and that the extinction crisis gives us nothing to worry about. But how are we to know which species are expendable and which are not? Since we can't tell which are the keystone species, it makes more sense to protect as many species as we can. Furthermore there is almost certainly an "insurance effect" at work, in that more biodiverse communities are more likely to contain species that can take over the work left vacant by any keystone species that disappear but are is difficult to predict.

Tilman's approach was extended by the BIODEPTH project, in which plots with different amounts of native grassland biodiversity were set up in eight European countries, from the cold north to the warm south (Hector et al. 1999). Despite the wide range of climatic conditions, high biodiversity in each country was strongly correlated with improvements in many key ecological functions, such as nutrient cycling, resistance to predators, and biomass production—once again evidence in favor of the organismic view (figure 7.2). Until now the analysis of the BIODEPTH data has focused on the impact of biodiversity on each ecosystem function in isolation from the rest, but a new analysis by Hector and Bagchi (2007) has shown that in fact each species contributes to a wide variety of ecosystem functions simultaneously, so that focusing on isolated

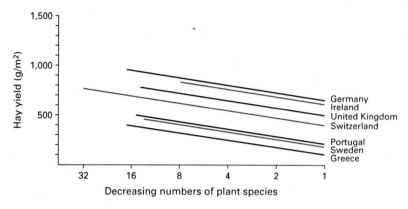

**Figure 7.2**
Key result from the BIODEPTH experiment.

ecosystem functions seriously underestimates the level of biodiversity needed to maintain the health of ecosystems.

Laboratory experiments also tend to support the idea that biodiversity improves the health of ecosystems. Scientists at Imperial College, London, have developed the "Ecotron," a series of chambers with controlled light, temperature, and humidity levels that house artificially assembled ecological communities, each with differing amounts of biodiversity. The main result of this research is that more diverse communities fixed more carbon dioxide from the air (Naeem et al. 1994). This may seem a fairly mundane finding, but it caused a stir in scientific circles by showing that biodiversity could have a key role to play in absorbing some of the vast amounts of the Earth-warming carbon dioxide gas that our economy is emitting into the atmosphere; terrestrial biodiversity may be major help combating global warming. New work in the Ecotron mimicked the elevated carbon dioxide and temperature that are expected with climate change. The surprising result was that climate change had little impact on the fauna and flora living above ground, but that the community of soil organisms was greatly altered. More carbon dioxide in the atmosphere stimulated photosynthesis among the plants, which then transported some of this carbon to their roots as sugars. The extra soil carbon changed the community of soil fungi, which in turn changed the community of fungus-eating spring tails (Jones et al. 1998). These changes in below ground ecology could, if writ large, have a massive impact on nutrient feedbacks and carbon storage in soils, but as yet no one knows whether this means that soils will be able to hold more or less carbon.

The fact that there was a change raises concern and could have an effect on future strategies for dealing with climate change.

In another series of experiments, scientists created artificial ecological communities by seeding glass bottles containing water and nutrients with diverse communities of bacteria and their larger protozoan predators. In these experiments greater diversity led to less variability in the flow of carbon dioxide in and out of the community (McGrady-Steed et al. 1997). The message here is that more diverse real-world communities could provide more predictable and dependable emergent ecological functions such as carbon capture and storage.

Mathematical modeling has also contributed to the new understanding of the relationship between biodiversity and ecological health. We now know from detailed fieldwork that ecological communities are replete with weak interactions with many predators focusing on eating a few individuals from a fairly wide range of species. Models that take account of these insights show that virtual communities with realistic feeding relationships and abundant weak interactions are more stable than previously thought possible (McCann et al. 1998). Another group of mathematical models known as community assembly models work by creating a pool of virtual plants, herbivores, and carnivores, each with its own body size and preferences for food and space. One species at a time is placed in an virtual arena where it interacts with other species that are already present. After a while, an astonishing thing happens—persistent communities self-assemble with a final membership of about fifteen species. As the number of species builds up, it becomes harder and harder for an invader to find a toehold in the nexus of interacting species. Communities that have existed for longer are harder to invade than newly established ones, strongly suggesting that communities develop an emergent protective network that becomes more effective as the community matures. Amazingly the challenge for an invader lies with the community as whole. An inferior competitor in a mature, well-connected community has a better chance of surviving an invasion from a superior competitor than it does as a member of a less well-connected more recently established community (Drake 1990).

The research we have considered—from field, lab, and computer modeling—tends to support Clements's idea that ecological communities can indeed be thought of a "superorganisms" that function more smoothly and predictably as their biodiversity increases. But perhaps Clements and Tansley were both right after all, each having perceived different sides of the same coin. If so, there is nothing inevitable about which species will colonize a bare patch of land, or indeed nothing inevitable about how a particular succession will progress (Tansley), but as soon as the species in a given place begin to web themselves together, the whole community becomes a unit with powerful emergent properties (Clements).

So far we have looked at the effects of biodiversity on ecological health at the local level, but could there be a relationship between biodiversity and the health of the planet as a whole? This question, considered absurd by the scientific community as recently as ten years ago, is now beginning to loom large in the minds of scientists trying to understand how humans are changing the Earth, which is increasingly recognized as a fully integrated system with life as a key player.

## Biodiversity and the Health of Gaia

It is now generally agreed that life affects climate in at least two fundamental ways: by altering the composition of the atmosphere; and by changing how solar energy heats up the Earth's surface and how this heat is distributed around the planet. But how could biodiversity be involved in making these globally important processes work more effectively? The Ecotron and BIODEPTH experiments have taught us that diverse ecological communities on the land can change the composition of our atmosphere by increasing the absorption of carbon dioxide. It is almost certain that biodiversity in the oceans also enhances this effect. Marine phytoplankton use carbon dioxide for photosynthesis much as land plants do, drawing it out of the air and into their tiny bodies. Dead phytoplankton sink, taking carbon that was once in the atmosphere with them to a muddy grave in the sediments below. This "biological pump" could also be more effective at removing carbon dioxide from the atmosphere if it is the case that larger phytoplankton are more often found in diverse communities, since it is known that these larger organisms increase the slow drift of carbon to the ocean depths (Fasham 2003).

Biodiversity may also improve the absorption and distribution of energy from the sun. It could be that more diverse communities on land and in the ocean are better at seeding clouds, possibly via the emission of more diverse cloud seeding chemicals, but this remains to be established. What is more certain is that a greater diversity of land plants could enhance cloud-making and energy distribution in two other important ways—by transpiring more water from the soil through roots and out into the air from pores on the undersides of leaves, and by providing more leaf surfaces from which rainwater can evaporate directly.

A big rainstorm has just finished watering several hundred square kilometers of Amazon forest. The leaves are all wet, and those at the top of the canopy glisten in the early afternoon sun. Some of the energy in the sunlight passes deep into the leaf where it fuels photosynthesis, but a fairly large portion is absorbed directly by the recently arrived film of water on the leaf surfaces. As the water molecules receive a influx of solar energy, they begin to gyrate like inspired dancers, and when sufficiently energized, they dance their way into the air as water vapor. This is evaporation. In the case of a leaf drying in the sun, the solar energy that might have heated the leaf is transferred to water vapor, and as this is swept away by the wind, the leaf is kept cool, just as we are when we sweat.

The energy held in water vapor can be released as heat whenever condensation converts it back into liquid water. This energy is called "latent heat" because it remains invisible until condensation happens. On the other hand, any solar energy absorbed by the surface of the leaf causes the molecules there to vibrate and to immediately reemit the energy as sensible heat, which one can detect directly with the skin or indirectly with an infrared sensor.

But it is not just rainwater that evaporates from the surface of a leaf, so does water that has traveled from the soil into the plant through tubes leading all the way from the roots to the thousands of microscopic pores beneath a leaf's surface. This water, carrying with it life-giving nutrients from the soil, eventually passes through the leaf pores into the air, a process known as transpiration. Amazingly plants keep the flow of water going without the kind of muscular contraction seen in animal circulatory systems. They do this by continually and deliberately leaking water through the pores, thereby creating a mysterious kind of suction that draws in new water all the way down at the roots. On warm days water entering a leaf from the soil is heated up by the sun's rays, and passes out of the leaf pores as water vapor. The summed effect of evaporation of water from leaf surfaces and transpiration of water from within the plant is considered to be a single process known as "evapotranspiration," which is vitally important for Gaia's climate. Because of it, a huge amount of solar energy is stored as latent heat in water vapor that can travel long distances before condensing to release its energy as heat, sometimes thousands of kilometers away. But evapotranspiration also has local effects. In the southern boreal forests of western Canada, where the deciduous trembling aspen (*Populus tremuloides*) is abundant, temperature rises steeply in the early spring when, unimpeded by aspen leaves, the sun's rays warm the ground. But as the aspen leaves unfurl and swell out to their full size, the rate of temperature increase drops dramatically because evapotranspiration cools and moistens the air (Hogg et al. 2000).

Foliage is thus very important in regulating the surface climate. In general, the more leafy a forest, the more evapotranspiration and so the more cloud production, local rainfall, local cooling, and plant matter production by photosynthesis (Bonan 2002). A more diverse flora could well improve transpiration by providing a bigger and more varied mat of below-ground root structures with better water-trapping abilities, and it could also enhance evaporation by providing a larger and more complex total leaf surface area from which rainwater can evaporate. Both of these effects would send more water vapor into the air for cloud-making. Some plants evapotranspire more than others. Because they have far fewer leaf pores, needle leaf trees pass less water into the air than their broadleaved cousins, thereby keeping themselves warmer—an advantage in the high latitudes (Bonan 2002).

Another climatically important characteristic of vegetation is its roughness, a measure of how much resistance plants give to the wind (Bonan 2002). When wind blowing over the land surface encounters plants such as trees, grasses, and shrubs, it transfers some of its energy to the leaves, making them dance about. This sometimes frenzied leafy dance mixes the air, making both evapotranspiration and the transfer of sensible heat from leaf to air much more effective than on a perfectly still day. The higher up the canopy you go, the more efficient are these transfers of energy from wind and sun to leaf. A dense rainforest canopy, with its high roughness, transfers much more energy to the air than the far less leafy, low roughness grasses in a savannah. The intricate leaf surfaces of a more diverse flora could create a rougher land surface that increases air turbulence, and this might well increase the transfers of heat and moisture to the air, influencing weather patterns on both local and global scales.

These impacts of biodiversity on local and global climates in turn feed back to influence biodiversity. Clouds seeded by biochemical substances emitted by the Amazonian vegetation keep the forest cool and recycle its water, thereby allowing the forest to persist and preventing the encroachment of the nearby drought-tolerant savannah (Artaxo et al. 2001).

The heat released when the clouds condense helps to configure the Earth's climate system as a whole into a state that favors forest growth in the Amazon region. Herein lies a great lesson for living in peace with Gaia, namely that the very structure of an ecosystem—which species are present, the depths of its roots, the extent of its leafiness, its albedo, and its release of cloud-seeding chemicals into the air—all have massive effects not only on climate both locally and globally, but also on the great cycling of chemical elements around the planet.

We have seen how biodiversity is a key player in creating habitable conditions on the Earth, including a climate that favors our own existence as well as that of the rest of biosphere. Biodiversity also provides the entire community of life with a host of other benefits, such the stabilization of soil, recycling of nutrients, water purification, and pollination. When they favor humans, these benefits have been called "ecosystem services" by a new breed of economists who are attempting to calculate how much these services are worth in financial terms. The results are staggering—in 1997 global ecosystem services were estimated to be worth one to two times the global GDP (Costanza et al. 1997).

Recently the results of the most comprehensive survey of the state of the world's ecosystem services were made public. The Millennium Ecosystem Assessment (2005), compiled by 1,360 scientists from 95 countries, deliberately took the approach of looking for the interconnections between human well-being and ecosystem health. The results make for sobering reading: 60 percent of the ecosystem services investigated have been degraded. Human activity has changed ecosystems more rapidly in the past fifty years than at any other time in human history. About 24 percent of the planet's land surface is now under cultivation; a quarter of all fish stocks are overharvested; 35 percent of the world's mangroves and 20 percent of its coral reefs have been destroyed since 1980; 40 to 60 percent of all available freshwater is now being diverted for human use; forested tracts have been completely cleared from 25 countries and forest cover has been reduced by 90 percent in another 29 countries; more wild land has been ploughed since 1945 than during the eighteenth and nineteenth centuries combined; demands on fisheries and freshwater already outstrip supply; and fertilizer runoff is disturbing or suffocating aquatic ecosystems.

The report makes it abundantly clear that the UN Millennium Development Goals of halving poverty, hunger, and child mortality by 2015 cannot be met unless ecosystems are nurtured and protected, since it is the poor who are most directly dependent on their services, particularly for freshwater and protein from wild fish and game. Furthermore it has become clear from a handful of successful projects that the way forward lies with encouraging local people to become involved in protecting their own ecosystems. This has worked well in Fiji, where local fishermen established restricted areas that reversed serious declines in fish stocks, and in Tanzania, where villagers now harvest food and fuel from 3,500 square kilometers of degraded land that they were allowed to reforest (Giles 2005).

All of this should be enough to convince the most hard-headed among us that it is very much in our own interest to maintain as much of our planet's biodiversity as possible. At the same time utilitarian arguments for protecting biodiversity may not prevent it from being seriously degraded, for ultimately we may not be able to save what we do not love. If we are to develop a worldview that has any chance of achieving genuine ecological sustainability, we will need to move away from valuing everything around us only in terms of what we can get out of it, recognizing instead that all life has intrinsic value regardless of its use to us (Naess 1990). Scientific and economic arguments such as those we have been exploring for protecting biodiversity can help a great deal, but on their own they are not enough. We need, as a matter of the utmost urgency, to recover the ancient view of Gaia as a fully integrated, living being consisting of all her life-forms, air, rocks, soil, oceans, lakes, and rivers if we are ever to halt the latest, and possibly greatest, mass extinction.

## References

Artaxo, P., M. O. Andreae, A. Guenther, and D. Rosenfeld. 2001. Unveiling the lively atmosphere-biosphere interactions in the Amazon. *Global Change Newsletter* 45: 12–15.

Bonan, G. 2002. *Ecological Climatology*. Cambridge: Cambridge University Press.

Costanza, R., R. d'Arge, R. de Groot, S. Farber, M. Grasso, B. Hannon, K. Limburg, S. Naeem, R. V. O'Neill, J. Paruelo, R. G. Raskin, P. Sutton, and M. van den Belt. 1997. The value of the world's ecosystem services and natural capital. *Nature* 387: 253–59.

Drake, J. A. 1990. Communities as assembled structures: Do they govern pattern? *Trends in Ecology and Evolution* 5 (5):159–64.

Fasham, M. J. R. 2003. *Ocean Biogeochemistry: The Role of the Ocean Carbon Cycle in Global Change*. IGBP Global Change Series. Berlin: Springer.

Giles, J. 2005. Ecology is the key to effective aid, UN told. *Nature* 437: 180.

Gulland, J. A. 1971. The effect of exploitation on the numbers of marine mammals. In B. T. Grenfell and A. P. Dobson, eds., *Dynamics of Populations.* Wageningen: Centre for Agricultural Publishing and Documentation, pp. 450–68.

Hector, A., B. Schmid, C. Beierkuhnlein, M. C. Caldeira, M. Diemer, P. G. Dimitrakopoulos, J. A. Finn, H. Freitas, P. S. Giller, J. Good, R. Harris, P. Högberg, K. Huss-Danell, J. Joshi, A. Jumpponen, C. Körner, P. W. Leadley, M. Loreau, A. Minns, C. P. H. Mulder, G. O'Donovan, S. J. Otway, J. S. Pereira, A. Prinz., D. J. Read, M. Scherer-Lorenzen, E. D. Schulze, A. S. D. Siamantziouras, E. M. Spehn, A. C. Terry, A. Y. Troumbis, F. I. Woodward, S. Yachi, and J. H. Lawton. 1999. Plant diversity and productivity experiments in European grasslands. *Science* 286: 1123–27.

Hector, A., and R. Bagchi. 2007. Biodiversity and ecosystem multifunctionality. *Nature* 448: 188–90.

Hogg, E. H., D. T. Price, and T. A. Black. 2000. Postulated feedbacks of deciduous forest phenology on seasonal climate patterns in the western Canadian interior. *Journal of Climate* 13 (24): 4229–43.

Huston, M. A. 1997. Hidden treatments in ecological experiments: Re-evaluating the ecosystem function of biodiversity. *Oecologia* 110: 449–60.

Jones T. H., L. J. Thompson, J. H. Lawton, T. M. Bezemer, R. D. Bardgett, T. Blackburn, M. K. D. Bruce, P. F. Cannon, G. S. Hall, S. E. Hartley, G. Howson, C. Jones, G. C. Kampichler, E. Kandeler, and. D. A. Ritchie. 1998. Impacts of rising atmospheric carbon dioxide on model terrestrial ecosystems. *Science* 280: 441–43.

Klein, B. C. 1989. Effects of forest fragmentation on dung and carrion beetle communities in central Amazonia. *Ecology* 70: 1715–25.

Lovejoy, T., and L. Hannah, eds. 2005. *Biodiversity and Climate Change.* New Haven: Yale University Press.

Lovelock. J. 2006 *The Revenge of Gaia.* London: Penguin Alan Lane.

Millenium Ecosystem Assessment. 2005. *Ecosystems and Human Well-Being: Current State and Trends.* Washington, DC: Island Press.

McCann, K., A. Hastings, and G. R. Huxel. 1998. Weak trophic interactions and the balance of nature. *Nature* 395: 794–98.

McGrady-Steed, J., P. M. Harris, and P. J. Morlin. 1997. Biodiversity regulates ecosystem predictability. *Nature* 390: 162–65.

Naeem, S., J. Lindsey, L. J. Thompson, S. P. Lawler, J. H. Lawton, and R. M. Woodfin. 1994. Biodiversity in model ecosystems. *Nature* 371: 565.

Naess, A. 1990. *Ecology, Community and Lifestyle.* Cambridge: Cambridge University Press.

O'Neill, B. C., F. L. MacKellar, and W. Lutz. 2004. Population, greenhouse gas emissions and climate change. In W. Lutz and W. Sanderson, eds., *The End of Population Growth in the 21st Century*. London: Earthscan, pp. 283–309.

Pontin, C. 2007. *A New Green History of the World*. New York: Vintage.

Primack, R. 2006. *Essentials of Conservation Biology*. Sunderland, MA: Sinauer Associates.

Reynolds, J. C. 1985. Details of the geographic replacement of the red squirrel (*Sciurus vulgaris*) by the grey squirrel (*Sciurus carolinensis*) in eastern England. *Journal of Animal Ecology* 54: 149–62.

Tilman, D., and J. A. Downing 1994. Biodiversity and stability in grasslands. *Nature* 367: 363–65.

Tilman, D., J. M. H. Knops, D. Wedin, P. Reich, M. Ritchie, and E. Siemann. 1997. The influence of functional diversity and composition on ecosystem processes. *Science* 277: 1300–02.

Wilson, E. O. 1992. *The Diversity of Life*. Cambridge: Harvard University Press.

Wilson, E. O. 2002. *The Future of Life*. New York: Little Brown.

Worster, D. 1994. *Nature's Economy: A History of Ecological Ideas*. Cambridge: Cambridge University Press.

# 8

## Global Warming, Rapid Climate Change, and Renewable Energy Solutions for Gaia

Donald W. Aitken

The Gaian system refers to the interconnected natural responses of the Earth to restabilize living and physical systems when perturbed beyond normal bounds. Gaia is, of course, not limited to providing only for human beings, but humans have a huge stake in the outcomes. Nowhere is the strange abandonment of the Gaia stabilization responsibilities of humans more evident or consequential than in our use of energy, in particular in the burning of fossil fuels and its resulting climate destabilization. While many other environmental problems may be deemed of equal or greater urgency, the interaction of anthropogenic climate destabilizations with all natural and human systems is leading many scientists today to identify climate change (and its cause, global warming) as the most pressing environmental issue that needs to be addressed by global cooperation.

### The Pivotal Energy Role of the Earth's Atmosphere

This chapter explores the nature of global warming and some of the impacts that are becoming evident today and appear to be heading the Earth system toward *tipping points* beyond which recovery by human actions will not be possible. I review the presently inadequate international response. The scientifically agreed-upon upper limits for the increase in global temperature and the concurrent maximum concentration of carbon dioxide in the atmosphere are presented to underscore the need for the adoption of stringent new policy and energy transition timetables for all nations. I conclude with a brief overview of energy solutions, focusing on the enormous potential of the same renewable energies that have been utilized in the Earth system during its entire history.

All of the solar energy that is absorbed, used, and reused by the Earth's physical and living systems is ultimately re-radiated out to space, only having resided for a time within the Earth's cycles and masses, including a tiny fraction parked in the Earth's vegetation and other life forms. If this outgoing radiation did not exactly equal the incoming energy from the sun, on average, we would either be a frozen, presently lifeless planet like Mars, or we would be an unbearable oven like Venus. After all, all three planets are within the "life zone" around the sun, the region in the solar system in which conditions for the emergence of life could arise. It is the physical properties of the Earth's atmosphere, then, that have been pivotal to making the Earth's temperature and climates livable and suitable for the development of life forms—including us. Tamper with those properties and we are tampering with all life-supporting systems on Earth.

Carbon dioxide along with the other greenhouse gases (e.g., water vapor, nitrogen oxide, methane, and anthropogenic chlorinated compounds) play a major role in the regulation of the flow of energy through the atmosphere from the sun to the Earth, and the counterflow of reradiated energy from the Earth back out to space. The flows must remain in balance to maintain thermal equilibrium. Even though these flows are individually substantial, if they are just slightly mismatched, the destabilizing effects can be great, for an altered energy balance requires that the entire energy equilibrium of the Earth and its physical and living systems must change.

The flows do get unbalanced from time to time. The sun goes through small oscillations in brightness. The Earth's axis periodically changes its relationship to its orbit over long periods. And volcanoes erupt, injecting huge amounts of both dust and carbon dioxide into the atmosphere. The 1991 eruption of Mt. Pinatubo in the Philippines, for example, injected between 15 and 30 million tons of sulfur dioxide gas into the air. In two weeks the cloud had gone around the Earth, and a two-year global cooling was launched. Volcanic eruptions come and go, as do other surface events on Earth. Over time they average out. But what humans are now doing with the burning of fossil fuels is not averaging out; carbon dioxide is accumulating in the Earth's atmosphere. From the standpoint of the Earth system, this has gone out of bounds leading to increasing destabilizations of the planet's energy, temperature, and climate systems.

## Fossil-Fuel Burning and the Atmospheric Carbon Dioxide Balance

On Mars virtually all of the carbon dioxide is trapped in its soils and rocks, and it is very cold with an average temperature of –50° Celsius. On Venus 96 percent of the atmosphere is carbon dioxide, creating a thermal blanket that surrounds and bakes that planet at an average temperature of +420°C. On Earth, on the other hand, only about 0.04 percent of the atmosphere is carbon dioxide, but even this minute amount plays a critical role in enabling our planet to stabilize at an ambient average temperature of about 14.6°C (58° Fahrenheit).

Until recently we did not know how sensitive the Earth's temperature and climates were to this small but evidently critical amount of carbon dioxide; we are now finding that out as we pour billions of tons of carbon into the atmosphere. When fossil fuels are burned, the carbon that has long dwelled in them is released as carbon dioxide into the atmosphere. This is now yielding an average of 6.1 billion tons of new carbon into the Earth's cycles (about 1 metric ton of carbon for every person on Earth). Even though much of this is absorbed into the oceans, and some is also absorbed by a general increase in the growth rate of photosynthetic plant life on the Earth's surface, about 3.5 billion metric tons of that new carbon is being added to the atmosphere each year. The consistent accumulation of these small amounts of excess $CO_2$ over the past 150 years, however, has added up to about 39 percent more carbon dioxide in the atmosphere than would otherwise be there from natural processes alone (IPCC 2007).

Aerosols (dust and small particles) from fossil fuel combustion reduce the flow of the incoming solar radiation while the carbon dioxide product of that combustion, along with the other greenhouse gases injected into the atmosphere by human actions, retard the flows of the outgoing radiation. The two flows are not equally affected, with the outgoing radiation flow impacted the most. The result has been a gradual *net* accumulation of excess energy on the Earth's surface and oceans at the rate of a little under 1 watt/meter squared, averaged over the entire Earth. While seemingly small in absolute amount, this imbalance rate, if it had existed and continued unchecked during the previous 10,000 years, would have raised the Earth's temperature more than 100°C and boiled the oceans.

The amount of carbon dioxide in the atmosphere, and its effect on the Earth's energy balance, are steadily increasing to levels not seen for perhaps a million years or more, and growing at an extraordinarily rapid rate compared with geological history during the development of the human species. Both the present *change* and the *rate of change* of atmospheric carbon dioxide are leading scientists to regard our predicament as ominous.

## Swiftly Unfolding Consequences

A number of research paths are converging on the history of atmospheric carbon dioxide content and Earth temperatures that can be traced for up to 750,000 years. The results show a straightforward relationship between carbon dioxide and temperature: as the one increased or decreased, so did the other. While the precise cause-and-effect details are not clear, the overall patterns are clear, and can reasonably be expected to continue to in the same relationship into the future (figure 8.1).

Equally clear from these various research results is that the burning of fossil fuels has taken the Earth's atmosphere, and hence energy flows and

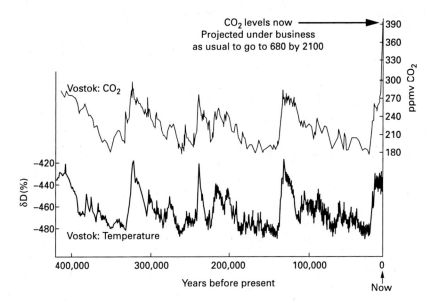

**Figure 8.1**
Present and projected future concentrations of carbon dioxide depicted on the 400,000-year Vostok ice core data sets for both Earth temperature and atmospheric carbon dioxide content. The IPPC is the Intergovernmental Panel on Climate Change. Figure courtesy of Dr. Robert Correll.

balances, into uncharted waters. Furthermore, since this has all happened in a short 150 years or so, the rise in carbon dioxide and temperature show up as sudden spikes at the end of charts of Earth's recent temperature and $CO_2$ histories. Both of those spikes are rising rapidly as the world's old and new fossil-fuel power plants (primarily the coal plants), as well as other sources (e.g., transportation) continue to increase levels of carbon dioxide in the atmosphere (figure 8.2).

So what can we expect to be the future results of our actions? Certainly a rapid warming of the Earth now appears underway. But since "climate" is the Earth's mechanism for the redistribution of its surface energies, it is equally inevitable that the Earth's climate must change as well.

How significant might that change be? For the past 10,000 years, following the exit out of the last ice age, the Earth's temperature has been remarkably mild and stable—nicknamed a "sweet spot" by climate scientist Robert Correll—not increasing or decreasing more than 0.5°C (see Lempinen 2007). This set the climatological stage for the evolution of great civilizations. The IPPC analyses, however, suggest that by the end

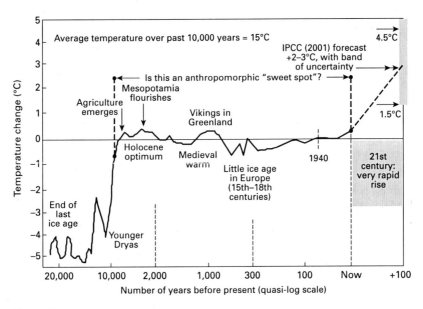

**Figure 8.2**
The last 10,000 years seems to have been ideal for the development of human societies. Is this an historic "sweet spot" that enabled humans to flourish? Figure and caption text by Dr. Robert Correll.

of this century, in the absence of stringent global control of greenhouse gas emissions, the Earth's temperature could climb as much as 4.5°C above that 10,000 year "sweet spot" average. This is roughly equal to the temperature difference between the last ice age and today, demonstrating how only a few degrees of warming or cooling can have extraordinary consequences for the biosphere and for human civilizations within it.

We do not need to wait for decades to discover how sensitive the Earth is to the impacts of our fossil fuel burnings, nor how quickly the Earth's energy systems can be unbalanced. The average temperature has only risen by 0.74°C (about 1.3°F) in the past 100 years. What is remarkable is the nature and pace of changes *already* taking place as the result of this small change, and how much more rapidly those changes are happening than even the projections of the best computer models. This would suggest that, while the Earth system is inherently robust, its balances are also very finely tuned.

Rather than offer a litany of all that *might* happen, then, I would rather lean on a few examples of what is happening *already* to underscore the urgency of the needed human response to the imperiling of the Earth systems on which both human beings and our contemporaneous species and ecosystems so vitally depend.

### Are Hurricanes Telling Us Something?

We remember all too well how Hurricane Katrina devastated New Orleans and other Gulf coastal regions in 2005, bringing untold suffering and damage. The previous year had seen four major hurricanes striking the US shores. Was global warming the cause? While it cannot be proved to be the cause of any particular storm, the accumulating evidence is showing a disturbing overall *pattern of climate change*—storms, floods, droughts—tracking the rate of carbon dioxide increase in the atmosphere.

For example, evidence shows that the number of the *most intense* tropical storms has increased by 80 percent over the past 35 years, and that the average intensity (which produces the damage) of storms created in the Atlantic Ocean has more than doubled during the 1983 to 2005 period (Webster et al. 2005; Kerr 2006). (Hurricane Wilma, on October 19, 2005, was for a while the strongest hurricane ever recorded with sustained winds of 170 miles an hour.) Hurricanes spawned in the North Atlantic further reveal an unambiguous correlation of increasing hurricane strength with human emissions, scientifically distinguishable from other possible natural causes (Mann and Emanual 2006). Recent theoretical models have improved to the point that some of these observations are predictable, although the models produce more complicated correlations in different ocean basins, and the models are not all in agreement. They do not, however, disagree with the statistical observations of hurricane intensity (Emanual et al. 2008). Hurricanes draw their energy largely from the surface waters over which they pass. Since research has shown unequivocally that the temperatures of the upper layers of the world's oceans are tracking the increase in lower atmospheric temperatures, the correlations of increasing tropical storm and hurricane strengths with increasing ocean surface temperatures is not surprising, but expected.

The damage caused by Hurricane Katrina has variously been estimated to be US$200 billion to $300 billion. Estimates for what it would have cost in coastal protection to prevent most of this damage are in the range of US$2 billion to $3 billion, or about 1 percent of the cost of the resulting damage. The message here is that prevention and mitigation of global warming impacts will be, in the long run, an economic bargain compared to the costs of inaction.

## Events in the Arctic Ocean

Perhaps even more alarming than the increasing global temperatures and intensities of storms worldwide are the unexpectedly rapid effects of global warming in the Arctic and Antarctic regions. In the Arctic the average temperature has been climbing for a number of decades twice as fast as elsewhere. The temperature of Alaska has been growing 6 to 8 times faster than the rest of the world.[1] The Antarctic peninsula is warming more rapidly than anywhere else on Earth (Bell 2008). Increased precipitation, shorter and warmer winters, and decreases in snow cover are already being documented in Arctic regions. The reduced solar energy reflection (or albedo) from the loss of ice and snow is moreover expected to accelerate the warming trend, generating a feedback loop that may be leading to runaway meltings and other far-reaching regional changes.

Summer sea ice in the Artic Ocean has been declining over the past several decades, reaching an all time low in 2005. The steady decline was replaced by a plummeting decline in 2007 when the summer Arctic sea ice decreased by 23 percent from the 2005 low. Further examination led to the startling discovery that more than two-thirds of the rapid sea-ice melting is now happening from below, an amount representing 5 times the normal summer loss, caused by the warming of the waters, rather than from above, as usually caused by the summer sun (Perkins

2007). The 2005 melting had already led climate scientists—who have been carefully modeling the potential impacts of global warming—to note that rather than earlier calculations of taking 100 years for the Arctic Ocean to become ice free in the summer, it could happen by 2050. The 2007 melting has now led some to note that summer sea-ice melting may be 20 years ahead of the theoretical projections; they have further revised their estimate of an ice-free Arctic summer to 2030. Climatologists are surprised at how much faster these consequences are unfolding than even the extreme scenarios of their models. Normally cautious scientists are now stating openly that 2007 may prove to have been the "tipping point"—the point at which the melting of summer Arctic sea ice became self-perpetuating (ibid.).

### Observations in Greenland and the Antarctic

The melting of floating sea ice alone does not raise sea levels. It is the melting of land-based glaciers, especially those covering Greenland and the Antarctic continent, that will contribute directly to sea-level increase. The Greenland ice sheet would raise the oceans by 24 feet if it melts; a melted west Antarctic ice sheet would increase sea level another 19 feet; and if the east Antarctic ice sheets were to melt sea levels would increase another 170 feet. These calculations give a total of 213 feet potential sea-level rise from melting ice on land. It has long been assumed that these meltings would take millennia.

But here again the effects appear to be accelerating at a pace well beyond those expected from computer modeling. The ice-melt rate in Greenland, for example, from 2004 to 2006 was *250 percent greater* than the ice-melt rate from 2002 to 2004. The record high Greenland ice sheet melt of the last 50 years, set in 2005, was trumped in 2007. Studies have demonstrated that, whereas the melting had been more related to regional climate changes between 1960 and 1990, the pattern of melting has since changed to reflect global temperature variations, putting the fingerprint of global warming firmly on Greenland's ice sheet losses (Hanna et al. 2008).

The disquieting increase in Greenland's ice melt is being matched on the west Antarctic continent. Scientific studies have shown that between 1996 and 2006 there was a 59 to 75 percent increase in annual ice loss from west Antarctica, accompanied by a 140 percent increase in ice losses from the Antarctic peninsula; the ice loss from Antartica now nearly equals that from Greenland. The most startling event was the complete collapse of the Rhode Island sized Larsen B iceshelf in a span of two months in 2002, following 10,000 years of stability. The floating ice shelves in front of glaciers and ice streams serve as dams to slow the ice stream motion. As they break up, the ice stream flow into the sea accelerates. In addition summer melt water (appearing more and more on the surface of glaciers) is apparently being conducted to the glacier bases, helping grease, and hence accelerate, the glacier flow rate (Bell 2008). The more rapid flow of glaciers is interpreted as the source of the accelerated ice loss in Antarctica.[2]

Whereas some scientists (see IPPC 2007) still conservatively project less than a 1 meter sea-level rise by the end of this century, the evidence of the rapid and accelerating changes now taking place has led other leading climatologists to suggest that it could be 2 meters or more—with a devastating potential flooding and storm erosion impact for many continental coastal lands, and a death knell for low-lying islands.

### The Oceans

Global warming has introduced two additional disturbing changes to the world's oceans. First, there is already a measured reduction in primary ocean productivity in nine of the twelve ocean basins. As life itself is a crucial participant in the Earth system, such a destabilization of life in the world's oceans can certainly be expected to ripple through all global systems. This will be particularly difficult for the future human populations that depend on fish protein. (Much is now drawn from aquaculture, but that is also introducing problems with chemicals introduced into fish feed and illnesses in fish raised in captivity.)

Second, drops in oxygen levels, observed throughout the Pacific Ocean and with oxygen-deprived "dead zones" appearing with ever-increasing frequency and intensity near shorelines, are being linked not just to the usual source of overfertilization from agricultural runoff (obviously directly human caused) but apparently also both to the warming of the waters and to changes in ocean circulations that are likely a result of global warming (and hence indirectly caused by human interventions in the atmosphere; Juncosa 2008).

Things can only get worse as, third, the increase in carbon being absorbed by the oceans is already leading to their acidification. At the minimum this will reduce the oceans' ability—as the major global $CO_2$ sink—to absorb the increasing $CO_2$. But of greater ecological consequence is that when that acidity reaches a critical level, phytoplankton can begin to calcify and disintegrate. Recent research suggests that primary production could increase from this, while others project that with increased calcification, phytoplankton disintegration would reduce primary productivity. Projections suggest that this could happen with atmospheric $CO_2$ concentrations in the range of 580 to 720ppm. Since the World Bank has projected a $CO_2$ concentration of 750ppm by 2100 (a level that analysts concur *must* be avoided), this could be a genuinely realistic scenario with potentially disastrous potential consequences for all of the ocean food chains, and thus obviously for people as well.

## First International Responses

International efforts have been underway to address global warming and climate change, but the results to date have been minor and continue to be mired in politics, leaving little room for the kind of response needed. But there is hope, and emerging evidence that an international response is possible.

The world's governments banded together to address a global atmospheric problem in March 1985 with the adoption of the Vienna Convention for the Protection of the Ozone Layer. This was followed up in September 1987, when twenty-four nations signed the "Montreal Protocol on Substances that Deplete the Ozone Layer," putting into action the aims of the 1985 Vienna Convention by setting legally binding controls and targets for phasing out the chlorinated chemicals that were causing the problem. The results have proved to be amazingly effective. International cooperation avoided a global atmospheric hazard with serious health consequences to humans and other species.

Less than three months after the implementation of the Montreal Protocol to protect the ozone layer, in December 1987 the United Nations General Assembly passed a resolution entitled "Environmental Perspective to the Year 2000 and Beyond," in which global warming first surfaced as an important policy area that should command the international attention.[3] Launched by UN resolution a year later, the "Intergovernmental Panel on Climate Change" (IPCC) was created jointly by the World Meteorological Organization and the United Nations Environment Programme (UNEP). The responsibility of this international assemblage of scientists was to assess the science and risks of "human-induced climate change," and to provide counsel to the United Nations, policy makers, and the public.

The first IPCC assessment report, published in 1990, was crucial in representing an intenational scientific convergence on the view that human activities are the main driving force behind global warming. That report, in turn, provided the empirical basis for the creation of the next policy framework by the United Nations—the United Nations Framework Convention on Climate Change (UNFCCC). In March 1994 the UNFCCC went into effect with 189 nations as signatories to this Treaty—including the United States. By their signatures, the countries that signed the UNFCCC declared that they were "determined to protect the climate system for present and future generations."[4] There have been three additional IPCC assessment reports since then, the most recent released in the summer of 2007. In addition to steadily refining the scientific conclusions and formalizing recommendations for international response, over the years the four assessment reports have been notable for their increasing certainty that human actions are now the primary cause of global warming and their growing concern regarding both the present and future probable consequences.

In December 1997 the famous Kyoto Protocol was adopted, delineating binding targets and a timetable for greenhouse gas emission reductions by the developed nations.[5] It was opened for signature in March 1998, but not ratified until early 2005, when 55 nations (accounting for 55 percent of the global carbon dioxide emissions) had signed. The goal of the Kyoto Protocol is to reduce the greenhouse gas emissions (in carbon dioxide equivalents) by the developed nations to an overall average of 5.2 percent below the 1990 levels, accomplishing this in the 2008 to 2012 window. Different signatory nations were assigned different targets, with the total to meet the overall goal. The United States refused to sign. Russian compliance put the total signatories over the top, enabling the Kyoto Protocol to become international policy.

The developing nations were given a pass in this agreement, a source of great US displeasure, to prevent economic hardship during development. But with China surpassing the United States in carbon dioxide emissions in 2008, it is clear that a next, post–Kyoto Protocol will need to differentiate between those developing nations (China, but also India) that are now on a par with developed nations in emissions and those that are not.

The thirteenth "Conference of the Parties" (COP13) in 2007 in Bali, which was to begin to develop standards for the post-2012 period, dissolved into political wrangling and rhetoric, with the United States continuing to stand in strong opposition to any mandatory targets. As this is being written, the hopes are being pinned on COP14, to take place in 2009, when, according to agreement by all nations, a threshold is to be firmly agreed upon regarding maximum global atmospheric concentration of greenhouse gases.

The Kyoto Protocol represents a beginning, standing as an important international acknowledgement of the need for cooperative action. However, even if all nations were to meet the Kyoto $CO_2$ reduction targets, emissions would *continue to grow*. Clearly, more stringent goals and reduction targets for greenhouse gas emissions need to be set and implemented. The question is how? Will it be technically and economically feasible to meet stringent reductions in greenhouse gas emissions in ways that can also facilitate the return to a Gaian balance of Earth processes?

## Setting Global Warming and Greenhouse Gas Targets for the Earth System—As We Know It

According to international scientific consensus, in order to avoid "dangerous climate change," the long-term temperature rise of the planet's surface lands, atmosphere, and waters should not exceed 2°C. We are almost 40 percent of the way to that point already. The maximum global temperature goal that scientists are converging on requires the concentration of $CO_2$ be no more than 450 ppm by 2050 (Meinshausen 2006). Since we are now at 390 ppm of $CO_2$—a rise in concentration by about 110 ppm from the pre-industrial levels—this target constitutes a formidable challenge to all nations.

The approximate figure of 450 ppm represents the concentration range of greenhouse gases at which there is a 50 percent chance that the temperature will not exceed the 2°C limit, and a nearly 70 percent chance it will not exceed a 3°C rise. It is the best available estimate of the upper limit for avoiding disastrous ice melting in the Arctic and Antarctic regions, unacceptable ocean level rise, and dangerous climate and ecological changes.[6] In other words, this is a goal for reestablishing a balance of the Earth's physical and living systems within acceptable bounds, and within a range that can again be stabilized and maintained by Earth system processes.

While there is no certainty that even this figure is safe, it sets a defensible target for action. The European Union has already adopted these figures—2°C and 450 ppm—as the basis for the EU emission reduction goals. These targets indicate that by 2050 global greenhouse gas emissions must be reduced by 50 percent over the base year 2000 levels. But in recognizing that the developing nations will have difficulty in meeting these reduction goals even if they try, the industrial nations—including a cooperating United States—will need to reduce their emissions by an average of 70 to 80 percent below year 2000 levels by 2050.[7]

In order to have a reasonable chance of accomplishing these deep cuts, industrial nations must have their emissions peak by 2010, and thereafter begin to decline at an average rate of 4 percent per year—that is, at a faster rate than the global emissions are presently increasing. If the emissions do not begin to decline until 2020, the decline rate will have to be an almost unachievable 8 percent per year. The inevitable conclusion is that a meaningful response by all nations must be set in motion immediately, if there is to be hope of success. While the path will not be easy, these aggressive targets can be met. A wide range of technology and policy options will be required, woven into international agreements that include significant support by the industrial nations for the developing nations. The starting point, however, must be the international acceptance of $CO_2$ concentration and temperature goals as binding global limits.

Governments opposing mandatory rules argue that voluntary goals by industries and governments should be adequate. However a national survey of 500 big industries in Britain, the United States, Germany, Japan, India, and China reported in early 2008 that climate change ranked only eighth in the concerns of the business leaders, behind, for example, increasing sales and securing growth in emerging markets—and would probably become of even lesser concern if climate change causes the global economy to deteriorate (David et al. 2008).

Some businesses have become environmentally responsible on their own initiative, recognizing that tackling energy efficiency and greenhouse emission reductions will place them in an advantageous position in the future, at a time when other businesses will have to scramble to meet inevitable new emission rules, including the payment of carbon taxes.[8] But, by a large margin, the businesses polled stated that setting goals for emission reductions was properly the role of governments. The unfortunate present result of that nonperforming "proper role" has been a nearly 2 to 3.5 percent annual *increase* in global $CO_2$ emissions with no immediate prospects of rates abetting. Voluntary controls seem fraught with competing interests and political agendas, and are not adequate in confronting the enormous technical and policy challenge that we face.

The nations that signed onto Kyoto now have the 2008 to 2012 period to deliver their committed emission savings, so this rate of global increase may indeed reduce for a time. On the other hand, the United States and China, neither signatories to the Protocol, and both staunch opponents of mandatory emission reduction standards, together account for nearly 50 percent of global greenhouse emissions. China is bringing online a new coal-fired power plant every one or two weeks, while the United States has over 150 new coal plants in various stages of planning or construction.

If we move our focus away from national government and corporate initiatives to voluntary actions by *local* governments, the story changes. In the United States, for example, on February 16, 2005, the day the Kyoto Accord went into effect, the mayor of Seattle, Washington, launched the US Mayors' Climate Protection Agreement. In June of that year, the US Conference of Mayors passed that agreement unanimously. Their target is modest, fashioned after the then vice president Al Gore's acceptance in Kyoto of a US target of a 7 percent reduction in greenhouse gas emissions from 1990 to 2012. While the United States did not sign the final agreement, by the end of January 2008, 780 cities in all 50 US states and Puerto Rico (representing more than one-third of the US population) had pledged to meet or exceed the terms of the unsigned US Kyoto Protocol obligation.[9]

This precedent was followed up at the December 2007 UN Climate Change Conference (COP13) in Bali. Whereas the participating national governments had great difficulty in securing significant agreements amid much squabbling and posturing, the World Mayors' and Local Governments' Climate Protection Agreement was launched at the same meeting on December 12.[10] A Gaia-appropriate target of reducing global emissions by 60 percent from 1990 levels by 2050, and by 80 percent for the industrial nations, was adopted. Signatories also agreed to offer annual reports on their greenhouse gas emissions and document their efforts to reduce them.

## Emission-Reducing Strategies: Considering the Alternatives

Energy efficiency in buildings provides the technically easiest, least expensive, and fastest measure for reducing global warming emissions. Both of the Mayors' Climate Protection Agreements include energy efficiency improvements to city facilities. The US Agreement goes further, in adopting the goals of the "Architecture 2030 Challenge," whereby the fossil fuel reduction standard for all new buildings is to be increased to 60 percent in 2010 all the way to 90 percent by 2025. The aim is to become carbon neutral by 2030.[11] This standard was adopted as policy by the American Institute of Architects and was included in the US Energy Independence and Security Act of 2007 (signed into law at the end of 2007) for all federal buildings.

Buildings in the United States are directly and indirectly responsible for 48 percent of the US greenhouse gas emissions,[12] and use 67 percent of all of the country's electricity. By 2035, however, the amount of new and refurbished buildings in the United States will approximately equal all of the buildings currently in place in the entire country.[13] A nationwide adoption of the Architecture 2030 standards has the potential therefore to displace up to 36 percent of the nation's greenhouse gas emissions by 2030. This efficiency measure alone could account for almost half of the 2050 target of an 80 percent reduction in US emissions.

Generally, the most impacting buildings are in or close to cities. About 50 percent of the world's population lives in cities, today also the locus of 75 percent of the worldwide consumption of energy and 70 percent of the world's consumption of electricity. By 2030, two-thirds of the globe's population is expected to live in cities, accounting for an even larger share of global energy use and emissions. National policy for the reduction of greenhouse gas emissions must consequently begin in the cities. That's where the people are—as well as the buildings, the vehicles, and the opportunities for the most environmentally and economically advantageous emission reductions. Alternatively, cities banding together to bring about change, as in the United States or Spain, could *de facto* begin to define national policy.

Over the next 50 years, a multitude of solutions to reduce global warming emissions will need to be pursued *simultaneously*, each involving serious and concerted effort. Some proposals are pie-in-the-sky, relying on grand unknown new technologies, or proposing giant sunshades in space, or other risky geoengineering with substances injected into the atmosphere or oceans. Others, on the surface, appear to many decision makers to have some merit, such as today's rising enthusiasm for the nuclear power option. There needs to be a way for decision makers to make early assessments of the reality and potential value of technology options that are going to be funded.

More realistic options, based on available technology, and appropriate for the first fifty years of this century's global energy transition, were outlined in 2004 in a well-known paper by Stephen Pacala and Robert Socolow.[14] They derived a series of wedge-shaped pieces of a carbon-emission pie chart, with each wedge worth a reduction in global carbon emissions of 25 billion tons by 2056. Seven such wedges taken together (representing 7 policy and technology tracks) could stabilize global $CO_2$ emissions at about today's level.[15] This in turn would lead to an ultimate $CO_2$ level about twice the pre-industrial level (around 560ppm). On present-day scientific assessments, this is still a dangerously high $CO_2$ level. The primary value of Pacala and Socolow's work, however, was to convert a set of numerical emission-reduction targets into understandable and feasible numerical requirements for the application of related technology options.

The value of this simple approach can be illustrated by returning to the nuclear power example. Just one wedge would be provided by a doubling of today's nuclear power capacity, and then only if it offsets coal-fired generation. With 435 nuclear power plants in operation worldwide today, and because almost all of those plants will have been retired and decommissioned within the next 50 years, 17 or 18 *new* nuclear power plants would be required to become operational every year for the next 50 years. Given the realities of the availability of materials, components, skilled labor, finances, and regulatory oversight, this may represent 3 to 5 times the maximum realistic rate for nuclear plant construction.[16] An alternative way of looking at this is that, within the present maximum rate for new nuclear power plant construction, by 2030 only 1/7 of a wedge will have been built, leaving about 98 percent of the remaining requirement for the reduction of carbon emissions to other policies and technologies.[17]

These simple calculations reveal that in all likelihood, nuclear energy could only provide a fraction of the required carbon-free energy by 2050, probably no more than a few percent. In today's seeming rush to revive nuclear energy, this should provide a note of realism to decision makers who will need to see to it that the other necessary technical options are also adequately pursued (Hultman et al. 2007).

Indeed, when one steps back and looks at future energy resource options, the naturally available renewable energy resources stand out, not only because of their enormous resource potential but also because they are the same ones that have powered Gaia processes for all Earth's history: solar, wind, ocean thermal, biofuel, hydroelectric, geothermal, and tidal. They are moreover the only energy sources that are guaranteed to be available in perpetuity.

Consider the following comparisons: The energy value of all sources stored in the Earth[18] is estimated to be about 9,100,000 terawatt hours.[19] The annual resource potential of renewable energy is about 350,400,000 terawatt hours per year—almost 40 times the total of stored energy in the planet. The annual energy from solar radiation accounts for 99.9 percent of this. Wind energy, while small compared to solar, is still about three times the total of all energy used by humans (which is about 80,000 terawatt hours per year). These figures invite the conclusion that about 0.075 percent of the land area on the planet receives energy equivalent to the total annual use of energy by all human civilizations. The renewable energy resource potential dwarfs all other long-range resource potentials—and it is renewed every year.

That's the good news. What is often considered to be the bad news is that the renewable energy resources are diffuse. You cannot pour a concentrated amount of any of them out of a can, or make a lot of money by drilling into a solid or liquid resource of it. To gather them requires constructing collection devices over large areas and paying for those collection devices up front, although after that initial investment the energy is free.

Renewable energy systems are still early in their global applications, so that they are more costly than the conventional resources. Price comparisons are highly distorted, however, by both direct and hidden subsidies for the conventional energy sources. Hidden subsidies include the costs of a military presence and wars to protect access to oil, the failure to include the costs of the environmental and human health impacts arising from their use, and the failure to adjust current prices for future risk in supply. The greatest of the latter risks will be the (apparently near-term) beginning of the period when global supply of fossil fuels cannot keep up with growing demand—followed by an absolute decline in supply (peak oil). Other risks include the effect of growing global nationalism and militarism on global energy supplies, accompanied by risks from terrorism to energy infrastructure and nuclear power plants, as well as risks of nuclear terrorism, using materials meant for the production of nuclear power.

Renewable energy applications are blessedly immune from these impacts and risks, providing greater security, human health and energy independence advantages, and lower impact on natural systems. Renewable energy creates local employment and converts fossil-fuel energy dollars that previously went out of the cities or the region into local and regional benefits with real economic advantages (also not included in fuel price comparisons). The advantage of fossil fuels, in terms of ease of use and extensive infrastructure, is short term and evanescing. The enormity and benefits of the solar resource, in particular, compared with the finite amounts and serious shortcomings of fossil resources, assures that renewable energy will be the long-range energy resource for human civilizations. Following the transition (to unfold in this century) the climate system will hopefully stabilize at familiar equilibriums—friendly to human life and extant species and ecosystems.

## Renewable Energy Rising

Something is happening with renewable energy worldwide: the use of those resources as carbon-free technology and policy solution to global warming is growing with astounding rapidity. Solar photovoltaics (PV) and wind generation are now the fastest growing energy sources in the world. Renewable energy supplies over 17 percent of the world's primary energy, if one includes traditional biomass and large hydropower.[20] In 2007 the "new" renewables (i.e., solar, wind, geothermal, small hydro, modern biomass, and biofuels) accounted for only 2.2 percent of the world demand for primary energy and 3.5 percent of global electricity production. Yet these small numbers mask the dramatic rates of growth for renewable energy capacities. The small percentages of the global total for these renewables also masks the greatly accelerating pace of annual investments in those technologies. In addition the renewable energy industries have been shown to create a proportionately larger number of employment opportunities for unit energy output than the conventional energy sources, in essence turning money previously paid for fuels and large central power plants into money paid for people.[21] The global stock markets have responded favorably to all this, with stock values for the publicly traded renewable energy companies often leading the way in annual growth rates and frequently growing in market value faster than the market as a whole.

The most concrete sign that renewable energy has come of age in the global transition to clean, noncarbon resources is the scope of the national standards now being set for the progressive adoption of renewable energy into the total energy mix. These are not in the 2 percent range, but rather they average in the 10 to 20 percent range. This policy growth has spurred and supported the market growth of renewables. While a complete overview of renewable energy policy developments is beyond the scope of this chapter, a few highlights are worth noting.

A recent survey[22] has revealed that, by 2007, 61 countries had set national renewable energy supply targets and mandatory timetables for achieving them. In the United States, renewable portfolio standards (RPS), defining the share of renewable energy in the electricity production mix required by certain dates, have been adopted in 50 percent of the states and the District of Columbia, ranging, for example, from a low of 8 percent by 2020 for Pennsylvania to a high of 20 percent by 2010 for California.[23] The European Commission adopted binding goals that extended their 2010 policies to new targets of 20 percent of final energy to come from renewable resources, along with 10 percent of transportation fuels, by 2020. Also included in the list of countries adopting binding renewable energy standards are 13 of the developing countries. China, among them, adopted their plan in September 2007, calling for 15 percent of primary energy from renewable resources by 2020.[24] In addition at least 57 countries, including 21 developing nations, have set binding targets for the share of electrical power to come from renewable energy.

The rise of renewables—from technological innovations to policy shifts—demonstrates that present and long-range directions for the development and implementation of renewable energy are now on a par with major conventional energy resources. Renewable energy has reached adolescence. It is also clear that renewable energy policies, as they emerge from this adolescence, require cooperative parenting skills at every level of government, from cities to states to nations. It is equally clear that society will need to transfer the enormously distorting and wasteful present funding from pathological applications like wars to the productive development and deployment of the energy technologies that will power the future human societies and industries. The funds are there, and available, if channeled by appropriately maturing societal priorities.

### Can Renewable Energy Meet the Needs of the Future?

Earlier in this chapter I examined the implications of accomplishing one wedge (25 billion tons of $CO_2$ displaced over 50 years) from nuclear power. While technically possible, it is probable that the realities of such a necessary pace of construction of new nuclear power plants would make that a difficult target. On the other hand, the nuclear industry feels that it is quite achievable. If it is, achieving this target would still leave 86 percent of the carbon free energy to come from other sources 50 years from now.

What would it take to deliver one wedge of wind energy or solar energy? The authors of the wedge formalism suggest that to displace one wedge of coal-fired generation by either wind or solar technology would require 2,100 GW (2.1 TW) of installed capacity, including sufficient conventional-fired backup to level the intermittent outputs of these two resources. This is a huge amount, representing 23 times the total global installation of wind power, and 210 times the global installation of solar power, at the end of 2007. How do these fare with the same kinds of realism tests applied earlier to nuclear power?

An analysis was undertaken in 2005 by Stanford University scientists to quantify the world's practical and realistic wind power potential, based on minimum necessary wind velocities for economic energy production and average turbine hub heights. Their research revealed 72 TW of potential (Archer and Jacobson 2005). Tapping into just 3 percent of this would produce one wedge of $CO_2$ reduction, so the wind resource is there in abundance. The present annual growth rate of wind energy (a doubling every three years) yields that this 3 percent could be achieved in 14 years. While this leads to unrealistic rates of installations, spreading it over 50 years could well be achievable.

As for solar energy for all practical purposes it is unlimited. The one wedge target for solar power systems could be reached in 16 years with a doubling time of half of what occurred during the 2002 to 2006 period. Again, this could become a realistic target if spread over 50 years. Adding another wedge produced from biomass used directly for heat and electricity, and to produce carbon-free fuels, would allow the world to meet over 40 percent of global energy needs from renewable energy resources after 50 years.

This is getting close to the 2050 global target of a 50 to 60 percent reduction in global $CO_2$ emissions to avoid the "dangerous climate change" threshold, although it is still well below the necessary 70 to 80 percent reduction target of the developed nations. But as earlier noted, another 40 percent reduction in $CO_2$ emissions could be accomplished in the developed nations from energy efficiency improvements, in particular with buildings.

One conclusion is that while adding the one wedge of nuclear power would further reduce the $CO_2$ emissions, that may not be necessary, for the combination of the renewable energy resources and energy efficiency may well be enough to avoid the dangerous climate change thresholds. This conclusion is supported by a number of published models and scenarios that appear to show a feasible renewable energy future meeting from 20 to 50 percent of global energy needs in 30 to 50 years, and up to 100 percent of global energy needs by the end of this century (Martinot 2008). For example, a 2007 analysis by scientists of the National Renewable Energy Laboratory in the United States concluded that 57 percent of the necessary reduction in year 2030 $CO_2$ emissions (for the United States) could be achieved by energy efficiency, with renewable energy technologies providing 50 percent of the residual energy required in that year.[25] These together would produce the necessary reduction in US $CO_2$ emissions by 2030, continuing to lower after that.[26]

The question, of course, is whether this will all be fast enough to prevent serious climate changes and enormously expensive consequences for human civilization. Recent publications have been sounding the alarm that even these kinds of legislative actions may no longer be adequate to prevent the most serious consequences of global warming, and that *leaps* in policies for much higher and more aggressive energy efficiency and conversion goals to carbon-free energies are now needed by *all* nations[27] (Hansen 2008).

## Gaia and Humanity: Reversing Dead-end Human Behavior

The essence of Gaian thinking is that the complex living and physical systems of the Earth work together as an integrated whole, with feedback mechanisms within and between systems resulting in stability or recovery from perturbations that are not excessive. Gaia theory expands on this to deduce that life is an *active* participant in this process, by contributing profoundly to the maintenance of dynamic stability of natural systems within ranges necessary for life itself.

Human beings have, within the past 100 years, become the greatest single impacting form of life on Earth. A 2007 analysis, for example, revealed that humans now consume almost 25 percent of the Earth's *total biological productivity*.[28] Consequently the active participation of the life element of Gaia is now largely being determined by the present and near-future actions of the human species. All available evidence today suggests that many human actions are introducing physical and chemical forcings that are too great for the Earth system to counter, thereby leading to ever-greater destabilizations.

The geological record, and especially climatic evidence, indicate that Gaia can be dangerously fast in destabilizing, and maddeningly slow in recovering. This adds an enormous urgency to all nations, religions, and cultures to redefine their most basic social institutions in terms of meeting our collective global responsibility toward a newly stable, biologically rich and diverse Gaia that can support human civilizations for millennia to come.

While the challenges of climate change have been the subject matter of this chapter, the broad picture of our adverse impact on the Earth system is more extensive: human overpopulation and unsustainable land-use patterns; depletion of freshwater resources; wars (which are terribly destructive of environments, to say nothing of people and cultures); rapidly expanding extinctions, forest destruction, and overall diminishment of biodiversity; loss of topsoil and decline in soil fertility; air, soil and water pollution; chemical contaminations of almost all living things, including human bodies; and overconsumption of the Earth's resources.

Ironically, this litany suggests that people are knowingly (albeit not willfully) disrupting the Earth system in such a way as to make human life more difficult, and perhaps even untenable in the long run. This does not contradict Gaia theory. Life on Earth can persist in the future in a new equilibrium of physical and living systems, featuring ecosystems and biomes that are different from those of today—without human beings. There is nothing in Gaia theory, after all, that says that life on Earth must include humans.

Humanity now bears the burden of determining whether the Gaia stabilizations necessary for the continued support of the human species can be protected and recovered. This awareness is only now dawning, driven by dramatic and disturbing changes in local and global climate and water systems, changes that are known to be largely the result of human decisions and actions. But humans have *elected to cause* these changes. Why are we being so obtuse?

Because, paradoxically, it seems that we—the ostensible pinnacle of evolutionary development—have invented religious, political, and economic institutions that have caused humans to adopt policies and to undertake actions based on preference, belief, or expediency. The impact on the Earth's living and physical systems, or on future generations, has rarely been considered. As a consequence almost all actions of societies and civilizations over the past couple of centuries—a blip in geological time—have conspired to destroy Gaian life-support systems. Only recently have some societies accepted a measure of this responsibility through emerging environmental protections.

Previously the scale of human impact was small in the great scheme of the Earth's interconnected living and nonliving systems. Now it is suddenly too great. Previously human social, political, cultural, and religious institutions had little effect on Gaia's global mechanisms. Suddenly the effects are dominant.

These same social institutions, however, can provide the tools that must be used to reverse these dangerous directions. This is fortunate, for while the Earth's physical processes possess great inertia, generally moving and changing slowly, humans can elect to change their social structures and adopt new global responsibilities on much shorter time scales. Restoring the Earth is now our responsibility. We need to understand that the path to global re-dependence on the naturally occurring renewable energy sources is not only desirable, it is entirely feasible. Furthermore it has become imperative. Just as we dream of a future without wars over resources, we can share a vision of a future powered by the sun. And what we can dream, we can accomplish.

## Notes

1. Arctic statistics quoted here are from the *Arctic Climate Impact Assessment*, drawn from an overview lecture by Robert Correll. The complete report can be downloaded at http://www.acia.uaf.edu/pages/scientific.html.

2. Inland glaciers are also melting. During the summer 2003 heat wave in Europe about 10 percent of the Alpine glaciers melted. At this pace 75 percent of the glaciers in Switzerland will be gone by 2050. By 2030 Glacier National Park in the United States. will only have memories where the glaciers once were.

3. United Nations General Assembly, 96th Plenary meeting, Resolution 42/186, December 11, 1987, http://www.un.org/documents/ga/res/42/ares42-186.htm.

4. The full text of the UNFCCC United Nations Framework Convention on Climate Change can be found at http://unfccc.int/resource/docs/convkp/conveng.pdf.

5. Kyoto Protocol to the United Nations Framework Convention on Climate Change, United Nations 1998, http://unfccc.int/resource/docs/convkp/kpeng.pdf.

6. An entire conference was devoted to the problem of "Avoiding Dangerous Climate Change" in 2005, producing a book of peer-reviewed papers that substantiate the reasonableness of these numerical targets. The 16.3 MB book can be downloaded from http://www.defra.gov.uk/environment/climatechange/internat/pdf/avoid-dangercc.pdf.

7. A short and readable summary of these points can be found in *Global Warming: A Target for U.S. Emissions* by the Union of Concerned Scientists, at http://www.ucsusa.org/global_warming/science/emissionstarget.html.

8. For many resources on this subject, see the Pew Center for Global Climate Change, Business and Climate reports, at http://www.pewclimate.org/companies_leading_the_way_belc/business_resource_portal/business_reports_res.cfm.

9. A running tally of cities that have signed onto the agreement is kept on the home page of the US Conference of Mayors, http://www.usmayors.org/uscm/home.asp.

10. http://www.cities-localgovernments.org/uclg/upload/news/newsdocs/World_Mayors_Local_Governments_Climate_Protection_Agreement.pdf.

11. See http://www.architecture2030.org/home.html.

12. US Energy Information Agency. See http://www.architecture2030.org/home.html.

13. Edward Mazria, founder of the Architecture 2030 Challenge, personal communication.

14. See also Pacala and Socolow (2006).

15. The world emits about 7 billion tons of carbon per day. This represents about 25 billion tons of $CO_2$ equivalent, with the ratio of 3.67 tons of $CO_2$ for each ton of carbon.

16. The US Nuclear Regulatory Commission is anticipating fast-track license applications by 2009 for 28 new nuclear reactors to be built at 19 sites in the United States.

17. The International Atomic Energy Agency (IAEA) in its October 2007 "high projection" suggests that world nuclear capacity could increase by 84 percent, up to 679GW(e), which would have nuclear on track to accomplish one full wedge by 2050.

18. Coal, uranium 235, petroleum, natural gas, and tar sands. Figures from Steven Heckeroth and Richard Perez.

19. One terawatt hour is one billion kilowatt hours, or 1,000 gigawatt hours. Global consumption of stored energy is about 80,000 terawatt hours/year.

20. The global statistics used in this section are from Martinot (2008).

21. In Germany the renewable energy industries, which produce about 6 percent of Germany's electricity, have created more jobs than the nuclear power industry, which produces about 30 percent of Germany's electricity, for about a 5:1 ratio in job-producing advantages by the renewables.

22. Martinot (2008).

23. This policy concept was co-introduced by the author of this chapter in California Public Utility proceedings in 1995, acting on behalf of the Union of Concerned Scientists, in partnership with Nancy Rayder, acting on behalf of the American Wind Energy Association.

24. But the carbon reduction benefits of this modest goal will be completely swamped if China continues to build the 800 coal-fired power plants they are anticipating.

25. See *Tackling Climate Change in the U.S.: Potential Carbon Emissions Reductions from Energy Efficiency and Renewable Energy by 2030*, www.ases. org/climatechange. A summary of the conclusions can be found in Kutscher (2007).

26. An even more ambitious plan for the United States was published in early 2008, showing a purely solar energy path that could provide 69 percent of the country's electricity and 35 percent of the total energy by 2050 (Zweibel et al. 2008). The total cost to accomplish this solar transition for the United States was estimated to be $400 billion over 40 years (the total 40-year cost represents two *years* of the Iraq war costs).

27. Former vice president Al Gore, co-recipient with the IPCC scientists of the 2007 Nobel Peace Prize for their joint work on climate change, and the publicizing of the enormity of the problem, announced in 2008 that a worthy world goal would be to strive for no fossil-fuel use within ten years.

28. *Science News* 172: 235.

# References

Archer, C. L., and M. Z. Jacobson. 2005. Evaluation of global wind power. *Journal of Geophysical Research* 110: D12110.

Bell, R. E. 2008. The unquiet ice. *Scientific American* 298 (2): 60–67.

David, T. H., G. Lean, and S. Mesur. 2008. Big business says addressing climate change "rates very low on agenda." *The Independent*, January 27.

Emanual, K., R. Sundararajan, and J. Williams. 2008. *Bulletin of the American Meteorological Society* 89 (3): 347–67.

Hanna, E., P. Huybrechts, K. Steffen, J. Cappelen, R. Huff, C. Shuman, T. Irvine-Flyng, and M. Griffiths. 2008. Increased runoff from melt from the Greenland ice sheet: A response to global warming. *Journal of Climate* 21 (2): 331–41.

Hansen, J. 2008. Global warming 20 years later: Tipping points near. Presentation before the National Press Club and the House Select Committee on Energy Independence and Global Warming, June 23.

Hultman, N. E., J. G. Koomey, and D. M. Kammen. 2007. What history can teach us about the future costs of U.S. nuclear power. *Environmental Science and Technology* 41 (7): 2088–93.

IPCC. 2007. http://www.ipcc.ch/.

Juncosa, B. 2008. Suffocating seas. *Scientific American* 299 (4): 20–22.

Kerr, R. A. 2006. A worrying trend of less ice, higher seas. *Science* 311 (5768): 1698–1701.

Kutscher, C. 2007. Tackling climate change in the U.S. *Solar Today* 21 (2): 26–29.

Lempinen, E. W. 2007. Science and environment: searching for climate change's "tipping points." *Science* 318 (5855): 1396–97.

Mann, M. E., and K. Emanual. 2006. Atlantic hurricane trends linked to climate change. *EOS* 87 (24): 233–44.

Martinot, E. 2008. *Renewables 2007—Global Status Review*. Washington, DC: Worldwatch Institute.

Meinshausen, M. 2006. What does a 2°C target mean for greenhouse gas concentrations? In H. Schellnhuber et al., eds., *Avoiding Dangerous Climate Change*. Cambridge: Cambridge University Press.

Pacala, S., and R. Socolow. 2004. Stabilization wedges: Solving the climate problem for the next 50 years with current technologies. *Science* 305 (5686): 968–72.

Pacala, S., and R. Socolow. 2006. A plan to keep carbon in check. *Scientific American* 297 (9): 50–57.

Perkins, S. 2007. Portrait of a Meltdown. *Science News* 172 (December 29): 387.

Webster, P. J., G. J. Holland, J. A. Curry, and H.-R. Chang. 2005. Changes in tropical cyclone number, duration, and intensity in a warming environment. *Science* 309 (5742): 1844–46.

Zweibel, K., J. Mason, and V. Fthenakis. 2008. A solar grand plan. *Scientific American* 299 (January).

# 9

# Gaia's Freshwater: An Oncoming Crisis

Barbara Harwood

Nothing is more critical to the functioning of Gaia than water, the life-giving elixir that flows in all living things. It facilitates the complex photosynthetic chemistry that draws upon the sun's energy to support cellular growth. It carries and distributes nutrients and removes wastes to support living tissues and ecosystems. Water is not an equal opportunity liquefier. On land, clean freshwater is required. Saltwater with unimpeded nutrient flows and supporting thermal conditions is required for oceans to thrive. When humans interfere with the myriad interconnected water-based processes, we risk creating imbalance between life and its environments, possibly to the detriment of all life.

In this chapter, I trace the international scope of the perilous changes introduced by humans that have led to serious risk for the Earth system as we know it, and perhaps even for the survival of humanity. I then turn to signs of hope in the ways a few individuals and groups have made great strides in finding sustainable solutions for the water environments of their villages or communities.

## Signs Are Troubling

It is paradoxical that on a planet covered with water, we worry about getting enough of it for our survival. But about 97 percent of our water is salty. Another 2 percent is locked up in ice and glaciers. A small portion exists as moisture in the Earth's atmosphere, and less than 1 percent is potable water available for humans and the other creatures with which we share the biosphere. But when water is measured in billions of cubic kilometers or quadrillion acre-feet, 1 percent is still a lot.

If we continued to live as humans have for millennia, our freshwater would be restored by natural cycles of respiration, evaporation, and rainfall—but we have not. Our population growth has surged beyond the ability of freshwater to sustain it in most parts of the world, and often that growth is greatest in the poorest and driest countries on Earth. Signs are troubling. It is projected that freshwater needs in the world will increase 40 percent by 2020, and yet water tables are dropping. Freshwater is being contaminated or wasted around the world, while 1.4 billion people already have little or no access to clean water. And not enough people are paying attention.

Fueling the impending crisis, multinational corporations have inserted themselves into the water cycle in one place or another. Global giants like Coca-Cola drain freshwater springs in India, leaving the surrounding population with easier access to Cokes than to freshwater. It takes four liters of water to produce a one-liter bottle of soft drink. Beyond the questionable nutritional wisdom of this switch, it is inadvisable to extract fresh groundwater used by India's poorest populations and pour contaminated wastewater back onto the ground, souring wells (Stecklow 2005).

Corporations appropriate freshwater for industrial purposes, from making chemicals to cooling power plants, often polluting the water downstream and rendering it unusable by populations dependent on it for drinking and crop irrigation. Waste sludge dumped on land from industrial processes seeps into groundwater. Drug and cosmetic production leads to the dumping of hormones and other chemical by-products, known to be endocrine disrupters, into water supplies. Nitrates from fertilizer, livestock manure, and septic tanks pollute groundwater and contribute to eutrophication and dead zones in coastal waters. Heavy metals from coal mine wastes, chlorinated solvents from manufacturing processes, and salts from seawater intrusion add contaminants. Pesticides from agriculture, golf courses, and backyard pest treatment, as well as petrochemicals from leaky underground tanks and refineries flow not only into rivers but also downward into precious aquifers, contaminating water sources that for centuries have been the purest on the planet.

Government policies and World Bank funds often skew water policy in other ways that benefit corporations. According to *Fortune* magazine, in a world fleeing the vagaries of technology stocks, water is the best investment sector for the century (Tully 2000). The World Bank places the value of the current water market at close to $1 trillion. Its frightening prediction is that with only 5 percent of the world's population currently getting its water from corporations, "the profit potential is unlimited."[1]

Worldwide there seems to be a silent recognition of the intractable conflict between municipalities or local water districts spending precious funds to treat contaminated water versus corporations and agriculture fighting to reduce pollution controls to maximize profits. Too often government sides with polluters, as in 2001 when the Bush administration put on hold the planned reduction in the arsenic standard for drinking water from 50 to 10 ppb. (Some epidemiological data suggest the standard should be zero.) The reason given was that it would force many cities to upgrade their water systems, and it would cost too much money.

In addition to polluting and draining rivers and lakes, we are overdrawing water from groundwater aquifers around the world. This remarkable depletion of ancient water is responsible for 200 billion cubic meters of nonsustainable water consumption annually. The most critical depletion of aquifers has occurred in vital food-producing areas of India, China, North Africa, the Middle East, and the United States. In the central United States, the life-sustaining Ogallala aquifer underlying eight states in the nation's breadbasket—an underground lake larger than Lake Huron—is dropping by about 100 feet per year in some areas. The amount of annual depletion in the Ogallala aquifer is equal to the annual flow of 18 Colorado Rivers (Nicklas 2005).

Adding to the problem of aquifer depletion is the recent discovery that using groundwater for crop irrigation or industrial purposes adds more planet-warming carbon dioxide to the atmosphere than volcanoes do. "On average, groundwater holds from 10 to 100 times as much carbon dioxide as the water in lakes and rivers," according to Gwen L. Macpherson, hydrogeologist at the University of Kansas (Perkins 2007).

Big dams may increase economic activity and provide local employment, recreation, and hydropower, but they also destroy fisheries, displace populations, ruin beautiful landscapes, and cause environmental havoc downstream. "Today, as a result of the dam-building era, we move huge quantities of water out of watersheds, often to cities hundreds of miles away," states Robert Glennon. "To accomplish this feat," he continues, "we have engineered pumping stations with pipes that go over mountains and drilled tunnels through them, delivering water miles from its source to cities built in the deserts" (Glennon 2002: 21). Sponsored and funded by government policy, we have created enormous growth in areas that lack adequate water resources, such as Los Angeles and Las Vegas. "We literally move water uphill to wealth and power," Glennon adds (ibid.).

All this adds up to daunting if not impossible problems around the world. And for me, the worldwide water crisis has become personal; it has already arrived in my own backyard.

## Water versus Papayas

I stared for what seemed like hours at the beautiful young papaya tree of our family garden. Four years old already, it had four giant papayas ripening at the base of its leaves, and several other smaller ones draping around its core above. But it had planted itself in the *wrong* place—the only logical corner in our garden to put the three rainwater collection tanks above ground so they could gravity-feed our organic garden.

Despite our living 200 yards from Mexico's largest inland body of water, Lake Chapala, we worry about water. The water table below us has dropped about 250 feet in the last five years from burgeoning development. We have an extensive rainy season each summer and fall—from four to six months long—and the rain pours off the mountains above and to the north of us and into the lake, replenishing it by several inches each season.

But there are still problems with the lake. First, it is long, wide, and very shallow. Because it is so shallow, more water evaporates from it each year than is removed by the city of Guadalajara for use as the sole water supply for its five million people. There is also tremendous waste of water as it is withdrawn. Guadalajara has such a leaky water system, experts say, that upward of 40 percent of the water it imports from the Lake Chapala drains away unused.[2] Add to that the chemical contamination of the lake from heavy metals in industries that border its primary river source, the Lerma, and the pesticide contamination from farmers on its borders, and you have an endangered, polluted lake, unfit to serve the freshwater needs of the communities that now line its shores. As I contemplated our papaya tree, I realized that this is the same sort of water supply predicament nearly everyone on this planet is either already facing or will face within the next few years.

## Water Challenges in China

On the other side of the planet from Mexico, in northwest China's Shanxi Province, a farmer named Qiao Sanshi sits on a low wooden stool near his rainwater collection "cellar," a tank barely below ground, patiently waiting for a visitor. He is clutching in his hand the most precious gift he can offer that person: a glass of water. Because over five thousand such cellars for collecting rain have been installed in his small Hequ County, he can provide to a guest something which most of us still take for granted. But in many parts of north China, where water tables are dropping by a meter or more every year, this is impossible. One in three people living outside cities in China have no access to safe drinking water.

Dropping water tables are also affecting China's food supply. Its wheat harvest, grown largely in the semi-arid north, has dropped precipitously in this century. From 2002 to 2004 China went from being essentially self-sufficient in wheat to being the world's largest importer (Brown 2005: 102). In a country where jobs created by industrial development barely stay ahead of population growth, farmers regularly lose the water battle with industry.[3] As long as the world wheat supply can provide imports to feed the Chinese, this is viable. But with water tables falling worldwide and rivers being drained, China's dependence on grain imports may not be sustainable over the long term. The country's water emergency is dire. With 22 percent of the world's population and only 6 percent of its water resources, China is among the world's thirstiest countries. According to China's own news organization,[4] over 400 of China's 699 major cities are water short and 50 of those are labeled "seriously threatened," including Beijing, whose depleted groundwater led a Beijing wit to send relatives an email invitation to the 2008 Olympics with B.Y.O.W. at the bottom: Bring your own water.

To remedy Beijing's situation, the Chinese government, rather than teaching water conservation to reduce per capita consumption, is building the world's largest dam. The water behind the Three Gorges dam on the Yangtze River in the south will be pumped northward to Beijing in the world's largest water diversion project, the $25 billion transfer of 50 billion cubic meters of water a year into the drying Yellow, Huai, and Hai rivers through three 800-mile long channels. The Three Gorges dam is one of the world's most controversial public-works projects. It has displaced at least 1.13 million people from the villages upstream to allow for the 660km long reservoir behind the dam, and it has visibly increased pollution upstream.

Prime Minister Wen Jiabao himself, as early as 1999, had warned that the very "survival of the Chinese nation" was threatened by looming water shortages.[5] Officials outside the country have now begun to fear that China's water crisis responses will profoundly affect their own countries. For example, the Chinese construction of several huge dams on the upper Mekong River threatens the downstream countries Cambodia, Thailand, Laos, and Vietnam, whose economies and food supplies are almost totally dependent on the natural flows of the Mekong.

One can imagine a scenario in which, as China's industrial development takes precedence, water is increasingly drained and polluted, leaving the country without adequate drinking water for its billion-plus people, and without water for crop irrigation, power-plant cooling, or hydroelectric production. In desperation, China may look around its neighborhood for water sources it can appropriate.

Perhaps China was already contemplating this predicament when it invaded Tibet almost half a century ago. The seven great Asian rivers originate high in the Hindu Kush of the Tibetan plateau, giving China, essentially, the power of life and death over everyone in every country living downstream. If China decides to "re-allocate" these rivers for its own needs, India, with 500 million dependent on the Ganges, and Pakistan, with millions dependent on the Indus, could consider military measures to regain control of their rivers. Both India and Pakistan have nuclear weapons. It is a grim picture, and no one really talks about it in such terms, but this kind of unthinkable confrontation is a real possibility if radical water efficiency measures and reorganization of water management are not instituted soon in China.

## Worldwide Stress: Divvying up the Rivers

It is not surprising that rivers are a source of contention that can escalate into serious conflict. They are often multinational, and the flow of the same river through different countries can vary based on climate, topography, and distance from the source.

In Egypt, a country with little rainfall, the Nile is the country's lifeline, so richly harvested by Egyptians that it is almost dry every year as it flows into the Mediterranean. In its upper reaches, it is two rivers. The White Nile originates in southern Rwanda and flows through Tanzania, Lake Victoria, Uganda, and into southern Sudan where it merges with the Blue Nile near Khartoum. The length of its flow provides water not only to Burundi, Uganda, and Rwanda but also to Sudan and Ethiopia. If populations in those upriver countries double by 2050 as projected, and Egypt continues to grow as well, who will be entitled to what amount of Nile water?

The Colorado River in the United States shares a similar quandary. The river's origins are high in the Colorado Mountains, fed by a snow pack that is decreasing annually. Whether or not global warming is exacerbating the problem, flow at its origins has been reduced for enough years that it is causing concern. The recent Nobel Prize winning IPPC team (Intergovernmental Panel on Climate Change) warned in their August 2007 report that "warming will decrease the snowpack blanketing the mountains in winter, increase the flooding in winter and spring as precipitation shifts from snow to rain, and reduce the water flows out of the mountains in summer." For lands and peoples sustained by the water from these mountains, the report predicts the shrinking snowpacks will "exacerbate competition for overallocated water resources" (Milius 2007).

It was the once mighty confluence of waters from the Colorado and its tributaries that created the Grand Canyon. Now, in addition to the upstream ice-melt, hydropower dams like Hoover and Glen Canyon have reduced the lower river's flow volume significantly, sacrificing much of it to Lake Power and Lake Mead, the two primary drinking water sources for rapidly increasing desert populations in Arizona and Nevada.

There is yet another wrinkle. The Colorado River Compact of 1922 provides a strict allocation of river water, based on flow-volume calculations made during its highest flows. This pact decreed that a portion of Colorado River water must be relinquished to California to provide water for the growing Los Angeles and San Diego cities and environs. As the flow has been reduced, Lake Mead and Lake Powell are already at historic low levels. Serious disagreements over access to the water of Colorado River threaten. Las Vegas has burgeoned into one of the fastest growing cities in the United States. Six of the ten fastest growing cities in the United States are in water-stressed areas of the southwest that require Colorado River water for survival. Two, Gilbert and Chandler, Arizona, are suburbs of Phoenix, and both Phoenix and Tucson are booming with immigrant retirees moving into the state. Three are in the Los Angeles basin. Southern California grows in population by millions every year. People flock there because of the warm, dry climate, but in these desert areas, where will more water come from? What of the issue of Mexico's water rights to the Colorado River—which barely trickles crosses the border, removing the source of livelihood of hundreds of farmers in the arid north of Mexico?

## Can We Learn What Rivers Teach?

A similar scenario exists in India and Pakistan. Regardless of what China may do in the future with the Ganges, the latter already has little water left after it flows through Bangladesh en route to the Bay of Bengal. India's population has reached a billion and shows few signs of slowing. Where will the people of India get more water? Pakistan's agriculture, providing sustenance for 157 million people, is completely dependent on the Indus River, which is now almost dry when it reaches the ocean. Pakistan's population is projected to more than double by 2050. Even if China doesn't take a drop from the Indus, where will the additional needed water come from?

Somehow one feels an even greater sadness about what could happen to smaller countries downstream of the great artery of Southeast Asia—the Mekong—precisely because this river is still one of the world's most pristine and undeveloped. Nicknamed "sweet serpent of southeast Asia," the Mekong winds 2,800 miles from its icy origins in Tibet, through southeast China's deep gorges, before flooding into the rainforests of Laos and Cambodia and emerging rich with sea life into the Mekong Delta in Vietnam. Wars in the area over the last century have largely prevented its economic development, the damming plight of most of the world's rivers. So it has retained its natural flow patterns, including the mysterious backward flow for part of the year of its major tributary, the Tonle Sap River. Since the Tonle Sap flows past the riverside palace of the king and queen of Cambodia, they preside, with great pomp and circumstance, over the colorful and festive celebration of the river's annual reversal each fall (Pearce 2006: 93).

In addition to its mystical significance in Cambodian culture, the backward flow has immeasurable ecological benefits. As the reverse flow surges upriver for over a hundred miles, it draws into a lake flooding vast areas of rainforest and providing a great nursery. When it flows outward again, this nursery floods the lower deltas with bountiful fish. It has been said that the people in this lower Mekong region all live like kings and queens, regardless of their incomes, because the primary catch in this mammoth river is the tiny trey reil, a small sardine that is rich in protein and calcium. Upward of sixty million people in Laos, Cambodia, and Vietnam depend on the two million fish a year caught in their nets from the Mekong (ibid.: 96).

The major societies of the world could choose to learn from this great natural river and leave it to flow unfettered toward the sea, or they could choose to begin, bit by bit, diverting it upstream so that downstream populations begin to suffer. One hopes that we have seen enough of river destruction in the world and that we can leave this one alone. This is the kind of lesson that the Earth's natural processes continue to offer to human societies that have the wisdom to pay attention. But the source of those lessons may vanish if the rivers continue to bow to human overuse and pollution.

### Breaking the Destructive Chains

It is not only the world's rivers that are threatened. Lakes and inland seas upon which millions depend are being polluted or drying up. In southern Russia and southwest Asia, the beautiful Aral Sea—once the world's fourth largest inland body of water—has virtually turned into a desert, adding dry land twenty times the size of Manhattan every year. The scale of what has happened here, according to the United Nations, represents the greatest environmental disaster of the twentieth century (Pearce 2006: 201).

The lake once famous for its beauty and recreational attraction is now split into three pools that contain only one-tenth its original volume. Fish can no longer survive in its high salinity. The accelerated growth of the world's thirstiest crop plant—cotton—in Uzbekistan has drained the lake's major river source, the Abu Darya. Uzbekistan's economy is dependent on cotton. Because the Russian government is the sole purchaser of its cotton—largely used to make military uniforms and clothing—meeting cotton production targets is a national obsession. Even though the collective farms have been privatized, individual landowners can still lose their land if they fail to meet cotton targets.

It is a war between economics and water use, just as water crises are in much of the world. One could be tempted to say that this interwoven chain of dependencies has no solution: the people must grow cotton to survive; the government must sell cotton to Russia to survive; the cotton must have water to grow, and the water must come from the Abu Darya, its only source. But thinking in such a manner must change. Instead of selling cotton to get currency to buy food, dryland crops like wheat, and backyard orchards, vineyards, and vegetable patches could feed the people and reduce water dependency, just as they did in generations before the Soviet Union demanded double the cotton production from Uzbekistan back in the mid-1960s. Uzbekistan is now independent from Russia and thus could think independently about solutions to its water crisis. Chains of habit and tradition seem extremely difficult to break, and whether or not Russia rules it, the giant nation on its northern border still wields enormous economic and military influence.

## Bulldozing the Garden of Eden

While I have focused most extensively on the repercussions of the freshwater crisis for people, it is not only humans who are suffering. Freshwater species from snails to fish to amphibians are dying out five times faster than terrestrial species. In fact freshwater animals are being destroyed as fast as rainforest species, which are generally considered to be the most imperiled on Earth. The first estimate of extinction rates of North America's freshwater animals shows that they are the most endangered group in the continent. A silent mass extinction is occurring in our lakes and rivers (Ricciardi and Rusmussen 1999). One in three European freshwater fish face extinction. Twelve distinct species, found mostly in Central Europe in the 1970s and 1980s, are already extinct, largely as a result of overextraction of river waters and dams that impede flow. Sea plants are also suffering. Ten species of coral and 74 species of seaweed from the Galápagos Islands alone are critically endangered.[6]

The largest mammal water dweller in China, the Yangtze River dolphin, has gone extinct in the last three years. "The baiji was a remarkable mammal that separated from all other species over 20 million years ago. This extinction represents the disappearance of a complete branch of the evolutionary tree of life," according to conservation biologist Sam Turvey of the London Zoo.[7] In addition to Yangtze River dolphin, the Yangtze paddlefish is (was) probably the largest freshwater fish in the world (at least 21 feet); it has not been seen since 2003. The huge Yangtze sturgeon breeds only in tanks now because it has no natural habitat; a large dam stands between it and its breeding grounds. "The whole Yangtze River ecosystem is going down the tubes in the name of rampant economic development," according to Robert L. Pitman of the NOAA Fisheries Ecosystem Studies Program. "The disappearance of an entire family of mammals is an inestimable loss for China and for the world. I think this is a big deal and possibly a turning point for the history of our planet. We are bulldozing the Garden of Eden, and the first large animal has fallen."[8]

## A Few Drops of Hope

When one hears of the world water woes, throwing up our hands in despair seems natural. But, of course, we cannot give up. Earth's inter-connected processes are now critically depend on the collective and individual decisions we make. Indeed there are some wonderful individuals around the world who are leading the way to a new relationship between humanity and freshwater systems.

One of the most famous individuals is India's "water man," Rajendra Singh. In over a thousand villages of the dry northern Rajasthan state, Singh's organization has mobilized rural communities to build and revive over 4,500 traditional water-harvesting structures to collect rainwater, thereby regenerating 6,500 square kilometers of land. Because of the renewed water flow into the water table from these so-called "johads," the rivers in Rajasthan, which were once drying up, are flowing again. Singh has effectively drought-proofed these villages even as poor monsoons have increasingly dried up water sources throughout India.

Wynnette LaBrosse, a Silicon Valley entrepreneur, has donated a million dollars to start an innovative program to bring clean drinking water to parts of the developing world. Through the organization Water-Partners International, LaBrosse's money will fund grants in India, Bangladesh, and Kenya to start a "water credit" program, a mix of charity and entrepreneurship where communities are given money to build pipes and wells and then pay back the costs into the loan fund with money they have saved.[9]

Ashok Gadgil, an inventor and scientist at the University of California, whose passion for getting water to the world is reflected in every word he speaks, has invented UVWaterworks, a low-cost, low-maintenance, energy-efficient solution to water decontamination. Four gallons of water per minute can be purified by flowing past a UV light at a cost of 2 cents per 250 gallons; one small solar panel can easily generate the 40 watts of electricity used by the unit. A single device can provide 15 liters of drinking water each day for 3,000 people. Three hundred devices have already been installed, serving 900,000 people in rural communities of Mexico, the Philippines, and India.

In the driest country in the world, Namibia, a clever engineer in 1968 realized that wastewater could be treated to supplement the capital city's potable water supply. At that time, it was used primarily for recharging groundwater aquifers, supplying industrial processes, and irrigating certain crops. Now, in drought years, the treated wastewater has supplied up to 30 percent of Windhoek's drinking water. Locations around the world are taking note (including southern California), as is Peter Glieck who questions why we treat water until it is pure enough to drink and then use it to flush toilets. Perhaps, he suggests, we need to provide pure water only for human and animal consumption, while using "gray water"—namely treated wastewater—for the rest (2001: 44).

Water conservation is also taking hold in unlikely places. In Mexico City, officials launched a conservation program that replaced hundreds of thousands of inefficient toilets. The improvements have already saved enough water for an additional 250,000 residents (ibid.). Numerous other options for water conservation in both industrial and nonindustrial nations are becoming increasingly available, including less wasteful washing machines, drip irrigation instead of pivot or flood irrigation for croplands, and "xeriscaping" in outdoor landscape.

These few examples represent the beginnings of what can grow into national and international efforts with similar-sized impacts. There is no way of knowing how much time is left before the human impacts on Earth's life-giving waters are beyond the point of redemption. But a vision for living in harmony and with dignity within the biosphere still frames our opportunities.

## Notes

1. Cited in http://www.thirdworldtraveler.com/Water/Water_Privateers_BG.html.

2. Guadalajara is not alone in the problem of aging infrastructure. It is estimated that the water lost from Mexico City's leaky pipes annually is enough to meet the water needs of a city the size of Rome.

3. According to Lester Brown, "A thousand tons of water can be used to produce a ton of wheat, worth about $200, or it can be used to expand industrial output by $14,000" (2005).

4. See www.china.org.

5. *The Economist*, May 21, 2005, Asia, p. 46.

6. See http://www.iucn.org/en/news/archive/2007/09/12_pr_redlist.htm.

7. See http://www.guardian.co.uk/environment/gallery/2007/nov/15/endangered species?picture=331277589.

8. See http://www.nytimes.com/2006/12/26/science/26field.html?partner=rssnyt &emc=rss.

9. See Buchanan (2004).

# References

Brown, L. 2005. *Outgrowing the Earth: The Food Security Challenge in an Age of Falling Water Tables and Rising Temperatures*. New York: Norton.

Buchanan, W. 2004. A first for charity—$1 million grant for drinking water. *San Francisco Chronicle*, October 9, p. A7.

Glennon, R. 2002. *Water Folliess: Groundwater Pumping and the Fate of America's Fresh Waters*. Washington, DC: Island Press.

Glieck, P. 2001. Making every drop count. *Scientific American* 292 (February): 44.

Milius, S. 2007. Wildfire, walleyes, and wine. *Science News* 171: 378–79.

Nicklas, M. 2005. National water crisis: Our aquifers. 2005 International Solar Energy Congress, Orlando, FL.

Pearce, F. 2006. *When the Rivers Run Dry*. Boston: Beacon Press.

Perkins, S. 2007. Groundwater use adds $CO_2$ to the air. *Science News* 172: 301.

Ricciardi, A., and J. B. Rasmussen. 1999. Extinction rates of North American freshwater fauna. *Conservation Biology* 13: 1220–22.

Stecklow, S. 2005. How a global web of activists gives coke problems in India. *Wall Street Journal*, June 7, pp. A1, A6.

Tully, S. 2000. Water, water everywhere. *Fortune*, May 15, p. 193.

Yardley, J. 2005. Spill in China brings danger and cover-up. http://www.nytimes.com/2005/11/26/international/asia/26china.html?pagewanted=2&_r=1&sq=â€œeSpill%20in%20China%20Brings%20Danger%20and%20Cover-up,â€&st=cse&scp=1

# 10
## Deep Time Lags: Lessons from Pleistocene Ecology

Connie Barlow

Scientists involved in Gaian research—also known as geophysiology, Earth systems science, or whole-Earth science—as a matter of course provision their global climate and chemical cycling models with their best understandings of time lags inherent in Earth's thermal and chemical reservoirs. For example, how long will it take the carbonic acid content of the world's oceans to equilibrate with today's (and tomorrow's) elevated concentrations of carbon dioxide in the atmosphere?

Time lags are just as important to understand for biodiversity preservation. New forms of population modeling help conservation biologists estimate the probabilities that a particular population (of any given size) of plant or animal will "wink out" owing to fluctuations in natural conditions—even if the population seems to be self-maintaining in the present. Such models have served as wake-up calls to conservationists that even stabilized populations of threatened species may be doomed to extirpation unless their numbers can be increased or corridors established to facilitate cross migration with neighboring populations.

Another kind of time lag also impinges on biodiversity preservation. This time lag has come to the attention of conservation biologists, thanks to the work of those who specialize in Pleistocene ecology. In the late 1970s ecologist Dan Janzen, working in Costa Rica, began to suspect that his studies of seed dispersal in the large-seeded, fruit-bearing plants had gone awry. The studies were flawed by the then-unexamined (and universal) assumption that dispersal candidates could include only those fruit- or seed-eating mammals that *currently* were native to the plant's home range—or that had likely been there just prior to the arrival of Europeans in the Western Hemisphere. Janzen had previously concluded that several large-seeded tropical plants were dispersed by rodents who extracted and buried the seeds for later consumption. But when he noticed the same seeds protruding from the dung of domestic horses, he realized it was time to invite Pleistocene ecologist Paul Martin to join his studies.

Martin advised Janzen that not only were horses native to North America until the close of the last episode of glacial advance (about 13,000 years ago), but there were lots of other now-extinct mammals that might also have coevolved with the plants in question: notably, giant ground sloths and elephant-like gomphotheres (Martin 1990). Janzen and Martin coauthored a now-classic paper in evolutionary ecology; published in *Science* in 1982, they titled it "Neotropical anachronisms: The fruits the gomphotheres ate."

I spent three years examining the genesis of that paper and exploring how its "deep-time" perspective has inspired subsequent research projects in evolutionary ecology and conservation biology. I worked my findings into a popular book, *The Ghosts of Evolution: Nonsensical Fruit, Missing Partners, and Other Ecological Anachronisms* (2000). One section of the book used the deep-time perspective to re-examine the circumstances of perhaps the world's most endangered species of conifer tree: the Florida torreya (*Torreya taxifolia*). It occurred to me that torreya's desperate plight owed to its failure to migrate north (perhaps for want of a seed disperser) from its Ice Age refuge in the Florida panhandle to habitat better suited to the tree's needs in peak interglacial times. That better habitat would likely have been the core of torreya's range during previous interglacials: the southern and central Appalachian Mountains.

As it turns out, I was not the first to make this suggestion. Bill Alexander, forest historian at the Biltmore Gardens of Asheville, North Carolina (in the central Appalachian Mountains), observed his garden's own grove of Florida torreya, and concluded that North Carolina seemed more conducive to the well-being of this conifer than was northern Florida (personal communication). In a 1990 article, botanist Rob Nicholson speculated, "Is *Torreya* an early victim of global warming and a precursor of a new wave of inexplicable extinctions?" How prescient he was! Thanks to a host of recent scientific papers (e.g., Barlow and Martin 2005; McLachlan et al. 2007; Hoegh-Guldberg et al. 2008) and popular articles (e.g., Fox 2007; Nijhuis 2008; Marris 2008), Florida torreya has become a "poster plant" for alerting the public and scientists alike to the lurking dangers of global warming and to the consequent need for what has come to be known as *assisted migration*. Assisted migration must not, of course, be promoted as an alternative to reducing greenhouse gas emissions. But it is decidedly unrealistic to assume that climatic change and its challenges to biotic diversity will vanish in the next decade or two. Again, time lags (melting polar and glacial ice) will take a long time to equilibrate even if the concentration of atmospheric $CO_2$ could politically and economically be stabilized at today's levels.

## Assisted Migration in a Time of Global Warming

It is easy to grasp that rapid and profound climate change will exacerbate the biodiversity crisis, especially in those regions where biological pre-serves are no more than islands of biotic richness encircled by a sea of civilization. As climate shifts regionally (and globally), where might threatened species be encouraged to go, and how will they get there? Conservation biologists are thus now supplementing discussion of geo-graphic corridors for connectivity with talk of assisted migration—that is, direct human involvement in choosing individuals to serve as founders for new populations deliberately transplanted to locations where that species does not currently exist (McLachlan et al. 2007; Hoegh-Guldberg et al. 2008).

Assisted migration is, of course, a less than ideal way for societies to ensure the continued existence of wild populations of plants and animals. The human mark on the future of biotic expression and evolution is already so overwhelmingly in the negative that we yearn for conservation practices that would allow nature itself to direct the recovery process. Yet in some instances, and increasingly so, massive human intervention will be essential for biodiversity preservation. We humans will deliber-ately choose the would-be immigrants (worse-case scenario: the stricken refugees) and provide the vessel for rapid and safe passage to the prom-ised land. And it is we who will decide where that land is to be found.

Florida torreya has attracted my attention (and increasingly that of others; e.g., Nijhuis 2008) for the simple reason that if there is any plant species for which assisted migration makes sense right now, it is surely America's most endangered conifer. Why? Because *Torreya taxifolia* has been struggling for half a century to persist in its current native range. Despite the best efforts by conservation scientists to nurture and coddle it in the wild, its numbers diminish each passing year. It is my contention that the combination of peak-interglacial climate conditions that the world is now in, elevated by human contributions to global warming, have for fifty years been urging this large-seeded (and charismatic) conifer tree to head north to cooler realms.

In the 1950s Florida torreya suffered a catastrophic decline, the ultimate cause of which is still unexplained. By the mid-1960s, no large adult specimens—which once measured more than a meter in circumference and were perhaps 20 meters tall—remained in the wild, felled by what seemed to be a variety of fungal pathogens. Today, the wild population persists as mere stump sprouts, cyclically dying back at the sapling stage, such that seeds are rarely, if ever, produced. *T. taxifolia* thus joins American chestnut in maintaining only a juvenile and diminishing presence in its present range.

Florida torreya is a yewlike conifer. Its large, single seed resembles that of a plum; it is encased in a fleshy packet (as is the seed of a yew). Historically, it has been found only along a short stretch of the Apalachicola River of northern Florida and the adjacent sliver of southern Georgia. It favors the cool and shady ravines that dissect the high bluffs of the river's eastern shore. Despite its current extreme endemism, the species was once a prominent mid- and understory member of its forest community, which even today includes an odd mix of northern and southern species: towering beech and hickory next to tall evergreen magnolia, and surrounded by stubby needle palm.

*Prehistorically*, the ancestral torreya species almost surely thrived as an understory tree on the slopes of the Appalachian mountains. As with its mountain-dwelling cousin to the west, California Torreya (*Torreya californica*), America's eastern torreya would have been shade-adapted, growing slowly while awaiting an opening in the canopy for the additional sunlight required to produce seed. The Appalachian torreya would have been similar to California Torreya in its supreme ability to re-sprout from rootstock after a fire, thus giving the plant a chance to mature and produce seeds (or pollen, as the genus is characterized by distinctively male or female trees), before the new recruits of rival species could shade it, once again, into a nonreproductive phase of survival.

Fundamentally, a deep-time perspective helps us see that the Apalachicola River of northern Florida is best understood as native habitat for eastern torreya only during a peak of glacial advance. After all, there is no dispute that the Apalachicola served as one of eastern North America's most important refugia during ice times (Delcourt 2002). There are still a few scattered beech trees lingering in the rich soils along that river, but the great bulk of the beech population long ago migrated and settled far to the north. A deep-time perspective thus opens up a new line of questioning: where would native range for species X have been during a peak interglacial—or during even more ancient times (species of genus *Torreya* coexisted with Cretaceous dinosaurs) when global climate was even warmer than it is today?

Assisted migration as a conservation tool is both fascinating and frightening for anyone focused on plants. It is fascinating because endangered plants can be planted by whoever so chooses, with no governmental oversight or prohibitions—provided that private seed stock is available and that one or more private landowners volunteer suitable acreage toward this end. This cheap-and-easy route for helping imperiled plants is in stark contrast to the high-profile, high-cost, and governmentally complicated range recovery programs for mobile animals, like gray wolf, lynx, and California condor.

Assisted migration frightens for precisely the same reasons it fascinates: anybody can do it, for good or ill, and with care or abandon. Its promotion could undermine decades of public education about the dangers of nonnative plants, as well as more recent efforts to promote the concept of wildlands corridors and connectivity. Still, in an age of deforestation, severe habitat fragmentation, and rapid global warming, assisted migration as a plant conservation tool should not be ignored. According to Peter Wharton, curator of the Asian Garden of the University of British Columbia Botanical Garden writes, the *Torreya* question is a door to immense issues relating to how we facilitate global "floraforming" of vegetational zones in a warming world. It represents another layer of responsibility for those of us who have a passion for forests and wish to promote the ecologically sensitive reforestation of so many degraded forest ecosystems worldwide (P. Wharton, personal communication).

The test case for assisted migration occurred in July 2008 when the citizen group I helped found (Torreya Guardians) undertook assisted migration for 31 seedlings of *Torreya taxifolia* purchased from a nursery in South Carolina. A handful of volunteers (and reporters documenting the action) gathered in the mountains near Waynesville, North Carolina, to spend a day planting the seedlings into wild forested settings on two parcels of private land. The Torreya Guardians' website documents that action.[1]

## Deep-Time Lags and the Imperative for Rewilding

The plight of the endangered Florida torreya tree is an exemplar of deep-time lags in which a species seems to have gotten "stuck" (perhaps for lack of its seed disperser) in once-suitable habitat that is no longer capable of supporting its survival. A second example of deep-time lags that is already informing the leading edge of conservation thinking involves plant–animal interactions at the landscape level. This is the proposal for "Pleistocene rewilding" (Donlan et al. 2005, 2006).

In 2005 a dozen leaders in conservation biology, led by Josh Donlan, coauthored a short advocacy piece (a "commentary") in *Nature* in which they contended that even the most biologically intact wilderness parks in America are missing key components of ecological interactivity. These components moreover had shaped American landscapes over millions of years. Notably, the zones of America too dry to support closed-canopy forests now lack the large mammalian plant browsers—as well as the large carnivores that had preyed upon those browsers—that had thrived in those areas throughout the Pleistocene epoch. Humans had brought back the large grazers (cattle and horses), but the browsers (camels, ground sloths, mammoths, and mastodons) were absent, and so were the large predators.

In consequence what ecologists had considered to be natural configurations of native vegetation were actually quite the contrary—at least from a deep-time perspective. Lacking capable carnivores and big browsers, much of the American west's grasslands, savannas, and deserts had been damaged by hoofed grazers, fostering soil erosion and selecting for the proliferation of shrubby plants (e.g., mesquite, creosotebush, and sagebrush). Cattle and horses eschew these shrubs—but such plants would have been eaten by big browsers native to North America during the ice times of the Pleistocene. Thus came the Pleistocene rewilding proposal to return close proxies of the lost browsers (Bactrian camels for America's extinct *Camelops*) and carnivores (the African lion for America's extinct lion) to carefully chosen test ranges of the American West.

In a longer paper published in 2006, the same set of authors elaborated on the half dozen reasons to undertake a test of the rewilding concept. One such reason is to offer Pleistocene ecologists a chance to witness and study how Pleistocene megafauna would likely have shaped the vegetational landscape of the arid and semiarid American west. Another is to provide the public with a chance to witness something similar to the pageant of American wildlife that would have greeted the first human immigrants to this continent (predecessors of the now native American peoples). What makes this radical proposal even possible is time lags. Communities of plant species have changed enormously since the end of the Pleistocene. But no once-dominant plant species of the savanna or grassland appears to have gone extinct (Delcourt 2002). Rather, it is the *patterning* of vegetation that is the character in question.

Unlike *Jurassic Park* fantasies of resurrecting the dinosaurs, the proposal to jump-start one or more Pleistocene parks is not only within the realm of possibility but arguably an ethical imperative. Humans are not responsible for the death of the nonavian dinosaurs. Yet the majority opinion in science is that humans are at least partly culpable for the huge loss of megafaunal species at the end of the ice times. Earth's "sixth mass extinction" began some 50,000 years ago when spear-toting, fire-wielding humans made their way to the once-isolated continent of Australia, and eventually into the Americas and onward to the islands of Polynesia, Madagascar, and New Zealand.

Deep-time lag, because of which continental vegetation has not yet fully adjusted to the loss of browsers, is the reason rewilding is a scientifically responsible proposal—even 13,000 years after America's "extinction of the massive" (Martin 2005). A deep-time perspective, penetrating far into the future, invokes a felt urgency for humans to engage in repopulating this continent with megafaunal stock that may eventually re-evolve species truly *native* to this land. This is the ethical ground from which the rewilding proposal ultimately springs. Here is how the dozen scientists and conservationists proposing "Pleistocene rewilding" concluded their call to action:

> In the coming century, by default or design, we will constrain the breadth and future evolutionary complexity of life on Earth. The default scenario will surely include ever more pest-and-weed dominated landscapes, the extinction of most, if not all, large vertebrates, and a continuing struggle to slow the loss of biodiversity. Pleistocene re-wilding is an optimistic alternative.
>
> We ask of those who find the objections compelling, are you content with the negative slant of current conservation philosophy? Will you settle for an American wilderness emptier than it was just 100 centuries ago? Will you risk the extinction of the world's megafauna should economic, political, and climate change prove catastrophic for those populations remaining in Asia and Africa? The obstacles are substantial and the risks are not trivial, but we can no longer accept a hands-off approach to wilderness preservation. Instead, we want to reinvigorate wild places, as widely and rapidly as is prudently possible. (Donlan et al. 2005: 914)

In conclusion, the deep-time perspective that comes naturally to those who work in the realm of geophysiology can now become the lens through which conservation biologists and other biodiversity activists go about their work. Specifically, the deep-time perspective encourages conservationists to revise the parameters we use for judging which species are native to a region. It also encourages us to be mindful of time lags in biological adjustments to shifts in climate, and thus in how we read the past and how we prepare for the future.

## Note

1. The Torreya Guardians' website has a page dedicated to providing citations and links to the classic and current scientific papers and news reports on the assisted migration debate and actions: www.TorreyaGuardians.org/assisted-migration.html.

## References

Barlow, C. 2000. *The Ghosts of Evolution: Nonsensical Fruit, Missing Partners, and Other Ecological Anachronisms.* New York: Basic Books.

Barlow, C., and P. S. Martin. 2005. Bring *Torreya taxifolia* north now. *Wild Earth* 1: 52–55.

Delcourt, H. 2002. *Forests in Peril: Tracking Deciduous Trees from Ice Age Refuges into the Greenhouse World.* Blacksburg, VA: McDonald and Woodward Publishers.

Donlan, J., H. W. Greene, J. Berger, C. E. Bock, J. H. Bock, D. A. Burney, J. A. Estes, D. Foreman, P. S. Martin, G. W. Roemer, F. A. Smith, and M. A. Soulé. 2005. Re-wilding North America. *Nature* 436: 913–14.

Donlan, J., J. Berger, C. E. Bock, J. H. Bock, D. A. Burney, J. A. Estes, D. Foreman, P. S. Martin, G. W. Roemer, F. A. Smith, M. A. Soulé, and H. W. Greene. 2006. Pleistocene rewilding: An optimistic agenda for twenty-first century conservation. *American Naturalist* 168: 1–22.

Fox, D. 2007. When worlds collide. *Conservation Magazine* 8 (1): 1–4.

Janzen, D. H., and P. S. Martin. 1982. Neotropical anachronisms: The fruits the gomphotheres ate. *Science* 215: 19–27.

Hoegh-Guldberg, O., L. Hughes, S. McTintyre, D. B. Lindenmayer, C. Parmesan, H. P. Possingham, and C. D. Thomas. 2008. Assisted colonization and rapid climate change. *Science* 321 (5887): 345–46.

Marris, E. 2008. Moving on assisted migration. *Nature Reports Climate Change,* August 28. http://www.nature.com/climate/2008/0809/full/climate.2008.86.html.

Martin, P. S. 1990. 40,000 years of extinctions on the "Planet of Doom." *Palaeogeography, Palaeoclimatology, Palaeoecology* 82: 182–201.

Martin, P. S. 2005. *Twilight of the Mammoth: Ice Age Extinction and the Rewilding of America.* Berkley: University of California Press.

McLachlan, J., J. Hellmann, and M. Schwartz. 2007. A framework for debate of assisted migration in an era of climate change. *Conservation Biology* 21 (2): 297–302.

Nicholson, R. 1990. Chasing ghosts: the steep ravines along Florida's Apalachicola River hide the last survivors of a dying tree species (*Torreya taxifolia*). *Natural History* (December): 8–13.

Nijhuis, M. 2007. Taking wildness in hand: Rescuing species. *Orion* (May–June): 43–47.

# IV

## Gaian Ethics and Education

# 11

## From the Land Ethic to the Earth Ethic: Aldo Leopold and the Gaia Hypothesis

J. Baird Callicott

Aldo Leopold is often called a prophet. This is mainly because he was warning of the onslaught of an environmental crisis—although not by that name—more than a decade before Rachel Carson (1962) and Stewart Udall (1963) sounded the alarm.[1] Heralding impending doom is, after all, one of the signal things prophets do. Further Leopold's seminal essay, "The Land Ethic," anticipated, by more than two decades, the emergence of formal, academic environmental ethics that came on the scene in the early 1970s.[2] And that too is prophetic—albeit of a good thing, not a bad. One can find scattered remarks in the works of Henry David Thoreau and especially John Muir that intimate a need for a new, non-anthropocentric moral relationship on the part of humans with nature. Leopold's "The Land Ethic," however, is the first sustained expression of such a need and the first outline of how such a relationship might be understood—at least in the Western intellectual tradition—and put into practice.

### Aldo Leopold's Land Ethic

The book in which "The Land Ethic" appears, *A Sand County Almanac*, is often called the bible of the contemporary environmental movement.[3] That is no accident. Leopold carefully crafted its style to give a subtle biblical ring and cadence to its words; and so—through his primary audience's Sunday-schooled ear—the book subliminally exudes an almost divine authority (Tallmadge 1989). Part I, the shack sketches, is a series of parables about meadow mice, pasque flowers, chickadees, and pines above the snow. Part II, "Sketches Here and There," is a series of homilies about the untoward ecological impact of the industrialization of agriculture, wetland draining, and wilderness desecration. Perhaps most transparently biblical in intent, the second most famous essay in *Sand County*, "Thinking Like a Mountain," recalls an occasion, a quarter-century prior, when Leopold and his companions, working as forest rangers in the southwest, murdered a mother wolf and wounded one of her pups. The moment is presented as a road-to-Damascus epiphany in which the wolf fixes Leopold (1949: 130) with her fading gaze— "a fierce green fire dying in her eyes"—and, in effect, mutely asks: Why persecutest thou me? Finally, the book ends with "The Land Ethic," a sermon on the moral obligations of people to land.

Trained at the Yale Forest School, Leopold eventually became an autodidact wildlife ecologist, so he grounded the land ethic in his principal intellectual heritage: ecology and Darwinian evolutionary biology (or more exactly, Darwinian evolutionary psychology). Leopold's land ethic has become the environmental ethic of choice among his fellow conservation biologists (Groom et al. 2006)—a field that Leopold also prophetically pioneered (Meine et al. 2006)—and other environmental and conservation professionals working in the field. This is doubtless because conservation biologists and other practicing conservationists and environmentalists feel a professional and intellectual affinity with Leopold, as well as because of the accessible and compelling way that he presents the land ethic. No subsequently developed environmental ethic has gained such wide currency and fond allegiance. However, resting something enduring—and one would surely hope that an environmental ethic would endure—on scientific foundations is risky because science is dynamic; it is self-correcting and thus ever changing. After 1949, when Leopold composed the land ethic, ecology underwent a major paradigm shift and Darwinian evolutionary biology (and psychology) became *neo*-Darwinian—and thereby putatively more rigorous and certainly more reductive. These changes in its scientific foundations are not fatal to the land ethic, but—if the land ethic is to continue to be relevant—its conceptual foundations must be reconstructed and its precepts revised accordingly.

In revised form, the land ethic remains applicable to the environmental concerns that it was conceived to address, but its applicability to other concerns is limited in two significant ways.

It is, after all, the *land* ethic. What about aquatic biotic communities and ecosystems? Leopold (1949) does state that "The land ethic simply enlarges the boundaries of the land community to include soils, *waters*, plants, and animals" (emphasis added). But by "waters" it is clear that Leopold has freshwater streams, rivers, and lakes in mind, for such waters—with soils, plants, and animals—constitute "collectively: the land." More specifically, then, what about *marine* biotic communities and ecosystems? Nearly three-fourths of the Earth's surface is ocean. Can the land ethic be expanded to embrace seascapes as well landscapes? In different essays published at different times and in different venues, I have defended opposite answers to that question (Callicott 1992; Callicott and Back 2008).

The land ethic is also limited by the spatial and temporal parameters to which it is scaled. It is spatially scaled to biotic communities, ecosystems, and landscapes; and its temporal scale is calibrated in decades. The spatial and temporal scales of Leopold's land-ethical thinking are clearly revealed in "Thinking Like a Mountain." There, he writes: "for while a buck pulled down by wolves can be replaced in two or three years, a *range* pulled down by too many deer may fail of replacement in as many *decades*" (1949, emphasis added). For most of the twentieth century such scales seemed large and long in comparison with ward politics and election cycles and commercial activities and economic cycles. Public concern, however, about global climate change began to be expressed in the last quarter of the twentieth century and has become increasingly acute and urgent as the twenty-first century unfolds. And the spatial scale of global climate change is planetary, while the temporal scale of the anticipated calamitous effects of global climate change—such as sea-level rise and desertification—is calibrated in centuries and millennia.

Wherever the truth may lie regarding the possibility of an oxymoronic marine land ethic, it is abundantly clear that the land ethic, even in revised form, cannot simply be scaled up to address the environmental-ethical challenge of global climate change. As noted, the scientific foundations of the land ethic are ecology and Darwinian evolutionary biology and psychology, while the sciences most centrally contributing to our understanding of global climate change are thermodynamics, geochemistry, evolutionary biology, and biological systems theory. Integrated into one interdisciplinary whole, those sciences constitute Gaian science. And the existence of a biospheric community—of which we can regard ourselves as plain members and citizens—is even more problematic than the existence of biotic communities. While the spatial dimensions of such a putative community are clear enough, how can we clearly define its temporal dimensions?

Given Aldo Leopold's cachet in environmental ethics, both among academic theorists and professional practitioners, the irrelevance of the land ethic to concern about global climate change is a pity. Fortunately, however, Leopold once more lives up to his posthumous reputation as a prophet—and, in this instance, quite amazingly. For back in 1923, he sketched, ever so sparingly, the Gaia hypothesis and a corresponding Earth—as distinct from a land—ethic. The Gaia concept coalesced in the 1970s. However, some of the leading exponents of Gaia theory, Lynn Margulis notable among them, attribute the earliest expression of Gaian thinking to Vladimir Vernadsky (1998). Vernadsky's proto-Gaian book, *The Biosphere*, was published in Russia in 1926 (Lazcano et al. 1998). While I do not suggest that the honor of being the first to broach the Gaia hypothesis should go to Leopold instead of Vernadsky—because of the faintness of Leopold's early sketch, because it was not published until 1979, and because his scientific and ethical thinking soon turned in a different direction—Leopold did commit his Gaia notions to writing, however briefly, three years before Vernadsky completed and published his. Of course, Vernadsky was no more aware of Leopold than was Leopold of Vernadsky, so there is no question of any influence of the thinking of the one on that of the other.

In what follows, I review the evolutionary and ecological foundations of the Leopold land ethic, how those foundations and the precepts that rest upon them are challenged by the subsequent paradigm shifts in evolutionary biology/psychology and ecology, and how the land ethic may be revised to accommodate its latterly shifting foundations. Then I turn to the conceptual foundations of a complementary Leopold Earth ethic that may address the contemporary acute and urgent concern about global climate change, which the land ethic is not able to do.

## The Conceptual Foundations of the Land Ethic

The very idea that an ethic can rest on scientific foundations flies in the face of one of the twentieth century's most disenabling shibboleths. Didn't the Logical Positivists, once and for all, divorce science and ethics, facts and values, *is* and *ought*? Aware of the pervasiveness of this schizophrenia in twentieth-century cognitive culture, Leopold (1933: 635) confronted it in "The Conservation Ethic," from which he borrowed heavily in composing "The Land Ethic": "Some scientists will dismiss this matter forthwith, on the ground that ecology has no relation to right and wrong. To such I reply that science, if not philosophy, should by now have made us cautious about dismissals." Without attempting here to exorcize the unholy ghost of Logical Positivism in the abstract, I will simply indicate how the Leopold land ethic is founded on evolutionary biology and ecology; you be the judge how cogently.

Darwin himself devoted a whole chapter to the "moral sense" in *The Descent of Man*. Many animals are social and of all the social animals— except perhaps for the haplodiploidal *Hymenoptera—Homo sapiens* is by far the most complexly and intensely social species. Among humans, ethics evolved to foster social integration and organization. For, as Darwin (1874: 120) colorfully put it, "No tribe could hold together if murder, robbery, treachery, etc., were common; consequently, such crimes, within the limits of the same tribe, 'are branded with everlasting infamy.'" Darwin's qualification, "within the limits of the same tribe" is important. Because ethics evolved to foster social integration, the effective limits of a society are also the effective limits of its ethics. The original human societies were small clans or "gens" (as nineteenth-century anthropologists called them), consisting of extended families. As the human population grew—so goes Darwin's account—and these small societies were forced to compete with one another for resources, the larger and better organized societies outcompeted those that were less so. Thus small clans merged to form tribes—in order to be larger and thus more successful competitors. Then, under the same competitive pressures, tribes merged to form nations. Darwin even anticipated a time—now upon us—in which all humanity would be united into a single global community: "As man advances in civilization, and small tribes *are* united into larger communities, the simplest reason would tell each individual that he *ought* to extend his social instincts and sympathies to all the members of the same nation, though personally unknown to him. This point being once reached, there is only an artificial barrier to prevent his sympathies from extending to the men of all nations and races" (1874: 126–27, emphasis added). As we see, Darwin has no qualms about inferring *ought* from *is* (or "*are*") —but maybe he just isn't as perspicacious as A. J. Ayer.

In "The Land Ethic," Leopold (1949) clearly alludes to Darwin's account of the origin and evolution of ethics by his choice of words— such as "struggle for existence," "social and anti-social conduct," "origin," and "evolve." Indeed he characterizes "this extension of ethics" over the last 3,000 years, from the time of Odysseus down to the present, as a process of "ecological evolution." To Darwin's sequence of societies—clan, tribe, nation, and global community—Leopold (1949: 203–204) adds the *biotic* community, our membership in which was discovered by twentieth-century ecology: "All ethics so far evolved rest upon a single premise: that the individual is a member of a community of interdependent parts"—which is Darwin's account of the origin and evolution of ethics in a nutshell; ecology "simply enlarges the boundaries of the community to include soils, waters, plants, and animals, or collectively: the land." When we recognize that we are simultaneously members not only of a hierarchy of human communities—family, clan, tribe, nation, global village—but also a biotic community, Leopold believes that we will acknowledge "a land ethic [that] changes the role of *Homo sapiens* from conqueror of the land-community to plain member and citizen of it" (ibid.).

## The Paradigm Shifts in Evolutionary Biology/Psychology and Ecology after 1949 and Their Implications for the Land Ethic

Darwin was notoriously unaware of the work of his contemporary, Gregor Mendel, and the latter's concept of the gene. What makes neo-Darwinian evolutionary theory "neo" is its integration of Darwinian evolution-by-natural-selection with Mendelian genetics. This so-called Modern Synthesis occurred in the 1930s (Dobzhansky 1937; Fisher 1930; Haldane 1932). In part because societies have no genes, which natural selection can cull and sort, and in part because of a more general Positivist and reductive fervor among twentieth-century evolutionary biologists, "group selection" became anathema by midcentury (Williams 1966). Darwin's account of the origin and evolution of ethics, however, explicitly turns on group selection. He wrote:

It must not be forgotten that although a high standard of morality gives but a slight or no advantage to each individual man and his children over other men of the same tribe, yet that an increase in the number of well-endowed men and an advancement in the standard of morality will certainly give an immense advantage to one tribe over another. A tribe including many members who, from possessing in a high degree the spirit of patriotism, fidelity, obedience, courage, and sympathy, were always ready to aid one another, and to sacrifice themselves for the common good, would be victorious over most other tribes; *and this would be natural selection.* At all times throughout the world tribes have supplanted other tribes; and as morality is one important element in their success, the standard of morality and the number of well-endowed men will thus everywhere tend to rise and increase. (Darwin 1874: 137, emphasis added)

Because Leopold apparently borrowed his account of the origin and evolution of ethics straight from *The Descent of Man*, it would seem to be vitiated by its implicit reliance on group selection. And so, for a while, it was indeed. As noted, however, science is dynamic and its paradigm shifts are often pendulum-like—or, perhaps better, gyre-like—in which an older concept returns to favor as part of a more comprehensive theory. And the more comprehensive theory, in this case, is "multilevel selection" (Wilson and Sober 1994; Sober and Wilson 1998). Even E. O. Wilson (2005) now acknowledges that the concept of kin selection, which adheres strictly to neo-Darwinian orthodoxy, is insufficient to account for the origin and evolution of ethics. Mainstream evolutionary theory is steadily gravitating toward Lynn Margulis's once unorthodox idea that natural selection favors symbiotic cooperation over ruthless competition (Margulis 1998). And, if anything, contemporary evolutionary psychology—which has succeeded the much maligned orthodoxly neo-Darwinian sociobiology—confirms Darwin's and, *eo ipso*, Leopold's general account of the origin and evolution of ethics.

In 1949, ecologists believed that biotic communities were more well defined, tightly integrated, and stable than they believe them to be today (Pickett and Ostfeld 1995). Today, ecologists recognize that biotic communities have no clear boundaries, nor are they capable of being clearly parsed into uniform types. Ecologists also now recognize that biotic communities have no *telos* toward which they develop, but are ever changing and routinely subject to natural disturbance—by fire, flood, drought, windstorm, and the like—as well as to human disturbance (Picket and White 1985). Thus the summary moral maxim of the Leopold land ethic, its golden rule, may seem obsolete: "A thing is right when it tends to preserve the integrity, stability, and beauty of the biotic community. It is wrong when it tends otherwise" (Leopold 1949: 204–205). Have biotic communities any robust integrity or stability to preserve? If not, only their beauty remains to be preserved—which is commonly understood to be "in the eye of the beholder"—a thin reed on which to rest an environmental ethic. By "beauty," however, Leopold meant not the scenic quality of landscapes, but the health of land. And by "health" he meant normal ecosystemic function—nutrient retention and cycling, soil building, hydrological modulation and purification, microclimate control, and so on. Contemporary ecologists recognize that such ecosystemic functions are both real and important, but they also recognize that ecosystems are not closed, self-regulating, and self-maintaining superorganisms—as the term "health" would imply—and thus that their functional components are not irreplaceable vital organs (Allen and Hoekstra 1992; Pickett and Ostfeld 1995).

As noted, the paradigm shift in ecology from "the balance of nature" to "the flux of nature" is not fatal to the Leopold land ethic. The human communities to which we belong—and which generate our humanly oriented duties and obligations—are no more sharply bounded or typological than biotic communities. And they too are constantly changing—growing or shrinking and altering in character with the ebb and flow of economic and demographic shifts. Moreover Leopold recognized that nature was dynamic, not static—at least at the evolutionary temporal scale. Thus we can scale down the dynamism of nature—which Leopold clearly recognized—to disturbance regimes and modify Leopold's land ethic in light of the paradigm shift in ecology that occurred between his day and ours. I have suggested that we rewrite the golden rule of the land ethic so as to reflect the more open, fuzzy, idiosyncratic, and dynamic nature of biotic communities as they are buffeted by myriad disturbances both local and remote: a thing is right when it tends to disturb the biotic community at normal spatial and temporal scales. It is wrong when it disturbs it otherwise (Callicott 1996). For in comparison with natural disturbances, human disturbances—such as industrial agriculture, rapacious forestry practices, urban, suburban, and exurban industrial and residential development—are often far more widespread and frequent than those that occur in nature.

## The Conceptual Foundations of the Earth Ethic

The paper in which Leopold faintly sketches the Gaia hypothesis and an associated Earth ethic is innocuously titled "Some Fundamentals of Conservation in the Southwest." The typescript is dated 1923 and remained unpublished until 1979 when it appeared in the first volume of *Environmental Ethics* (the journal)—the same year, coincidentally, in which James Lovelock's *Gaia: A New Look at Life on Earth* appeared.[4] It is divided into three sections, the first of which classifies the resources of the southwest as "minerals," "organic," "climatic," "historic," and "geographic." Leopold then expressly confines his discussion to the first two, and of these, only one paragraph is devoted to minerals. As to organic resources—farms, forests, and ranges—Leopold focuses on their water and soil underpinnings. And finally his focus narrows to the alarming erosion of the region's soil—which he correctly attributed to overgrazing and fire suppression—as the ultimate concern. Soil erosion negatively affects water resources by silting reservoirs and widening formerly narrow and clear streams and making them turbid and "flashy." Leopold (1979: 138) concludes the second section of "Some Fundamentals" with these words: "Erosion eats into our hills like a contagion, and floods bring down the loosened soil upon our valleys like a scourge." This, by the way, is a good example of Leopold's early experimentation with biblical phrasing and cadence.

The third section of "Some Fundamentals" is titled "Conservation as a Moral Issue." To the philosopher's eye, Leopold appears to suggest three distinct ethical modes for addressing the moral issue that conservation represents: (1) classical virtue ethics, (2) consequentialistic duties to future generations, and (3) Kantian deontological respect for the Earth per se as a living being.

> [1] Who cannot feel the moral scorn and contempt for poor craftsmanship in the voice of Ezekiel when he asks: "Seemeth it a small thing unto you to have fed upon good pasture, but ye must tread down with your feet the residue of the pasture? And to have drunk of the clear waters, but ye must foul the residue with your feet?"...Ezekiel seems to scorn waste, pollution, and unnecessary damage as something unworthy—as something damaging not only to the waster, but to the *self-respect* of the craft and the society of which he is a member. (Leopold 1979: 138–39, emphasis added)[5]

Plato, Aristotle, and their Greek contemporaries conceived of ethics in terms of virtue, which is centrally about *self*-respect. Concern or respect for others is only a benign side effect of personal *aretê*. In addition to reflecting badly on the waster himself, waste, pollution, and unnecessary damage also reflect badly on the polluter's society and the damager's craft—farming, ranching, and forestry in this case—notes Leopold, quite consistently with Greek thinking in this regard. Technically put, virtue ethics is not exclusively individualistic but also has at least two holistic dimensions: the professional self-respect of the guild as such and the self-respect of the collective society per se:

> [2] We might even draw from his [Ezekiel's] words a broader concept—that the privilege of possessing the earth entails the responsibility of passing it on, the better for our use, not only to immediate posterity, but to the Unknown Future, the nature of which it is not given us to know. (Leopold 1979: 139)

Here Leopold does suggest that in addition to self-respect—individually, professionally, and socially—we should also be concerned about the effect of waste, pollution, and damage on the welfare of other humans who will follow us. This adumbrates an anthropocentric utilitarian mode of ethical thought. Here too we have both an individualistic and holistic aspect. Immediate posterity is an aggregate of identifiable individuals: our existing children and grandchildren and those who will exist in our own lifetimes or in the lifetimes of those whom we know and care about. Distant future generations are conceivable only collectively, not as a set of identifiable individuals. Perhaps to bring this holistic aspect out, Leopold refers to them as the *Unknown* Future, and goes on, redundantly, to stress our irremediable ignorance regarding just who they might turn out to be.

[3] It is possible that Ezekiel respected the soil, not only as a craftsman respects his material, but as a moral being respects a living thing. (Leopold 1979: 139)

Having suggested three distinct ethical modes for conceiving of conservation as a moral issue all in one paragraph, Leopold devotes the rest of the essay to elaborating the third: respect for the soil—which immediately becomes "the earth"—as a living being. We cannot coherently imagine that our waste, pollution, and damage will adversely affect the *welfare* of the Earth—its happiness or, as contemporary utilitarians would express it, the Earth's degree of "preference satisfaction." But we can coherently imagine that waste, pollution, and damage violate our duty to *respect* the Earth—as a moral being respects a living thing. Respect for beings with intrinsic value is a hallmark of Kant's deontological (or duty-oriented) ethic. Kant, however, was a militant ratiocentrist, insisting that being rational is a sine qua non of an entity's intrinsic value, dignity, and respect-worthiness. But—excepting, perhaps, God and the heavenly host—the only rational beings that Kant recognized to actually exist were human beings. Hence Kant's ethics is also, *in effect*, militantly anthropocentric.

Leopold takes as his first task persuading his audience that the Earth is indeed a living being, and it is by that effort that Leopold anticipates the Gaia hypothesis. His second task is to undermine his audience's knee-jerk anthropocentrism, a legacy of Western religion and philosophy. I here review only the first.

Leopold (1979: 139) presciently notes that "[t]he very words *living thing* have an inherited and arbitrary meaning derived not from reality, but from human perceptions of human affairs." As Gaian theorists have indicated, life may better be understood in terms of its recursive formal properties—its autopoiesis and autonomy—rather than in the classical biological terms of morphology and reproduction. Leopold goes on to quote and paraphrase *Tertium organum*—by another Russian, coincidentally, P. D. Ouspensky (1922). Its title may suggest that that book is all about superorganisms, or third-order organisms like Gaia. First-order organisms, one may suppose, are single-celled; second-order organisms are multi-celled; and third-order organisms are to multi-celled organisms as multi-celled organisms are to single-celled organisms. But that's not what *Tertium organum* is about—not at all—nor is that the meaning of the title. Aristotle's corpus of treatises on logic is called the *Organon*. Francis Bacon wrote a *Novum organum*, published in 1620, a new epistemology for the nascent modern science. Ouspensky hubristically intended his *Tertium organum* to supersede Aristotle's *Organon* and Bacon's *Novum organum*. And despite its sensational popularity when Leopold got a hold of it, Ouspensky's book is really as vacuous as it is pompous and pretentious. Nor can I find any evidence that Ouspensky was acquainted with the work of Vernadsky, and vice versa. So what Leopold (1979) goes on to attribute to Ouspensky are really his own speculations:

[I]t is at least not impossible to regard the earth's parts—soils, mountains, rivers, atmosphere, etc.—as organs or parts of organs, of a coordinated whole, each part with a definite function. And, if we could see this whole, as a whole, through a great period of time, we might perceive not only organs with coordinated functions, but possibly also that process of consumption and replacement which in biology we call metabolism, or growth. In such a case we would have all the visible attributes of a living thing, which we do not now realize to be such because it is too big, and its life processes too slow. (1979: 139)

That this is indeed an early—and maybe the very first—instance of Gaian thinking is suggested by Leopold's attribution of a metabolism to the Earth as a whole. Metabolism is a defining characteristic of living organisms—which are in a far-from-equilibrium thermodynamic state, open to energy flows but closed in regard to their own processes. Earth's putative "organs with coordinated functions" represent its operational closure, while its putative "metabolism" is constituted by its openness to solar energy, gravitational influences from the sun and moon, and ambient cosmic materials.

Leopold goes on to suggest that if the Earth has both organs with coordinated functions and a metabolism that "there would also follow that invisible attribute—a soul or consciousness—which...many philosophers of all ages ascribe to all living things and aggregations thereof, including the 'dead' earth" (ibid.). This may immediately raise suspicions that Leopold's Gaian speculations are as fraught with teleological tendencies as the early Lovelock's—that Gaia was consciously planning and directing her own autopoiesis and evolution (Bormann 1981). But as each of us should well know simply by introspection, being conscious, as we are, by no means implies that we plan and direct our own ontogeny, physiology, and metabolism. Gaia's consciousness, like our own, might well be emergent and epiphenomenal and oriented toward perception and reaction to environmental changes—such as changes in the amplitude of solar radiation—rather than toward its own internal organization and processes. Leopold (1979) concludes his brief propaedeutic to a future Gaia hypothesis with a coda, fusing science, ethics, and poetry:

> Possibly in our intuitive perceptions, which may be truer than our science and less impeded by words than our philosophies, we realize the indivisibility of the earth—its soil, mountains, rivers, forests, climate, plants, and animals, and respect it collectively not only as a useful servant but as a living being, vastly less alive than ourselves in time and space—a being that was old when the morning stars sang together, and, when the last of us has been gathered unto our fathers, will still be young. (1979: 140)

## An Earth Ethic Rising

So what shape should an Earth ethic assume? As noted, Leopold hints that it should be essentially Kantian, based on deontological respect. About this possibility, however, he seems unduly sanguine: "Philosophy, then, suggests one reason why we cannot destroy the earth with moral impunity; namely, that the 'dead' earth is an organism possessing a certain kind and degree of life, *which we intuitively respect as such*" (Leopold 1979: 140, emphasis added). But do we? The indifference most people seem to show to what Holmes Rolston III (1994) calls "super-killing"—anthropogenic species extinction—let alone ordinary killing (anthropogenic destruction of individual organisms) makes one wonder if we really do intuitively respect life as such. Maybe that's why Leopold immediately turns to ridiculing Western anthropocentrism.

Perhaps an adequate and effective Earth ethic should be pluralistic. And indeed, Leopold seems to agree. While he devotes most of "Conservation as a Moral Issue" to his call for respecting the Earth as a living being, he begins, as noted, by invoking two other ethical modes: individual, professional, and social virtue; and concern for the welfare of future generations, both immediate posterity and the unknown future. I would add a fourth. Global climate change will not be felt equally by all members of the present generation. People living in low-lying deltas, such as that of the Ganges, and on oceanic atolls will suffer disproportionately, first from more incidental flooding and saltwater intrusion and finally eviction, as sea level rises and storm surges intensify (IPCC 2007). People living in almost all tropical and temperate coastal areas will suffer from the increased frequency and intensity of cyclones (IPCC 2007). Indigenous peoples of the Arctic will find their environments and thus their traditional cultures more radically altered by warming than peoples living at lower altitudes (IPCC 2007). In respect to all these untoward phenomena and many others, the more affluent will be better able to cope than the less affluent. Adding insult to injury, those least responsible for generating greenhouse gases, living in non-industrialized societies, will suffer, on average, more than those living in industrialized societies who are most responsible. Such injustices demand redress and that's what environmental-justice ethics is all about (Bauer 2006).

The spatial scale of global climate change requires a profound shift in our moral ontology. Ethics, as we know it, has assumed an atomistic ontology—moral agents and patients are thought to be individuals. But as these considerations of environmental justice suggest, the only effective moral agents and patients at the global scale are societies, not individuals. What compensation, we are asking, in effect, do less-at-risk-but-more-to-blame industrialized *societies* owe more-at-risk-but-less-to-blame non-industrialized *societies*? No environmental-justice ethicist insists that affluent American individuals should write personal checks to individual impecunious Micronesians in order to compensate them for sea-level rise and to enable them to cope with that.

The temporal scale of global climate change makes the necessity for a shift from an individualistic to a sociocultural moral ontology even more obvious. We used to marvel at the temporal horizons of moral deliberations among the Iroquois—who considered the effect of present choices out seven generations (Erdrich 1996). But, do the math. Assume a twenty-five-year interval for each generation—a reasonable age-span between human parents and their offspring:

$7 \times 25 = 175 + 75$ (the life span of the seventh generation) $= 250$ years.

The serious adverse effects of global climate change, however, are predicted to *begin* to be registered only by mid-century and not to kick in full tilt for several centuries hence and then to last for millennia (IPCC 2007; Solomon et al. 2009). And the lag time for realizing the effects of remedial actions we take now will occur on the same temporal scale—calibrated in centuries and millennia. Does one care—can one care—about the welfare of *individual* human beings living in the twenty-third century or the twenty-fourth?

To that question, one may at least reply that we do in fact care, and care deeply, about individual human beings who lived 2,500 years ago, even 3,000 years ago. Who can read the *Apology* and not share Plato's bitter condemnation of Anytos, Lycon, and Meletus and the kangaroo court of Athens that unjustly convicted Socrates of impiety and corrupting the youth and sentenced him to death? Who can read the *Odyssey* and not seethe with the same righteous indignation that consumed Telemachus because of the outrageous behavior of the profligate and calumnious suitors? But that's because we know Socrates and Telemachus as individuals. We care far less about the unknown past beyond the pale of history and, as Leopold points out, beyond immediate posterity, lies the "unknown future." Can we muster up the motivation to make sacrifices now that will benefit anonymous individual people living two or three thousand years hence?

At the very beginning of "The Land Ethic," Leopold points out that Western *civilization* goes back three thousand years to the time of Homer. If our moral ontology is socioculturally scaled, as opposed to individually scaled, and the survival and continuity of human civilization is what we're concerned about, not the welfare of unknown future individuals severally, what would the math look like then? We conventionally divide the history of Western civilization into three periods—ancient, medieval, and modern—each roughly a millennium in duration. So seven sociocultural generations come out to be seven thousand years.

So much for a moral-*patient* ontology scaled to the temporal parameters of global climate change; what about a moral-*agent* ontology? I was appalled by what I saw at the end of Al Gore's otherwise excellent documentary, *An Inconvenient Truth*: a list of things that each of us, *individually and voluntarily*, can do to reduce our carbon emissions. I myself do many of those things: replace halogen light bulbs with compact fluorescents, make my home-to-office-and-back commute by bicycle, and so forth. But I live in Denton, Texas—not Ashland, Oregon, or Boulder, Colorado. So I am painfully aware that my individual efforts to lessen the size and impress of my own personal carbon footprint are swamped by the recalcitrance of the overwhelming majority of my fellow citizens. Many of them have never heard of global climate change. Many of those who have are convinced that it's a hoax cooked up by self-righteous environmentalist elites who can't stand to see common people have their mechanized fun. And many of those who think that it's for real welcome it as a sign that the end times are near, the horrors of which they will be spared by the rapture. It will not suffice therefore simply to encourage people *individually and voluntarily* to build green and drive hybrid. But what's worse is the implication that that's all we can do about it, that the ultimate responsibility for dampening the adverse effects of global climate change devolves to each of us as individuals. On the contrary, the only hope we have to temper global climate change is a collective sociocultural response in the form of policy, regulation, treaty, and law. What is required, in the closing words of "Tragedy of the Commons," is "mutual coercion mutually agreed upon" (Hardin 1968).

## Notes

1. The prophet epithet appears traceable to Roberts Mann (1954); it was picked up by Ernest Swift (1961); Roderick Nash (1967) institutionalized it in *Wilderness and the American Mind*, titling one chapter in that classic, "Aldo Leopold: Prophet."

2. The world's first college course titled "environmental ethics" was offered in 1971 at the University of Wisconsin-Stevens Point. The first articles by academic philosophers were published two years later (Routley 1973; Naess 1973).

3. See, for example, Wallace Stegner (1989).

4. "Some Fundamentals of Conservation in the Southwest," was slightly edited and reprinted in Flader and Callicott (1989). One noteworthy change was made. The typescript spelled Ouspensky's name as "Onpensky." But there is no doubt that Leopold referred to Ouspensky because he accurately quotes from *Tertium organum* verbatim. The published article in *Environmental Ethics* did not correct the spelling of Ouspensky's name. Feeling that the error was introduced by Leopold's secretary as she typed up his manuscript, I corrected her error for republication in *RMG*. Leopold composed in pencil in a tight cursive script and so, quite excusably, his secretary read his "us" as an "n."

5. Leopold quotes from Richard G. Moulton, *Modern Reader's Bible* (New York: Macmillan, 1907). The editor and translator of Leopold's bible, Richard G. Moulton, was also author of *The Literary Study of the Bible: An Account of the Leading Forms of Literature Represented in the Sacred Writings* (Lexington, MA: Heath, 1899). This fact would seem to confirm that Leopold's interest in the bible was less devotional than rhetorical.

## References

Allen, T. F. H., and T. W. Hoekstra. 1992. *Toward a Unified Ecology*. New York: Columbia University Press.

Bauer, J. 2006. *Forging Environmentalism: Justice, Livelihood, and Contested Environments*. New York: Sharpe.

Bormann, F. H. 1981. The Gaia hypothesis [review of *Gaia: A New Look at Life on Earth* by James Lovelock]. *Ecology* 62: 502.

Callicott, J. B. 1992. Principal traditions in American environmental ethics: A survey of moral values for framing an American ocean policy. *Ocean and Shoreline Management* 17: 299–308.

Callicott, J. B. 1996. Do Deconstructive Ecology and Sociobiology Undermine Leopold's Land Ethic? *Environmental Ethics* 18: 353–72.

Callicott, J. B., and E. Back. 2008. The conceptual foundations of Rachel Carson's sea ethic. In L. Sideris and K. Dean Moore, eds., *The Philosophy of Rachel Carson*. Albany: State University of New York Press, pp. 94–117.

Carson, R. 1962. *Silent Spring*. Boston: Houghon Mifflin.

Darwin, C. R. 1874. *The Descent of Man and Selection in Relation to Sex*, 2nd ed. London: Murray.

Dobzhansky, T. 1937. *Genetics and the Origin of Species*. New York: Columbia University Press.

Erdrich, L. 1996. Read their lips! Three novel ideas for a Clinton speech. *Washington Post*, June 23, p. C-1.

Fisher, R. A. 1930. *The Genetical Theory of Natural Selection*. Oxford: Clarendon Press.

Flader, S. L., and Callicott, J. B. ed., 1989. *The River of the Mother of God and Other Essays by Aldo Leopold*. Madison: University of Wisconsin Press.

Groom, M. J., G. K. Meffe, and C. R. Carroll. 2006. *Principles of Conservation Biology*. Sunderland, MA: Sinauer Associates.

Haldane, J. B. S. 1932. *The Causes of Evolution*. London: Logman, Green.

Hardin, G. 1968. Tragedy of the Commons. *Science* 162: 1243–48.

IPCC. 2007. Intergovernmental Panel on Climate Change. *4th Assessment Report*.

Lazcano, L., D. Suzuki, C. Tickell, M. Walter, and P. Wesybroek. 1998. Foreword to the English translation. In V. I. Vernadsky, *The Biosphere*. New York: Copernicus, pp. 14–19.

Leopold, A. 1933. The conservation ethic. *Journal of Forestry* 31: 634–43.

Leopold, A. 1949. *A Sand County Almanac and Sketches Here and There*. New York: Oxford University Press.

Leopold, A. 1979. Some fundamentals of conservation in the southwest. *Environmental Ethics* 1: 131–41.

Mann, R. 1954. Aldo Leopold: Priest and prophet. *American Forests* 60 (8): 23, 42–43.

Margulis, L. 1998. *Symbiotic Planet: A New Look at Evolution*. New York: Basic Books.

Meine, C., M. Soulé, and R. F. Noss. 2006. A mission-driven discipline: The growth of conservation biology. *Conservation Biology* 20: 631–51.

Naess, A. 1973. The shallow and the deep, long-range ecology movements: A summary. *Inquiry* 16: 95–100.

Nash, R. 1967. *Wilderness and the American Mind*. New Haven: Yale University Press.

Ouspensky, P. D. 1922. Tertium organum: *The Third Cannon of Thought; A Key to the Enigmas of the World*. N. Bessaraboff and C. Bragdon, trans. New York: Knopf.

Pickett, S. T. A., and P. S. White, eds. 1985. *The Ecology of Natural Disturbances and Patch Dynamics*. Orlando, FL: Academic Press.

Pickett, S. T. A., and R. S. Ostfeld. 1995. The shifting paradigm in ecology. In R. L. Knight and S. F. Bates, eds., *A New Century for Natural Resources Management*. Washington, DC: Island Press, pp. 261–78.

Rolston, H. 1994. *Conserving Natural Value*. New York: Columbia University Press.

Routley, R. 1973. Is there a need for a new, an environmental ethic? In Bulgarian Organizing Committee, ed., *Proceedings of the Fifteenth World Congress of Philosophy*. Varna: Sophia Press, pp. 205–10.

Sober, E., and D. S. Wilson. 1998. *Unto Others: The Evolution and Psychology of Unselfish Behavior*. Cambridge: Harvard University Press.

Solomon, S., P. G.-K. Pattner, R. Knutti, and P. Friedlingstein. 2009. Irreversible climate change due to carbon dioxide emissions. *Proceedings of the National Academy of Sciences*. doi: 10.1073/pnas 0812721106.

Stegner, W. 1989. The legacy of Aldo Leopold. In J. Baird Callicott, ed., *Companion to* A Sand County Almanac: *Interpretive and Critical Essays*. Madison: University of Wisconsin Press, pp. 233–45.

Swift, E. 1961. Aldo Leopold: Wisconsin's conservation prophet. *Wisconsin Tales and Trails* 2 (2): 2–5.

Tallmadge, J. 1989. Anatomy of a classic. In J. Baird Callicott, ed., *Companion to* A Sand County Almanac: *Interpretive and Critical Essays*. Madison: University of Wisconsin Press, pp. 110–27.

Udall, S. L. 1963. *The Quiet Crisis*. New York: Holt, Rinehart and Winston.

Vernadsky, V. I. 1998. *The Biosphere*. New York: Copernicus.

Williams, G. C. 1966. *Adaptation and Natural Selection: A Critique of Some Current Evolutionary Thought*. Princeton: Princeton University Press.

Wilson, D. S. and E. Sober. 1994. Reintroducing group selection to the human behavioral sciences. *Behavioral and Brain Sciences* 17: 585–615.

Wilson, E. O. 2005. Kin selection as the key to altruism: Its rise and fall. *Social Research* 72: 159–66.

# 12

## Principles of Gaian Governance: A Rough Sketch

Karen Litfin

Characterizing the Earth holistically as a self-regulating system, Gaia theory represents a creative synthesis that simultaneously builds upon and transcends reductionist science. At least four general readings of Gaia theory are possible. First, by substantially impacting (or "regulating") the planet's geochemistry, living organisms contribute to sustaining the habitability of the biosphere. In particular, the presence of life tends to alter a planet's climate and atmospheric chemistry in ways that favor the continuation of life (Resnik 1992). Stated in this way, Gaia theory qualifies as a scientific hypothesis in that it is, in principle, falsifiable. Second, and more radically, Earth itself may be understood as a complex, bounded, self-organizing, adaptive living system (Margulis and Sagan 1995). Third, the evolving biosphere is self-organizing with emergent properties, but not necessarily along a homeostatic track (Volk 2003). Fourth, and most radically, Gaia is conceived by some as an evolutionary planetary intelligence that operates to the benefit of the whole, or perhaps even according to some larger purpose (Russell 2008). Of the four readings, the first is the most broadly accepted and has helped to spawn a paradigmatic shift in the natural sciences, most evident in the new integrative field of Earth system science. This chapter begins with the first interpretation, articulating Gaia theory in the more general language of systems theory. In particular, it highlights some implications of three essential characteristics of living systems—holism, autopoiesis, and symbiotic networks—for global governance. The chapter then proceeds to inquire into some of the ways that Gaia might find expression in political culture, with special attention to how Gaian thinking might elucidate questions of global justice and kindle the political imagination.

## Gaia and Political Culture

Gaia's cultural allure was evident at the outset. James Lovelock (2000) was astonished to receive twice as many letters in response to his first book on Gaia from people interested in its religious aspects as from those with a more scientific bent. Had Lovelock named his theory something along the lines of "homeostatic Earth systems theory," he surely would not have received such a response. Gaia, Earth goddess and mother of creation, simultaneously evoked the ancient transcultural theme of a living Earth while tapping into a contemporary hunger for a sense of connection and wholeness. While the literary, religious, and philosophical dimensions of Gaia theory have been widely explored, the political implications have received little attention. What concepts, metaphors, and promptings might Gaia theory offer us as we explore modes of governance commensurate to the task before us? This chapter embarks on an exploration of this question.

From a Gaian perspective our blue planet functions much like an organism—a self-contained living system embedded in the larger solar system, with internal metabolic systems of temperature and chemical modulation and an atmospheric membrane that separates it from outer space. Interestingly this holistic perspective of the Earth has emerged just at a time when the twin phenomena of globalization and environmental destruction call us to adopt a planetary perspective. Gaian thinking awakens us to the crucial fact that human systems are embedded in and utterly dependent on this greater whole. Because the Earth system is the wider context in which our social, political, and economic systems operate, and because our actions now have planetary consequences,[1] we are increasingly compelled to develop forms of governance that are compatible with the larger system that environs and sustains us. This monumental task will most likely to occupy generations to come. Therefore this discussion will be suggestive, tracing only the broadest implications of Gaia theory for contemporary social and political theory.

## Systems Theory and Gaian Governance

Gaia theory, which views the Earth as a complex and bounded system, draws upon the more general field of systems theory, the basic tenets of which open up fresh possibilities for considering questions of global governance. Much contemporary political discourse is grounded in an atomistic, reductionistic model that sees the world as constituted by discrete institutional entities and problems, approaching these problems largely in isolation from one another. For instance, international environmental law, which consists of myriad separate regimes for hundreds of issues ranging from toxic waste exports to fisheries management, is itself rooted in an atomistic demarcation of the planet into sovereign nation-states. By highlighting the embeddedness of human systems in the living Earth system, Gaian thinking fosters a kind of meta-position from which a systemic perspective on global environmental governance might emerge. In broad terms, global environmental problems represent a collision of human systems with the larger Gaian system. In contrast to the mechanical billiard-ball metaphors that inform much of modern political discourse, the Gaian image of a living Earth may be more amenable to the problems at hand. Moreover, as a scientific alternative to modern reductionism, Gaia provides important concepts and metaphors that can help move us toward a viable future.

Systems theory, which has been adapted to a range of fields including engineering, education, finance, health, psychology, and natural science, postulates three broad types of systems.[2] *Hard systems* include many of the technologies associated with industrial life such as electrical grids, transport systems, and telecommunications. Because of their mechanical character and their linear logic, hard systems are known for their short-term efficiency, predictability, and performance. *Living systems*, of which Gaia is the largest known instance, are nested systems of biota and their environments. These complex systems cannot be understood in terms of the linear, reductionist logic of nonliving systems. They require a more dynamic, holistic approach. *Human systems*, like living systems of which they are a part, are nested and complex, evolving, reproducing themselves, and dying. Purpose, which is not an obvious property of hard or living systems, is essential to human systems.

Human systems problems, because they have multiple interacting causes and involve many actors with differing perspectives, are generally exacerbated when addressed in terms of hard-systems logic and methods. When hard-systems thinking looks at human systems, it is partially blind because it ignores purpose, subjectivity, and complex interdependence. Imagine, for instance, the consequences of approaching familial relationships with the orientation of a car mechanic. On a societal scale, technological fixes and other hard-systems solutions to human systems problems often generate more intractable problems down the road.

Examples include arms races, the so-called Green Revolution, and tall smokestacks that reduced local pollution but created the new problem of transboundary acid rain. The global repercussions of hard-systems thinking are becoming evident just as living-systems thinking, including Gaia theory, counsels an alternative model.

## Holism

Living and human systems are bounded entities, distinguishable yet never entirely separate from their environments. A cell, the simplest living system, is a "membrane-bounded, self-generating, organizationally closed metabolic network" (Capra 2002). That network includes complex macromolecules, such as proteins, enzymes, RNA, and DNA. The permeability of the cell's membrane gives it access to the nutrients and waste depositories it needs to survive, while also making it vulnerable to incursions from outside. In Gaia theory the atmosphere functions as a membrane that simultaneously separates Earth from outer space, while being porous enough for both sunlight and meteors to enter the system. All living systems, including human systems and Gaia, maintain a degree of structural integrity without ever being fully independent.

This radical interdependence stands in sharp contrast to prevailing political and psychological notions of independence. Just as modern psychology valorizes the autonomous ego, modern political thought is premised upon individual rights and state sovereignty. Systems theory calls all of these into question. In the words of V. I. Vernadsky, the Russian systems scientist, "human independence is a political, not a biological concept" (quoted in Primavesi 2000: 6). The autonomous individual, reliant upon billions of bacteria, is a fiction. Human well-being is also utterly dependent on local ecosystems and the Gaian system, including the ceaseless decompositional and generative work of plants, phytoplankton, bacteria, fungi, earthworms, and other organisms. Current economic and political institutions reflect a mode of consciousness that is essentially oblivious to this radical embeddedness. The apparent incompatibility of the dominant human systems with Gaian equilibrium suggests a need for modalities of governance rooted in a systemic understanding of interdependence—with other nations, species, and Gaia itself.

Serendipitously, Gaia theory's planetary perspective enters the scene just as the global impacts of human systems are becoming evident. For the first time humanity has become a geophysical force with planetary effects. The rate of species extinction is between 1,000 and 10,000 faster than in the pre-industrial era, rivaling the last great wave of extinctions that wiped out the dinosaurs 65 million years ago (UNEP 2002). Climate scientists predict that global temperatures will rise between 1.5 and 6 degrees Celsius in the coming century, a warming on the order of the shift from the depths of the last ice age to the present interglacial period (IPCC 2007). Most key resources—including forests, minerals, petroleum, fresh water, topsoil, and fisheries—are being depleted at unsustainable rates.

Like life itself, human beings have evolved the capacity to inhabit virtually every corner of the Earth. Globalization of some form therefore seems to be part of our destiny. Yet, as part of a greater whole, we must harmonize our social, economic, and political systems with Gaia. International environmental law represents only a piecemeal movement in this direction. Because it sidesteps the crucial questions of purpose and process that give rise to the destruction, green diplomacy cannot offer a systemic solution. By rigorously explicating the networks of systemic interdependence that underpin the Earth's functioning, Gaia theory challenges us to locate human systems with the living whole. How might Gaia theory inform our search for modes of globalization that are compatible with the larger Gaian system?

Some environmentalists are uncomfortable with the notion of Gaia because they believe it encourages complacency. They worry that people will assume that Gaia, like any good mother, will simply clean up their mess. Yet such an assumption would be a tragic and shortsighted misreading; Gaia theory is concerned with the systemic functioning of the planet, not the welfare of any particular species. From a Gaian perspective we are far more expendable than bacteria. While some may find solace in Gaia's capacity for adaptation over the aeons, any future equilibrium state will almost certainly be far less favorable to our species than the present one. For most of Gaia's 3.8 billion years, glacial periods were frequent and species diversity far lower than at present. So a healthy dose of prudence is in order—not for Gaia's sake, but for our own. In Lovelock's words (1990: 212), Gaia is "stern and tough, always keeping the world warm and comfortable for those who obey the rules, but ruthless in her destruction of those who transgress." Gaia theory helps to reveal those rules to us.

Living systems are maintained through the dynamic interaction of their subsystems. In the Gaian system the main chemical subsystems involve the cycling of carbon, nitrogen, oxygen, and sulphur (Lovelock 1990). The largest human system, the global political economy, involves the dynamic interaction of corporations, governments, international organizations, banks, and nongovernmental organizations. In both cases the systems are self-making: they generate high degrees of order through complex relationships among their parts and with the environment, rather than as a consequence of external agency. Gaia theory tells us that viable human systems must function as a nested subsystem within the Earth system. While this insight may seem absurdly obvious, the mounting global ecological mega-crisis suggests that sometimes the obvious bears repeating.

**Autopoiesis**
Gaia theory suggests that the Earth system has been *autopoietic*, or self-making, over the course of billions of years. Autopoiesis, a term coined by Maturana and Varela (1998) from the Greek words for "self" and "making," highlights the self-generative nature of metabolic networks in living systems. The system continually makes and remakes itself, maintaining its structural integrity and organic functioning through exchange with its environment. The minimal autopoietic entity is a bacterial cell, and the largest one known is Gaia (Primavesi 2000). An autopoietic system undergoes unceasing change while preserving its web-like pattern of organization. During the first two billion years bacteria ruled the planet and devised all of life's essential processes: reproduction, photosynthesis, fermentation, nitrogen fixation, respiration, and locomotion (Capra 2002).

Despite the proliferation of life forms over the millennia, many of Gaia's essential characteristics have remained relatively stable. *Homeostasis*, the tendency toward constancy, is another property of living systems. Lovelock's Gaia hypothesis predicts that Earth's climate and chemical composition remain in homeostasis over long periods of time until some internal contradiction or external force causes a jump to a new stable state (Lovelock 1990). Most external forces have been asteroid or comet impacts, to which Gaia responds, whether gradually or rapidly, by moving into a new stable state. Human activity since industrialization represents a new kind of internally induced planetary crisis, one that seems to call for a conscious autopoietic response. Given the looming global eco-crisis and our utter dependence on Gaia, we might be curious to learn about how Gaia has handled past planetary crises.

Earth's first internally induced environmental crisis probably occurred with the invention of photosynthesis, when the consumption of carbon dioxide by bacteria threatened to destabilize Gaia's homeostatic balance. The ensuing accumulation of oxygen, one of their waste products, however, opened up a tremendous niche for oxidizing consumers, and the subsequent growth of more complex organisms (Margulis and Sagan 1995). These complex life forms in turn replenished the atmosphere's most important greenhouse gases: carbon dioxide and methane. Thus Gaia responded to the "crisis" of photosynthesis by generating a new atmospheric homeostasis.

Another crisis is suggested by the recent "snowball Earth" discovery, which coincidentally represents the strongest geological evidence against Gaia theory (Hoffman and Schrag 2000). Around 600 million years ago, just before the appearance of recognizable animal life, the entire Earth, including the tropics, apparently froze over for 10 million years or more. Geothermal flux from radioisotope decay in the Earth's mantle and the buildup of atmospheric carbon dioxide from volcanism prevented the oceans from freezing to the bottom, but ice grew to a depth of several kilometers as global temperatures plummeted to –50 degrees Celsius. Eventually carbon dioxide from volcanoes accumulated to record levels, warming the Earth and melting the ice, causing an extreme climate reversal that brought about a fierce greenhouse effect. This event seems to conflict with Gaia theory's predictions of homeostatic stability.[3]

Yet some interpretations of snowball Earth are consistent with Gaia theory. First, Gaia theory predicts that crises will be followed by long periods of stability; ten million years would qualify as a long period of stability, even if life didn't exactly flourish during that time. Snowball Earth thus offers a cautionary tale, reminding us that Gaian homeostasis does not always provide a comfortable home. Second, the extreme glaciations of snowball Earth occurred just before a rapid diversification of multicellular life, culminating in the Cambrian explosion of biodiversity around 550 million years ago. Paradoxically the long periods of isolation and extreme environments on a snowball Earth could have stimulated genetic change (Hoffman and Shrag 2000), and this crisis and apparent anomaly to Gaia theory could have facilitated the evolutionary lineage of our own species. Rather than falsifying Gaia theory, snowball Earth may be an odd and extreme chapter in Gaia's self-making.

The nature of feedback among tightly linked networks means that very small causes can quickly amplify into large effects; complex systems therefore have only limited predictability. Nonlinearity is therefore not just a mathematical concept linked to exponential change as a consequence of feedback mechanisms, but an ontological property of living systems with important implications for global environmental politics. A dramatic example occurred with the discovery of the Antarctic ozone hole in 1985. Scientists' predictions of incremental global ozone loss as a result of chlorofluorocarbons (CFCs) were overturned with the dawning recognition that these so-called miracle compounds could catalyze a chain reaction in the stratosphere, with one CFC molecule destroying as many as 100,000 molecules of ozone (Litfin 1994). In this case, because the human systems were relatively simple with only a small number of firms producing CFCs, international negotiators found it comparatively easy to address the problem via the Montreal Protocol and its subsequent amendments. Yet even with full compliance, this shining success story of environmental diplomacy will not return the ozone layer to pre-1985 conditions for another century. Moreover Gaia's long-term responses to the effects of ozone depletion, such as the massive death of phytoplankton, remain unclear. The potential for irreversibility is a corollary to nonlinearity.

If rapid planetary change is possible, as geologic history suggests, then an attitude of prudence and humility is appropriate. Gaian thinking therefore supports the precautionary principle: if the risk is high, then we should act to prevent harm even in the absence of scientific certainty. Because Gaia is the systemic vessel of all living and human systems, our actions should be especially constellated toward ensuring the stable functioning of Gaian systems. Nonlinearity means that in the presence of systemic perturbations, surprises are likely. Recognizing that surprises, by definition, cannot be anticipated, the wisest perspectives and policies will be those that enhance the resiliency of human and other living systems (Janssen 2002). This means understanding, as much as possible, the nature of those systems, attuning our systems to the larger Gaian system and taking precautionary action to limit harm.

Like living systems, human systems are autopoietic, tending to reproduce themselves and evolve new equilibrium states in response to changing conditions. In his essay on social autopoiesis, sociologist Niklas Luhmann (1990) describes human systems as self-generating communicative networks. These networks have both material and cultural effects, generating external social structures like corporations and states as well as internal structures of meaning like rights and roles. For example, the global economy is continually reproduced through networks of communication involving advertising, production, entertainment, financial transfers, and education, each of which has structural correlates in subjective consciousness and intersubjective experience. What happens when autopoietic human systems disrupt the Gaian equilibrium?

According to Gaian scientists, when the activity of an organism favors both Gaia and itself, it will tend to spread; eventually, both the organism and the associated environmental change may become planetary in scope (Lovelock 1990). We may therefore be tempted to optimistically infer from humanity's relatively rapid globalization that this trend is favorable to (or at least compatible with) Gaia. This logic, however, ignores the vastly different time scales associated with human vs. Gaian processes. A period of 100,000 years, for instance, is many times longer than all of human history, yet represents less than 0.003 percent of Gaia's lifetime. Only in the last part of the twentieth century did the human species become a geophysical force operating on a planetary scale; only in the last decade was human-induced climate change conclusively observed. The sobering fact is that we cannot know exactly when or how the Gaian system will respond to these rapid changes. The geological record evinces a pattern of punctuated equilibrium, with long periods of homeostasis followed by sporadic catastrophes, which in turn spark intense periods of innovation leading to new stable states. Gaian theorists believe that once a system shift gets underway, it moves into a new and very different state quickly, taking as little as a century to establish into a new geochemical equilibrium. Species diversity, however, will take millions of years to rebound. After the Cretaceous-Tertiary asteroid impact biological diversity is believed to have taken between five and ten million years to recover. Therefore it is prudent to bear in mind the converse of the optimistic inference mentioned above: any species that impairs Gaia's functioning may precipitate not only its own demise but that of many others, even as the web of life innovates toward a new homeostasis.

The concept of autopoiesis raises an important philosophical question. If a living system somehow "makes itself," does it do so purposefully? Because it hinted at such a possibility, Lovelock's original formulation of the Gaia hypothesis met with intense scientific criticism, especially from neo-Darwinists. Some critics interpreted him as proposing a sentient Gaia able to consciously control the Earth with foresight and planning. In his later formulation Lovelock illustrated the principle of homeostasis through a simple model that involved dynamic interaction but not intentionality. For instance, the automatic self-regulation of the carbon cycle, which has stabilized atmospheric concentrations of carbon dioxide at 0.03 percent, requires no foresight and planning. Yet this number is very different from the 95 to 98 percent concentrations of carbon dioxide on Venus, Mars, and pre-life Earth (Lovelock 1990). The Earth's improbable atmosphere is a consequence of feedback between biota and nonliving systems.

The feedback mechanisms that generate Gaian homeostasis require neither intention nor altruism, but rather only a reciprocal flow of influence.[4] Whenever the rate of change in a system is getting faster, positive feedback is at work. This kind of reinforcing feedback is important when a new equilibrium is getting established, but it can also lead to a pernicious spiraling effect, as in avalanches, stock market booms, and cattle stampedes. On a Gaian scale, for instance, because water vapor is itself a greenhouse gas, increased evaporation on a warming planet may increase the temperature further. When positive feedback gets out of control, the resulting runaway system can only be stopped if either the external environment or the internal negative feedback halts the positive feedback loop. Damping, or negative feedback, prevents the system from running away with itself. For instance, the absence of predators in an ecosystem will lead to an overpopulation of their prey; their numbers in turn will not be able to subsist on the given food supply, so they will fall to a sustainable level. In each of the cases above the feedback is an automatic function. The system is responsive, yet no purposeful agent is postulated as responsible; Gaia theory does not entail teleology.[5] Questions of larger purpose and intention in living systems are simply beyond the bounds of scientific methodology.

Purpose, however, is essential to human systems. It consists of the most cherished values that inform and orient our systems. While a human system's purpose might be unexamined, misunderstood, ignored, debated, and even disguised, reconfiguring it from its base requires identifying its drive and implicit values. The global economy is a self-reproducing network of networks, but can we point to a basic purpose or set of purposes that drive it? Growth, development, prosperity, and wealth—these are different words for what many would agree is the underlying purpose of the system. Growth as systemic purpose is evident in its almost universal acceptance—across the political spectrum from left to right, and around the world from to North to South.[6] While disagreement on *how* to pursue economic expansion abounds, there is a striking consensus on the fundamental objective itself. Under the prevailing capitalist ideology, the alternative to growth is economic collapse, both at the level of the firm and the state.[7] Yet systems thinking tells us that the growth imperative is a positive feedback mechanism, and therefore runs the risk of creating runaway processes that can only be stopped when either the external environment or an internal instability halts them. Systems theory does not predict exactly when or how that might happen, but it does warn us about positive feedback loops in general.

If human systems are to persist as a global subsystem of Gaia, then we will need to align our purposes with the functioning of Gaia. The longer we wait, the greater the risk. If economic growth is the purpose of the global economic system, then reconfiguring the current system means first and foremost rethinking our purposes. For human systems to be harmonious with the wider Gaian system, ecological sustainability must become a core human purpose.

## Symbiotic Networks and Gaian Governance

Gaia theory depicts the Earth system as a many-layered isomorphism, a vast autopoetic network of nested communities. Thus, when biologist Lewis Thomas asked himself, "What is the Earth most like?" he answered, "It is *most* like a single cell" (Thomas 1974). Living systems, from the cell to Gaia, are constituted through symbiotic networks whereby dissimilar entities coexist in a mutually beneficial arrangement. Contrary to the neo-Darwinist view of life as a harsh competition for survival, Gaia theory upholds cooperation as much more the rule than competition. Bacteria, the most long-lived class of organisms and the basis of all subsequent life, are inherently social; "they live by collaboration, accommodation, exchange, and barter" (Thomas 1974: 6–7). Gaia theory tells us that life did not colonize the planet by combat but by networking (Margulis and Sagan 1995).

Like other living systems, human systems consist of networks. On a global scale the human system comprises innumerable networks of production and consumption, diplomacy and warfare, advertising and entertainment, and education and ritual. Many (if not most) social systems are more rooted in cooperation than competition: for example, the family, global transportation, and postal networks. Yet the overarching premise of the global economy is competition. Firms compete with one another for resources and markets, workers compete for jobs, and countries compete for investment. Both capitalism and traditional Darwinian biology also presume the natural environment as a stable background in which individuals compete and to which they must adapt. Even Marxist political economy, an ostensibly more cooperative approach, depicts history as class struggle and nature as valueless material to which human labor brings value. In contrast to modern approaches to political economy, Gaia interprets life as the ability of cooperative networks to simultaneously adapt to, alter, and enhance their environments to their mutual benefit. Both the unrelenting drive to compete, an intrinsic consequence of the growth imperative, and the notion of environment as inert backdrop are therefore at odds with Gaia theory. A sustainable global economy, by way of contrast, would consist of symbiotic networks acting in harmony with Gaia.

In living systems, networks continuously reconstitute their elements in cyclical processes. In ecosystems, and in Gaia as a whole, recycling is the rule; one species' waste is always another species' source of nourishment. The Earth's major nutrients—carbon, hydrogen, oxygen, and nitrogen—are cycled and recycled. Cyclical exchanges of energy and resources in a living system are sustained by pervasive cooperation. Neither for Gaia, nor for any local ecosystem, is there an "out there" into which "waste" can be dumped. Such concepts as garbage and pollution are foreign to Gaia. Yet existing political approaches to waste (whether solid, atmospheric, toxic, biomedical, or nuclear) pursue safer technologies and disposal practices without ever questioning the very concept of waste itself. This is true for all levels of mainstream "waste management," from municipal policies to international treaties. Indeed industrial societies are based upon "the toilet assumption"—the implicit belief that waste can be simply "flushed away" (Slater 1970). Gaia theory reiterates the message of global environmental degradation: there is no "away."

What might Gaia tell us about principles of governance? Contrary to the fears of some that a Gaian politics would be reductive (to biology or the planet) and therefore invite abuse by demagogues, a thoroughgoing Gaian politics would be radically democratic, a nested system of governance from the neighborhood to the global level (Madron and Jopling 2003). There are no authoritarian regimes in Gaia, only mutually enhancing symbiotic networks. Unlike the current system, premised upon the growth imperative, Gaian democracies would be oriented toward purposes of sustainability and justice, and modeled on a network vision of participatory governance and forms of leadership that empower people. The prevailing command-and-control culture in business and politics would be replaced by a culture of dialogue. Autopoiesis, or self-making, would take on new meaning with the globalization of democracy as people organize themselves according to Gaian principles.[8]

The rise of network society, from global civil society to the Internet, coincides with the decline of the sovereign nation-state. State sovereignty is being radically reconfigured, by global networks of communications, finance, crime, terrorism, disease transmission, ecology, and transnational activism (Litfin 1997; Hawken 2007). From a Gaian perspective the nation-state is neither large enough to inspire a planetary identity nor small enough to nurture the kinds of local identity and civic involvement that are essential to participatory governance. This does not mean that the nation-state will cease to exist, but only that it may be incorporated into broader cross-cutting networks of supranational, regional, and local forms of governance.

Yet we are wise to remember that while Gaia theory can be helpful in reorienting our thinking about human systems, it is not a panacea. Systems language and concepts offer an integrative way of understanding current problems and redirecting our actions down a more sustainable path, but they do not lay the stones along that path. For this reason Lovelock warns that we need to be wary of opportunists whose use of ecological language is merely a mask for ulterior aims (1979). Gaia theory can help us with the essential task of seeing the big picture, but it does not resolve the thorny problems of practical politics. In this sense Gaia may be more important for its broader contribution to our ethical and political imagination than for its direct policy effects. Gaia theory does not so much represent a holy grail as a powerful corrective wind to reorient our sails.

## Gaia and Global Justice

In displacing humans from the center or apex of creation, Gaia theory offers both scientific and metaphysical support for an alternative to modernity's anthropocentric outlook. For some environmental theorists this is its most significant contribution to green politics (Dobson 1990). Yet this anti-anthropocentric message does not translate easily into a strategy for social and political change for the simple reason that human action is unlikely to harm Gaia's overall health. As Lovelock states, "On a planetary scale, life is near immortal" (1986: 28). If Gaia is a self-making system, then there is no palpable need for human action. If "nature is in control," as some ecological thinkers infer from the Gaia hypothesis (Spretnak and Capra 1985, cited in Dobson 1990), then apathy or even environmentally destructive behavior may be morally acceptable inferences from Gaia theory.

Along these lines, some anti-environmentalists cite Gaia theory in support of their views. Ron Arnold, for instance, whose *Ecology Wars* has been called "the bible of the wise-use movement," takes great solace in "Lovelock's clear-sighted vision of a self-protecting Earth managed for ages by self-knowing human stewards" (cited in Botkin 2001: 41–42). He cites the autopoetic resilience of Gaia in order to justify both current industrial practices and his own caustic attack on environmentalism. Rather than decentering humanity, "wise-use" proponents understand Gaia theory as giving free reign to human systems. The primary flaw in this logic, however, is that Gaian time scales are on the order of aeons, whereas human systems rarely consider anything longer than a generation. Contrary to anti-environmentalists' wishful thinking, Gaia's resilience says nothing about the resilience of human systems.

A core message of Lovelock's theory is that even if human-induced perturbations to Gaia were good for the biosphere—which seems highly unlikely—they could be disastrous for us. As Andrew Dobson articulates, "While the Gaia hypothesis might indeed lead us to contemplate our humble place in the grand scheme of things and thus to a 'decentring' of the human being, we quickly return to center stage as humility turns into fear for survival" (1990: 45). I would add that humility could be born not only of fear, but also from a sense of awe and gratitude for the larger systems that environ us. While undercutting anthropocentrism, Gaia theory has the paradoxical effect of highlighting, rather than diminishing, our place in the Earth system.

Yet Gaia theory raises some disconcerting ethical questions. If value in the Gaian system is related to the continuance of life in general, then must our ethical concern extend beyond humans to other creatures? To the planet? In some ways our concern for Gaia comes not so much from ethical obligation but from an enlarged sense of pragmatism: we want to save our own skins. Gaia will survive, but our interference may catapult her into a new and less hospitable state. Thus Gaian pragmatism evokes some general ethical principles. "Is" may not dictate "ought," but it can be suggestive. If, for instance, species diversity and a stable concentration of greenhouse gases *are* critical for a healthy functioning of the Gaian system, then we *should* prevent species extinctions and reduce our use of fossil fuels. If risks *are* high, then action to prevent harm *should* be taken, even in the absence of full scientific certainty. If current practices *do* risk destabilizing the Earth's climate and life support systems, then we *should* take precautionary action and change them. Thus Gaia theory, when combined with a commitment to viable human systems, seems to call into question the traditional fact/value distinction.

At a minimum, Gaian thinking supports a precautionary approach: if we value human life, to wait for full scientific certainty before curtailing behavior that might destabilize planetary life support systems would be foolhardy to say the least.

If Gaia focuses our attention on the Earth, what happens to our generally accepted ethical commitments to other people? What, for instance, of questions about justice under conditions of extreme global inequality? At first, we might think that if Gaia is the object of our concern, then we must sidestep thorny questions like North–South inequity and get onto the business of "saving the planet." Because Gaia's big-picture perspective challenges anthropocentrism, we might be tempted to ignore the comparatively small questions of justice and equity. And because the Gaian scale of global heating seems to dwarf other environmental concerns, we might be tempted to ignore the comparatively minor problems of pollution and conservation of local ecosystems.

Ironically, James Lovelock falls prey to both temptations in his latest book, *The Revenge of Gaia* (2006). Were he not a reputable scientist and were his prognostications not also supported by a host of peer-reviewed research on the rapidly unfolding crisis of human-induced climate change, his book would read like an alarmist science fiction fantasy. While he has done a great service in sounding the alarm, his policy prescriptions reflect an uncanny combination of a profound understanding of Gaian-scale natural systems alongside a disturbing insensitivity to social, ethical, psychological, and even smaller scale ecological questions. Lovelock's foremost policy recommendation is a rapid and large-scale transition to nuclear energy, ignoring the highly problematic questions waste disposal, weapons proliferation, terrorism, and affordability for developing countries. Some of his technological fixes, like pumping aerosols or launching gigantic mirrors into the atmosphere to reflect incoming solar radiation, are astonishing in their deviation from the precautionary thinking that seems to follow from a thoughtful application of Gaian thinking to human systems. Lovelock's proposal for a "sustainable retreat" into cities and a large-scale transition from agricultural land-use patterns to forests as carbon sinks is premised on a far-fetched scheme for laboratory-based food production by high-tech chemical firms. In the context of concerns about democracy and global justice Lovelock's most troubling proposal is for an enforcement body for restoring Gaia's health that would be controlled by the wealthy countries. Here he exemplifies the worst fears of developing countries:

that the rich minority, who caused the climate catastrophe in the first place, will use their power and wealth to preclude or block their own economic development. Like a good doctor Lovelock has diagnosed the immediate causes of the patient's (Gaia's) fever, but his prescriptions would benefit enormously from a good injection of holistic medicine.

Human systems are now too deeply intertwined with the Earth system for one-sided engineering panaceas and technocratic elitism to solve Gaian-scale problems: ethics, politics and psychology have become integral aspects of the Gaian system. While Gaia's planetary perspective may undercut humanism in the big picture, the pragmatic requirements of moving toward sustainability have the paradoxical effect of highlighting questions of justice and equity. Under the unacknowledged assumption that infinite growth on a finite planet was possible, we could anticipate that economic growth would eventually "trickle down" to everyone. But the recognition is dawning: the overconsumption of the North cannot be globalized without Gaian-scale consequences.

Even so, trends are at odds with this recognition. With 80 percent of the human population, developing countries represent the wave of the future. They are not going to change their development trajectories in the absence of a compelling moral and practical exemplar, nor without financial and technological assistance from the wealthy countries. Justice therefore becomes a matter of "geoecological realism" (Athanasiou and Baer 2002). As Dobson suggests, a return to a weak anthropocentrism is required in order to transform Gaian thinking from either pure science or mystified philosophy into a practical worldview with a strategy for social change (1990). Even if human beings are constituted by bacterial colonies, and even if human history represents only a tiny fragment of Gaia's lifespan, politics and economics are ultimately about people. Gaia's planetary perspective reminds us that we are all in this together. Therefore importing Gaian insights into the social and political arena requires that we pay attention to the needs and aspirations of other people—especially those who represent the wave of the global future.

Being in this together, however, does not mean that our ethical and political dilemmas have an easy Gaian answer. If politics is about who gets what, where, when, and how, then Gaia theory offers little guidance except to say that as a species, our well-being depends on Gaia's well-being. Marcel Wissenburg, who finds in Gaia theory no redeeming political value, offers the following assessment:

"Gaianism" is a modern variant of philosophical determinism. It supposes that everything depends upon everything else, that all of nature is a whole, that the whole rules itself, and that no part of it is autonomous. Gaianism necessarily leads to one of three courses of action: quietism, totalitarianism, or "anything goes." (1993: 9)

As suggested earlier, quietism and "anything goes" are only options if we are not concerned with the future of our species. Similarly we see that the risk of a Gaian totalitarianism, which would subject all human interests to the nonhuman interest of Gaia, is not so real once we recall that Gaia per se is not at risk. What *are* at risk, however, are the relatively comfortable Gaian conditions under which extant human cultures have evolved. Preserving those conditions is a monumental task, one that will demand sincere and focused attention to questions of distributive justice. Gaia theory therefore does not preclude questions of ethics and distributive justice. Indeed, since human systems are themselves living systems and subsystems of Gaia, Gaia theory can elucidate our search for equitable and sustainable modes of governance.

Science, a key source of both legitimation and conceptual models, has always provided grist for the political mill. As Theodore Roszak observes, "It is one of the glories of science that it can give back to the culture from which it grows" (1992: 30). Until recently the scientific metaphors that dominated the modern Western political imagination were drawn from an atomistic, mechanical, and reductionistic worldview. Nation-states, firms, and people were conceived as independent, acquisitive individuals competing for resources, power, and wealth; nature was either a backdrop to our human dramas or a source of wealth to be exploited by industrious humans. No doubt, importing science into political life can have pernicious effects, as it did with social Darwinism. Yet, just as the Enlightenment application of Newtonian science fostered democracy and a particular vision of human progress, so too does Gaia theory bring an emancipatory potential to our current situation.

Gaian concepts of holism, autopoiesis, symbiotic networks, and nonlinearity offer a very different language from mechanistic science and neo-Darwinian biology for understanding human systems. As David Abram suggests, Gaia theory has powerful implications for virtually every realm of human endeavor because it calls for a new way of perceiving our world. Whereas the modern separation of mind and matter upholds "a cultural program of environmental spoilage without hindrance of ethical restraint, [Gaia theory] shifts the locus of creativity from the human intellect to the enveloping world itself" (Abram 1990: 79). Human creativity, in its most harmonious and sustainable expression, would be a co-evolutionary impulse coherent with and evoking the larger Gaian creativity from which it emanates. If we take seriously the implications of our embeddedness in Gaia, we recognize that "we exist *in* this planet rather than *on* it" and we recover a sense of the Earth as "the forgotten ground of all our thoughts and sensations" (ibid.). In the most literal sense we exist within Gaia because we live within her atmosphere. A Gaian mode of perception engenders a radical critique of the instrumental, exploitative relationship with the world that informs prevailing human systems. Yet, as Abram remarks, Gaia is not merely an abstract theory; we experience Gaia only in particular places and through the medium of our embodied sensual awareness.

Gaia theory is emerging just as the challenges of globalization are becoming acute. As Vaclav Havel (1997) observes, we experience a sense of helplessness before these challenges because "our civilization has essentially globalized only the surface of our lives." Our external lives—our communication, transportation, financial exchanges, agriculture, and medicine—are globalized, but our inner lives orbit inside the myopic constraints of egoism and parochial identities. Gaia theory revitalizes our vision of the human condition, calling us back from our isolation, connecting us to the wondrous whole of creation and evoking a greater sense of responsibility. Gaian thinking provides one channel through which our inner identities and modes of perception can develop to meet the challenges of our externally globalized world. Embracing our embeddedness in the whole of creation and "trusting [our] own subjectivity as the principle link with the subjectivity of the world" (Havel 1997: 93), we claim our *responsibility* as *an ability to respond* to planetary challenges.

Gaia theory not only provides new ideas for understanding natural and human systems, it also introduces new concepts and metaphors to the political imagination. Symbols can be powerful sources of motivation, and the image of the Earth as a living, self-regenerating being is an especially powerful one. If affect precedes cognition, as many psychologists claim, then the emotional appeal of Gaia theory may surpass its more practical contributions to sustainability.

## Gaia and the Political Imagination

An Internet search for "Gaia theory" yields nearly two million websites, while a search for "Gaia" turns up nearly twenty million. Of the latter, most are about environmentalism and various forms of spirituality, but their topics also include the arts, urban planning, tourism, feminism, and even sporting goods. Gaia is ubiquitous among environmental activists and spiritual seekers. The popular embrace of Gaian imagery in rationalistic and technologically advances societies may seem surprising. Yet, because Gaia has deep mythopoetic roots, perhaps we should not be so surprised. Gaian thinking represents a return to a cosmology of human embeddedness rather than human exceptionalism. Indeed the philosophy of human exceptionalism—a premise of modernity's secular faith—is an extreme historical aberration. While the Enlightenment offered tremendous advances in science, technology, and human rights, it was also a great forgetting of different ways of knowing. Gaia theory is consonant with the organic worldview and the Great Chain of Being of premodernity, a worldview that carried with it certain moral taboos about how to treat a living Earth (Merchant 1983).

Gaia theory is cotemporaneous with the dawning recognition of the nexus between globalization and ecology, but also with the rise of feminism and Earth-based spirituality. Gaia theory, by virtue of its namesake and its content, resonates with strands of both of these movements. Some see in Gaia the rebirth of paganism, others the return of the goddess, and still others an ally in the politics of ecofeminism (Spretnak 1982; Hardin 2004). These approaches to Gaia have wrought intellectual and political fractiousness around questions of essentialism, gender equality, and historical interpretation (Biehl 1991; Merchant 1995). Yet, perhaps more important, these strands of the environmental movement have given Gaia theory "the emotional and moral force it may need to become politically relevant" (Roszak 1992).

Gaia theory at once revives an ancient symbol and endows it with scientific legitimacy, synthesizing empiricism with poetic inspiration. In much the same way that the image of the Earth as seen from space inspires environmentalism, Gaia is a symbol of wholeness, interdependence, and dynamic complexity. For many, Gaia also evokes awe and reverence, restoring a sense of connection to the cosmos that Western culture abandoned when it displaced the medieval conception of the

Great Chain of Being with a mechanistic worldview. By evoking a sense of the sacred, Gaia challenges modernity's utilitarian orientation while simultaneously leaning on its appeal to science. Yet in the context of a rational, technological culture, a simplistic revival of this ancient symbol runs the risk of shallowness. A spiritual symbol is not merely cognitive or sentimental, but rather it must stir us in the deepest parts of our being and reconstitute our actions and relationships.

Vaclav Havel, former president of the Czech Republic, finds in Gaia theory his inspiration for an alternative discourse of human rights, one that is rooted in a Gaian spirituality rather than secular anthropocentrism. While he upholds the discourse of human rights as an integral part of a new world order, he distances it from the language of the departing era. For Havel (1997: 181), the authority of a world democratic order can only be built on "the revitalized authority of the universe." Gaia theory, he suggests, offers an important source for this revitalized authority because it brings us to "the awareness of being anchored in the Earth and in the universe, the awareness that we are not here alone or for ourselves alone but are an integral part of a higher, mysterious entity....Only someone who submits to the authority of the universal order of creation, who values the right to be a participant in it, can genuinely value himself and his neighbors, and thus honor their rights as well" (1997: 171–72). Gaia need not only inform questions of environmental governance; it can also inspire a wider context for envisioning human rights.

Thus far our analysis has been largely conceptual, yet the gravity of the situation calls for pragmatic solutions. Where, if anywhere, are the vibrant experiments in Gaian governance? No doubt, small groups of people everywhere are taking up the challenge of revising the purposes and functioning of human systems in light of Gaia. In his recent book *Blessed Unrest*, Paul Hawken (2007) likens the decentralized global movement for ecological sustainability and socioeconomic justice to Gaia's immune system. While Hawken's geophysiological metaphor may be too literal for some, it is possible that the movements he explores, ranging from organic agriculture to fair trade to indigenous rights to recycling, represent embryonic experiments in Gaian governance.

Perhaps the most radically holistic of these experiments is the global ecovillage movement, which seeks to establish socially and ecologically viable alternatives on the grounded understanding that current systems cannot last. Ecovillages have taken root in every climate and on every continent; their belief systems are rooted in every major world religion, plus paganism and atheism. Underlying this diversity is a shared commitment to a holistic worldview consistent with systems theory (Litfin 2009). While the movement remains small—comprising several hundred recent communities in industrialized countries and perhaps 15,000 traditional villages with ecovillage design principles in the developing world—it is growing rapidly. If these communities were isolated experiments, disconnected from one another and from larger global processes, they might not be of interest to the study of international politics. Since 1995, however, with the formation of the Global Ecovillage Network (GEN), they have come together to share and disseminate information about sustainable living practices.

Both as conceptual underpinning and imaginal metaphor, Gaia circulates widely in the ecovillage movement. There are ecovillages with "Gaia" on several continents, and "whole Earth" images are popular in ecovillages everywhere. In the 1990s Ross and Hildur Jackson, founders of GEN working in Denmark, started three "Gaian" entities: Gaia Trust, which funnels financial assets from investments into seed grants for ecovillages; Gaia Technologies, which develops sustainable technologies, and Gaia Villages, which conducts research on the global ecovillage network. More recently the leadership of GEN has formed a collective—Gaia Education—that has developed a four-week comprehensive course on Ecovillage Design Education. A related venture is Gaia University, which began offering undergraduate and graduate degree programs in 2006. If any icon can be said to elicit universal appeal in the global ecovillage movement, it is Gaia. More important, the basic principles of Gaian governance outlined in this chapter are most evident in the global ecovillage movement: holism, symbiotic networks, participatory democracy, cyclical processes tending toward zero waste, and local experience grounded in planetary awareness.

The growth imperative has become a planetary malady, calling into question the viability of prevailing human systems. As we stand perched between hope and despair in our search for new models of governance, Gaia theory offers good pointers. First, our well-being is utterly contingent on the equilibrium of the larger Gaia system, along with its constitutive symbiotic networks and cyclical processes. Second, Gaian-scale crises can precipitate systemic shifts that dwarf human time frames. Third, we share a common bacterial ancestry with all other species, yet our capacity for *conscious* autopoiesis seems to be our distinguishing mark. In a time when fear and despair threaten our capacity for positive action, Gaia can serve as a source of faith, humility, and inspiration, reminding us that we are an integral part—and an astonishing result—of an evolutionary process that has been unfolding on our home planet for four billion years. We are the means by which Gaia is growing into self-awareness, and current conditions may be the labor pains of that birth of consciousness.

Gaia enlarges our vision of human purpose beyond the growth imperative, and reorients our action beyond the personal and local onto a planetary spatial and temporal scale. And because Gaia acts locally as well as globally, a Gaian awareness makes us more, not less, intimate with the particular landscapes of our dwelling. Yet, as David Spangler (1993) rightly warns, invocations of Gaia run the risk of becoming empty slogans if we do not allow them to *inhabit* us. If we sincerely want to reinvent our relationship with the Earth, we cannot simply deploy images of Gaia to meet emotional, religious, political, or commercial needs without allowing them to transform us in unexpected and radical ways. Both as scientific theory and cultural image, Gaia has the potential to become an intensely fertile idea for our time.

## Notes

I thank Richard Gammon, Jason Lambacher, David Schwartzman, Chris Uhl, Paul Wapner, Stephen Warren, Lauran Zmira, and two anonymous reviewers for their diverse and helpful comments on earlier drafts of this chapter.

1. Until the last decades of the twentieth century, the environmental consequences of human activity were almost entirely local and regional. Beginning in the 1980s, people became aware of a new category of problems whose causes and effects are both local and planetary. The most obvious of these global environmental problems are global climate change and stratospheric ozone depletion. Others, like deforestation, desertification, and loss of biodiversity are clearly local and regional but are global in their full impact. Losing species, for instance, is a local matter, whereas the current wave of mass extinction is a global crisis for the Earth's biosphere as a whole.

2. This threefold typology is adapted from Madron and Jopling (2003) and Checkland (1981).

3. I am grateful to Stephen Warren, a glaciologist and atmospheric scientist at University of Washington, for pointing out to me this apparent anomaly to Gaia theory. For a comprehensive exploration of Snowball Earth, see http://www.snowballearth.org.

4. The discussion in this paragraph is drawn from Madron and Jopling (2003).

5. But neither can Gaia theory rule it out. The question of purpose informs the observation that Gaia theory is a spectrum of ideas, ranging from the axiomatic to the speculative. At one end of the spectrum is the undeniable claim that life has dramatically shaped the Earth system. Moderate views understand Gaia as a self-organizing system or, more radically, a single planetary being. More speculative Gaian thinkers believe that an underlying intelligence is directing the co-evolution of Gaia's physical and living systems.

6. One might argue that economic growth is actually a means to more deeply held values of convenience and efficiency, values that have been so taken for granted that they have only recently received the serious analysis they deserve (see Tierney 1993; Princen 2005). While a focus on growth as a systemic human purpose should not preclude such an analysis, because economic growth is almost universally held as the primary means to these cherished values, it warrants special consideration.

7. An important counterpoint to the growth consensus is being promoted by a new generation of ecological economists who are elaborating upon John Stuart Mill's classic arguments for a steady-state economy. See Daly and Farley 2003 and http://www.ecoeco.org/, the website for the International Society for Ecological Economics, which was founded in 1989.

8. Even if Gaia theory were proved false, these political changes would be beneficial. Thus, in terms of human action and well-being, Gaia may be more important as a galvanizing metaphor than as a scientific theory.

## References

Abram, D. 1990. The Perceptual Implications of Gaia. In A. Hunt Badiner, ed., *Dharma Gaia: A Harvest of Essays in Buddhism and Ecology*. New York: Parallax, pp. 75–92.

Athanasiou, T., and P. Baer. 2002. *Dead Heat: Global Justice and Global Warming*. New York: Seven Stories Press.

Biehl, J. 1991. *Finding Our Way: Rethinking Ecofeminist Politics*. Montreal: Black Rose.

Botkin, D. B. 2001. *No Man's Garden : Thoreau and a New Vision for Civilization and Nature*. Washington, DC: Island Press.

Capra, F. 2002. *The Hidden Connections: Integrating the Biological, Cognitive, and Social Dimensions of Life into a Science of Sustainability*. New York: Doubleday.

Checkland, P. 1981. *Systems Thinking, Systems Practice*. New York: Wiley.

Dobson, A. 1990. *Green Political Thought: An Introduction*. London: Unwin Hyman.

Hardin, J. W. 2004. *Gaia Eros: Reconnecting to the Magic and Spirit of Nature.* Franklin Lakes, NJ: New Page Books.

Havel, V. 1997. *The Art of the Impossible: Politics as Morality in Practice*, transl. by P.Wilson. New York: Knopf.

Hawken, P. 2007. *Blessed Unrest: How the Largest Movement in the World Came into Being and Why No One Saw It Coming.* New York: Viking.

Hoffman, P. F., and D. P. Schrag. 2000. Snowball Earth. *Scientific American* 282: 68–75.

IPCC 2007. Intergovernmental Panel on Climate Change. *4th Assessment Report.*

Janssen, M. A. 2002. A future of surprises. In L. H. Gunderson and C. S. Holling, eds., *Panarchy: Understanding Transformations in Human and Natural Systems.* Washington, DC: Island Press, pp. 241–60.

Litfin, K. 1994. *Ozone Discourses: Science and Politics in Global Environmental Cooperation.* New York: Columbia University Press.

Litfin, K. 1997. Sovereignty in world ecopolitics. *Mershon International Studies Review* 41(2): 167–204.

Litfin, K. (forthcoming). Reinventing the future: The global ecovillage movement as a holistic knowledge community. In G. Kutting and R. Lipschutz, eds., *Environmental Governance, Power and Knowledge in a Local-Global World.* London: Routledge, pp. 124–42.

Lovelock, J. 1979. *Gaia: A New Look at Life on Earth.* Oxford: Oxford University Press.

Lovelock, J. 1986. Gaia: The world as living organism. *New Scientist* (December 18).

Lovelock, J. 1990. *The Ages of Gaia: A Biography of the Living Earth.* New York: Bantam.

Lovelock, J. 2000. "Foreword" to Anne Primavesi, *Sacred Gaia: Holistic Theology and Earth System Science.* London: Routledge.

Luhmann, N. 1990. The autopoiesis of social systems. In *Essays on Self-Reference*, New York: Columbia University Press.

Madron, R., and J. Jopling. 2003. *Gaian Democracies: Redefining Globalisation and People-Power.* Totnes, UK: Green Books.

Margulis, L., and D. Sagan. 1995. *What Is Life?* New York: Simon and Schuster.

Maturana, H., and F. Varela. 1998. *The Tree of Knowledge.* Boston: Shambhala.

Merchant, C. 1983. *The Death of Nature.* New York: Harper and Row.

Merchant, C. 1995. *Earthcare: Women and the Environment.* London: Routledge.

Primavesi, A. 2000. *Sacred Gaia: Holistic Theology and Earth System Science*. London: Routledge.

Princen, T. 2005. *The Logic of Sufficiency*. Cambridge: MIT Press.

Resnik, D. B. 1992. Gaia: from fanciful notion to research program. *Perspectives in Biology and Medicine* 35(4): 572–82.

Roszak, T. 1992. *The Voice of the Earth*. New York: Simon and Schuster.

Russell, P. 2008. *The Global Brain: The Awakening Earth in a New Century*. Edinburgh: Floris Books.

Slater, P. 1970. *The Pursuit of Loneliness: American Culture at the Breaking Point*. Boston: Beacon Press.

Spangler, D. 1993. Imagination, Gaia, and the sacredness of the Earth. In F. Hull, ed., *Earth and Spirit: The Spiritual Dimension of the Environmental Crisis*. New York: Continuum.

Spretnak, C. 1982. *The Politics of Women's Spirituality: Essays on the Rise of Spiritual Power within the Feminist Movement*. Garden City, NY: Anchor Books.

Thomas, L. 1974. *Lives of a Cell: Notes of a Biology Watcher*. New York: Bantam.

Tierney, T. F. 1993. *The Value of Convenience: A Genealogy of Technical Culture*. Albany: SUNY Press.

United Nations Environment Programme (UNEP). 2002. *Global Environmental Outlook-3 (Geo-3)*. London: Earthscan.

Volk T. 2003. *Gaia's Body: Toward a Physiology of Earth*. Cambridge: MIT Press.

Wissenburg, M. 1993. The idea of nature and the nature of distributive justice. In A. Dobson and P. Lucardie, eds., *The Politics of Nature: Explorations in Green Political Theory*. London: Routledge, pp. 3–20.

# 13

## In the Depths of a Breathing Planet: Gaia and the Transformation of Experience

David Abram

By providing a new way of viewing our planet—one that connects with some of our oldest and most primordial intuitions regarding the animate Earth—Gaia theory ultimately alters our understanding of ourselves, transforming our sense of what it means to be human. For much of the modern era, earthly nature was spoken of as a complex yet mechanical clutch of processes, as a deeply entangled set of objects and objective happenings lacking any inherent life, or agency, of its own. Such a conceptual regime helped sustain the cool detachment that was generally deemed necessary to the furtherance of the natural sciences. Yet the thorough objectification of nature also served to underwrite the sense of human uniqueness that has permeated the modern era. As long as the Earth had no unitary life, no agency, no subjectivity of its own, then we humans could continue to ponder, analyze, and manipulate the natural world as though we were not a part of it; our own sentience and subjectivity seemed to render us outside observers of this curious pageant, overseers of nature rather than full participants in the biotic community. The thoroughgoing objectification of the Earth thus enabled the old, theological presumption—that the Earth was ours to subdue and exploit for our own, exclusively human, purposes—to survive and to flourish even in the modern, scientific era.

Gaia theory, however, gradually undoes this age-old presumption.

### Minding Earth

By demonstrating that organic life is reciprocally entangled with even the most inorganic parameters of earthly existence, Gaia theory complicates any facile distinction between living and nonliving aspects of our world. By showing that Earth's organisms collectively influence their environment so thoroughly that the planet's oceans, atmosphere, soils, and surface geology together exhibit behavior more proper to a living physiology than an abiotic system, Gaia theory suggests that the biosphere has at least a rudimentary kind of agency. It suggests that like any living entity, the biosphere is not just an object but also, in some curious sense, a *subject*.[1]

To the extent that we take seriously the ongoing disclosures of Gaian science, we cannot help but feel a transformation in our own relation to the planet. If agency is an attribute of the biosphere as a whole, then the felt sense of *our own* agency need not isolate us from the material world that surrounds us. Just as our life is now recognized as part of a vast, planetary metabolism, so human sentience can now be felt as an extension, an elaboration, even an internal expression of the organic sentience of the biosphere itself.[2] Rather than the sole carriers of awareness within an essentially inanimate or mechanically determinate world, we now find ourselves fully embodied and embedded within a nature that has its own wild intelligence, and our own subjectivity seems no longer entirely ours.

Of course, awareness—or consciousness—is an exceedingly amorphous and ephemeral phenomenon, one that is notoriously difficult to pin down. Numerous scientific papers and books have been published in recent years trying to account for the emergence of consciousness, or to explain how awareness is constituted within the brain. Yet many of these explanations are dramatically at odds with one another, for there exists no clear agreement as to just what this enigma that we call "consciousness" actually *is*. Part of the difficulty stems from the intransigence of old notions—in particular, our age-old assumption that consciousness, or mind, is a uniquely human property, an utterly intangible substance that resides somewhere "inside" each of us.

It may be far more parsimonious, today, to suggest that mind is not at all a human possession but rather a property of the breathing Earth—a property in which we, along with the other animals and plants, all participate. The apparent "interiority" that we ascribe to the mind would then have less to do with the notion that there is a separate consciousness located *inside me*, and another, entirely separate and distinct consciousness that resides *inside you*, and more to do with a sense that you and I are both situated *within it*—a recognition that we are corporeally immersed in an awareness that is not ours but is rather the Earth's.

Our experience of awareness actually has much in common with our experience of the planetary atmosphere that circulates around and between earthly organisms, and that circulates *within* them as well.[3] Like the quality of awareness, we are steadily informed by the fluid air, and yet it is very difficult to catch sight of: we glimpse it only indirectly, as it bends the branches of an oak or rips a note from our hand and sends it tumbling along the street. We partake of the air ceaselessly, yet seem unable to fully bring it to our attention. Itself invisible, the atmosphere is that element *through which* we see everything else—much as consciousness, which we cannot see or grasp, is that through which we encounter all other phenomena. We are unable to step apart from consciousness, in order to examine it objectively, for wherever we step it is already there.

Consciousness, or awareness, is in this sense very much like a medium in which we are situated, and from which we are simply unable to extricate ourselves without ceasing to exist. Everything we know or sense of ourselves is conditioned by this atmosphere. We are intimately acquainted with its character, ceaselessly transformed by its influence upon us. And yet we're unable to characterize this medium from outside. We are composed of this curious element, permeated by it, and hence can take no real distance from it.

To acknowledge this affinity between air and awareness, however, is to allow this curious possibility: that the awareness that stirs within each of us is continuous with the wider awareness that moves around us, bending the grasses and lofting the clouds. Every organism partakes of this awareness from its own angle and place within it, each of us imbibing it through our nostrils or through the stomata in our leaves, altering its chemistry and quality within us before we breathe it back into the surrounding world. Consciousness, in this Gaian sense, may be ineffable, but it is hardly immaterial, for it is a quality in which we participate with the whole of our breathing bodies. Hence just as your body is different from mine in many ways, so your sensations and insights are richly different from mine. The contrasting experience of a praying mantis or a pileated woodpecker—or of a field of wild lupines, for that matter—is as different from our experience as their bodies are different from ours. Each being's awareness is unique, to be sure, yet this is not because an autonomous mind is held inside its particular body or brain but because each engages the common awareness from its own extraordinary angle, through its particular senses, according to the capacities of its flesh.

Such a Gaian way of articulating the mind—one that speaks of awareness as an attribute of the living biosphere, rather than a discrete property unique to privileged entities within that biosphere—offers an audacious and unexpected resolution to the mind-body problem that has long plagued Western philosophy. Yet it offers much more besides. By shifting the locus of intelligence from the human interior to the encompassing biosphere, such a way of speaking offers a corrective to contemporary assumptions that dramatically overlook the thorough dependence of human culture upon the continued creativity and flourishing of the more-than-human natural world.

In this chapter I would like to explore just a few of the experiential shifts and insights that might follow from such a transformed way of speaking. I hope to suggest, by these explorations, something of the way that Gaia might come to be experienced not merely as an objective set of facts but as a felt reality—as a vast and enigmatic presence whose life both pervades and exceeds our own.[4]

### Place and Awareness

When we allow that mind is a luminous quality of the Earth, we swiftly notice this consequence: each region—each topography, each uniquely patterned ecosystem—has its own *particular* awareness, its unique *style* of intelligence. After all, the air, the translucent medium of exchange between the breathing bodies of any locale, is subtly different in each terrain. The atmosphere of the coastal northwest of North America, infused with salt-spray and the tang of spruce, cedar, and fir needles, tastes and feels quite different from the air shimmering in the heat rising from the soil of the southwest desert—hence the black-gleamed ravens who carve loops through the desert sky speak a very different dialect of squawks and guttural cries than the cedar-perched ravens of the Pacific northwest, whose vocal arguments are filled with the liquid tones of falling water. Likewise the atmosphere that rolls over the great plains, gathering now and then into vorticed tornados, contrasts vividly with the mists that advance and recede along the California coast, and even with the blustering winds that pour through the Rocky Mountain passes. The specific geology of a place yields a soil rich in particular minerals, and the rains and rivers that feed those soils invite a unique blend of grasses, shrubs, and trees to take root there. These in turn beckon particular animals to browse their leaves, or to eat their fruits and distribute their seeds, to pollinate their blossoms or simply to find shelter among their roots, and thus a complexly entangled community begins to emerge, bustling and humming within itself. Every such community percolates a different chemistry into the air that animates it, joining whiffs and subtle pheromones to the drumming of woodpeckers and the crisscrossing hues of stone and leaf and feather that echo back and forth through that terrain, while the way that these diverse elements blend with one another is affected by the noon heat that beats down in some regions, or the frigid cold that hardens the ground in others.

Each place has its rhythms of change and metamorphosis, its specific style of expanding and contracting in response to the turning seasons, and this, too, shapes—and is shaped by—the sentience of that land. Whether we speak of a whole range of mountains, or of a small valley within that range, in every case there is a unique intelligence circulating among the various constituents of the ecosystem—a style evident in the way events unfold in that place, how the slow spread of the mountain's shadow alters the insect swarms above a cool stream, or the way a forested slope rejuvenates itself after a fire. For the precise amalgamation of elements that structures each valley exists nowhere else. Each place, that is to say, is a unique *state* of mind, and the many powers that constitute and dwell within that locale—the spiders and the tree frogs no less than the humans—all participate in, and partake of, the particular mind of the place.

Of course, I can hardly be instilled by this intelligence if I only touch down, briefly, on my way to elsewhere. Only by living for many moons in one place, my peripheral senses tracking seasonal changes in the local plants while the scent of the soil steadily seeps in through my pores—only over time can the intelligence of a place lay claim upon my person. Slowly, as the seasonal round repeats itself again and again, the lilt and melody of the local songbirds becomes an expectation within my ears, and so the mind I've carried with me settles into the wider mind that enfolds me. Changes in the terrain begin to release and mirror my own, internal changes. The slow metamorphosis of colors within the landscape; the way mice migrate into the walls of my home as the air grows colder; oak buds bursting and unfurling their leaves to join a million other leaves on that tree in agile, wind-tossed exuberance before they tumble, spent, to the ground; the way a wolf spider weaves her spiraling web in front of the rear porch light every spring—each such patterned event, quietly observed, releases analogous metamorphoses within myself.

Without such tunement and triggering by the earthly surroundings, my emotional body is stymied, befuddled—forced to spiral through its necessary transformations without any guidance from the larger Body (and hence entirely out of phase with my neighbors, human and nonhuman). *Sensory perception here is the silken web that binds our separate nervous systems into the encompassing ecosystem.*

Human communities too are inflected by the particular sentience of the living lands that they inhabit. There is a unique temperament to the bustling commerce and culture of any old-enough city, a mental climate that we instantly recognize upon returning after several years, and that we mistakenly ascribe solely to the human inhabitants of the metropolis. It is a result, we surmise, of the particular trades that the city is known for, or the dynamic mix of ethnicities that interact there, or the heavy-handed smugness of the local police force. Yet all such social dynamics draw nourishment from the elemental energies of the realm—from the heavy overcast that cloaks the sky for weeks at a time, or the profusion of flocking birds that nest on the ledges of apartment buildings, or the splashing speech of the river that rolls through downtown, tossing glints of sunlight into the eyes of all who walk near, or from the way the greasy exhaust from fifty thousand commuting cars interacts with the humidity of the summer air. The dismal social ills endemic to certain cities have often been stoked by the foolishness of urban designers who ignored the specific wildness of the place, the genius loci, the unique intelligence of the land now squelched and stifled by local industries. A calloused coldness, or meanness, results when our animal senses are cut off for too long from the animate Earth, when our ears—inundated by the whooping blare of car-alarms and the muted thunder of subways—no longer encounter the resonant silence, as our eyes forget the irregular wildness of things green and growing behind the rectilinear daze.

Each land has its own psyche, its own style of sentience, and hence to travel from Manhattan (in the Hudson River estuary) to the upper Rio Grande valley is to journey from one state of mind to another, very different, state of mind. Even to travel by train from New Haven to Boston, or simply to walk from one New England town to another, is to transform one's state of awareness. Traveling on foot makes these variations most evident, as the topography gradually alters, easing the stress on one's muscles as mountains give way to foothills and foothills become plains, and as the accents of the local shopkeepers transform in tandem with the shifting topography. The very texture of the air changes, as the moisture-laden atmosphere of the highlands, instilled with the exhalation of roots and decaying leaves and the breath of cool granitic caves, opens onto the drier wind whirling across the flatlands, blending the scents of upturned soil with hints of exhaust from the highway, and—especially strong in some places—the acrid smell of processed fertilizer.

Such alterations in the unseen spirit of the land are mostly hidden to those who make the journey by car, since then all the senses other than sight are held apart from the sensuous Earth, isolated within a capsule hurtling along the highway too fast for even the eyes to register most changes in the texture and tone of the visible. Still, subtle clues drift into the cabin, now and then—the insistent stench of those fertilized fields, or the reek from an unfortunate skunk, finding its way even into nostrils well-insulated by air-conditioning. And the ears can engage some aspect of the shifting psyche of the land if we turn on the radio—the percussive hip-hop and blues of the city opening onto the lilting voices and plucked strings of country music (laced with funk in some regions and more plaintive in others). Along certain stretches of highway the wave-lengths give way to a saturated array of Christian stations, with smooth or gravel-voiced preachers citing chapter and verse. This too is a register of the mind of that locale. Yet how much more thoroughly the land would feed our thoughts if we were not driving but rather strolling on foot across this land—or even pedaling a decent bicycle, the gusting wind swelling our lungs as our muscles work themselves against the slope.

If the automobile isolates our speeding senses, by and large, from the land around us, the airplanes in which we fly abstract us almost entirely from the breathing Earth. After checking in our bags at the airport, we tighten our buckles and loudly levitate up out of the ecosystem, shaking our senses free from the web of relationships that comprise the specific intelligence of that place. Only to plunk down some time later in an entirely different ecology—in an entirely different state of mind—without experiencing any of the transitional terrain between them, without our nervous system being tuned and tutored for this change by the gradual changes in the sensorial topography as we move across it. The sudden strangeness is jarring to our animal bodies, and especially shocking when we're compelled to adapt to the new circumstance in a matter of minutes.

Yet for those who have managed to keep their animal senses awake, the journey from one ecosystem into another is precisely a journey from one state of mind into another, strangely altered, state. From one mode of awareness, flavored by saltspray and the glint of sunlight on waves, to a different mode of awareness wherein the sorrowful cries of the coastal gulls become only a vague, half-remembered dream.

## The Land's Elemental Moods

Alterations in the very texture of the mind, however, are not only brought about by traveling from one geography to another. Those who dwell steadily in a single terrain, whether by choice or necessity, also experience profound shifts in the collective awareness. The psychological qualities of a particular place steadily metamorphose as the powers that comprise that place shift among themselves—as the first rays of morning, for instance, spill their warmth across the fields—or as that place exchanges specific constituents with other places. The migratory flight of certain birds, for example, may bring large flocks to settle for days or months in a particular region, and their arrival will alter the psyche of that land even as those birds enter and partake of that very psyche. Seasonal shifts in the collective sentience are especially obvious, of course, while other, more continuous transformations go mostly unnoticed by us. Yet we can be sure that there *are* manifold changes unfolding in the local mindscape, changes that imperceptibly—but inevitably—affect our emotions, our thoughts, and our actions.

I have suggested that the subtle intelligence of a particular place is akin to the medium of air that circulates invisibly within and between the inhabitants of that locale, nourishing their breathing bodies even as it bends the grasses and lofts the clouds. The most dramatic modulations within the collective psyche of a place are often those that alter the sensuous quality of this medium, changes that we commonly ascribe to the "weather"—those transformations in the collective atmosphere that often confound our conscious plans, sometimes curdling the unseen medium into a visible fog that slows our steps and clogs our thoughts, or suddenly congealing the depth around us into a thicket of slanting raindrops.

Changes in the weather transform the very feel of the world's presence, altering the medium of awareness in a manner that affects every breathing being in our vicinity. We sometimes refer to such phenomena, collectively, as "the elements," a phrase that suggests how basic, how primordial, these powers are to the human organism. The ephemeral nature of weather phenomena—the way such modulations in the atmosphere confuse the boundaries between the invisible and the visible, between inner and outer, between "subjective" and "objective"—ensures that the weather holds a curious position in the civilized world of modernity. We refer to it constantly; inquiring after or commenting on the weather establishes the most basic ground on which any social communication can proceed. Although it rarely occupies our full attention, the weather is always evident on the periphery of that attention, an ever-present reminder that the reality in which we live is ultimately beyond our human control.

For the activity of the atmosphere (presumably a strictly objective matter) remains the most ubiquitous, the most intractable, the most *enigmatic* of practical problems with which civilization has daily to grapple. Despite the best efforts of science and the most audacious technological advances, we seem unable to master this curious flux in which we're immersed, unable even to glean a clear comprehension of this mostly invisible field of turbulence and tranquil eddies so fundamental to our existence. The difficulty is compounded, today, by the abrupt warming of the global climate due to the industrial-era accumulation of greenhouse gases in the atmosphere; how this accelerating trend will affect the weather in particular regions is all but impossible to predict. We rely on satellites to monitor the atmosphere's unruly behavior from outside, hoping to gain from such external data a rudimentary sense of its large-scale patterns, so we can better guess at its next moves.

But suppose we were to analyze this turbulent dimension from within—from our perspective as sensing and sentient organisms thoroughly permeated by this flux? How then would we articulate its manifold modes of activity, its storms and its calms, its clarities and its condensations as they resound in our bodies and move through the terrain? We would need a term that suggests the subjective quality of these elemental phenomena, the way in which they subtly alter the palpable mind of the place, transforming the awareness of all who dwell there.

For our own species, at least, it's clear that such changes in the ambient weather do not *force* a change in our conscious thoughts, but rather alter the felt context of those thoughts—the somatic background, or mood, within which thinking unfolds. From our own creaturely perspective, then, we might say that shifts in the weather are transformations in the *mood* of the land. Different atmospheric conditions—different kinds of weather—are, precisely, different *moods*.

Wind, rain, snow, fog, hail, sunshine, heavy overcast—each element, or mood, articulates the invisible medium in a unique manner, sometimes rendering it (partly) visible to the eyes, or more insistently palpable upon our skin. Each affects the relation between our body and the living land in a specific way, altering the texture and tonality of our dreaming.

## Humidity

During the summer, near the coast, one sometimes wakes to a day that seems like any other, although as one goes through the motions of dressing and preparing breakfast, one notices one's thoughts lagging behind, as though they have yet to fully separate themselves from the state of sleep. A slowness attends all one's cogitations—the newspaper, today, seems written with less zip, reporting the same old thing with the same stale phrases, and when you put it aside you wonder if a minute's rest on the couch would be in order before tackling the day's work. The immediate tasks to be accomplished are after all somewhat vague and unfocused; it's hard to remember just what they are.

Only on stepping outside, and surveying the world from one's stoop, does the material cause of this mental lethargy become apparent. For the leafy trees, the electric wires, and the other homes are all bathed in a humid atmosphere that renders their outlines fuzzy and imprecise, while the mountains that usually rise from the far edge of town have dissolved, or are wholly shrouded, somehow, in the moisture-thick air. There's no cloud to be seen in the washed-out sky, only the too-big sun, hovering in the east, sweating like a spent tennis ball mouthed by too many dogs; its dull heat presses in from every direction.

## Lucidity

Then there are those rare days—not entirely unknown in any region of the Earth—that dawn with a clarity that muscles its way into every home and office, lending a crispness and cogency to almost every thought. One feels uncommonly good on such days, and others do too; deliberations move forward with unaccustomed ease. Ambiguities resolve themselves, or render themselves more explicit, the choices more defined and clear-cut. There's a delicious radiance that seems to come from the things themselves, from even the tables and the plush rug, and when we step outside we can taste it in the air and the way a few fluffed clouds rest, almost motionless, in the crystal lens of the sky. How far our vision travels on such days! When I climb to the top of the street I can see clear to the mountain range that rises from the plain in the neighboring state! And how sharp that horizon is. Long-term goals abruptly become evident; possibilities far in the future seem more accessible, lending perspective to the present. Hence planning goes more smoothly, with a marked absence of the usual friction—no sweat.

Although, to be sure, we're not always in sync with such felicitous weather—with the strangely clarified transparence that lifts the weight of the whole suburb on such unpredictable days, or that wraps the aspen branches outside our cabin with such a pellucid and form-fitting cloak of blue. Sometimes we're still carrying the strains and stresses of recent weeks, struggles that followed us into our dreams and now cling to our face and our feet, or we're still in the dank doldrums due to the wreck of a relationship we'd trusted our hearts to. These are the worst days for depression, when everyone we meet moves so smooth through the world. Even if we're off on our own, well away from the human hubbub, the despondence can be darker on such days when we feel that the stones and the singing sky and the green blades of grass are all tuned to another frequency. For there's an insistent and eager harmoniousness to things, an ease that we sense on the periphery—the hillside itself humming with pleasure for a whole afternoon—yet the mood cannot penetrate through the thick pellicle of our pain. The mismatch of the world with our own traumatized state feels distressing, even terrifying, shoving us deeper into the pit.

Of course I am writing of these earthly elements, or moods, from an entirely human perspective. Indeed I'm writing from the quite subjective perspective of a single human creature—myself. Nonetheless, I write with the knowledge that there cannot help but be some overlap between my direct, visceral experience and the felt experience of other persons— whose senses, after all, have much in common with my own. Moreover I've confidence that my bodily experience is a variation—albeit, in many cases, a very distant variation—of what other, nonhuman, bodies may experience in the same locale in a common season, at a similar moment of the day or night. For not only are our bodies kindred—all mammals, for instance, sharing a common ancestry, and hence still enacting differ- ent variations of what were once common sensibilities—but also we are all of us, here and now, interdependent aspects of a common biosphere, each of us experiencing it from our own angle, and with our own specific capabilities, yet nonetheless all participant in the round life of the Earth, and hence subject to the same large-scale flows, rhythms, and tensions that move across that wider life.

The world we inhabit is not, in this sense, a determinate or determinable set of objective processes. It is flesh, a densely intertwined and improvisational tissue of experience. It is a sensitive sphere suspended in the solar wind, a round field of sentience sustained by the relationships between the myriad lives, the myriad *sensibilities* that compose it. We come to know more of this sphere not by standing apart from our bodily experience but by inhabiting our felt experience all the more richly and wakefully, feeling our way into deeper contact with other experiencing bodies, and thus with the wild, inter-corporeal life of the world itself.

## Stillness

The pencil whirls above my scribbling fingers, letters arranging themselves on the paper as I list the matters I must attend to in the next couple days. It's too much stuff. Between getting the wood stacked for winter and my daughter's dental appointment, between repairing the steady roof leak and dropping a clutch of packages at the post office, I've no idea when I'll ever compose the lecture I agreed to deliver Thursday evening.

Something catches my peripheral vision, and I turn toward the window. My eyes widen in surprise: snowflakes! A great crowd of snowflakes floating down, a deep thicket of slowly tumbling white. How long has this been happening? I stand and stare for a few moments, then pull on a sweater and step out the door into a landscape transformed as if by a spell. My steps make no sound—the white blanket already plush upon the ground and layered in tufts upon the juniper and pine branches, as flakes drift down like loosened stars. A hundred of them swerve into my face, melting cold against my skin as I walk slowly through a world utterly transfigured by this silent grace cascading through every part of the space around me.

The surge and press of the week's worries has simply vanished. When I try to call those concerns back to mind, I cannot find them behind the teeming multitude of slowly falling flakes—past and future have dissolved, and I am held in the white eternity of a moment so beautiful it melts all my words. All weight has lifted; the innumerable downward trajectories have convinced my senses that I myself am floating, or rather rising slowly upward, and the ground itself rising beneath me—the Earth and I now rising weightless through space.

A sound—the flutter of a bird's wings, and a small explosion of snow from a branch the bird launched from. Then, just silence. Not silence as an absence of sound but as a fullness, as the very sound ten thousand snowflakes make as they meet the ground. A thick silence, muffling the whole valley, and for all I know the whole cosmos. I cannot imagine that any bird, squirrel, coyote, or hare is not similarly held in the visible trance of this slowly cascading silence. The clumps on the branches deepen.

The snow falls through the night, the porch light illuminating a charmed space through which powder floats steadily down. I turn it off before sleeping, then step outside to breathe the darkness: now even the house, and the car asleep in the driveway, have fallen under the spell.

By morning the snowfall has stopped. Yet the enchantment holds; when I step outside and snap my boots into my skis, there is a soft stillness everywhere. I glide between the trees and onto the dirt road, whose many ruts are now invisible; unbroken goodness extends from the tips of my skis in every direction. There is a hushed purity to the world, and to awareness itself as I glide across the snowy fields. The dentist will wait, and the post office will get its packages when the roads are clear. Thursday's lecture is forming itself, easily, as I glide over the white expanse, my body writing its smooth script across the unbroken pages.

Now and then a high limb releases its too-heavy mound of snow, and a spray of powder drifts down in sheets, glittering, scintillating, then vanishing into the clarified air.

## Wind

Of all the elements, wind is the most versatile and protean, offering in each region a different set of aspects, varying itself according to the season, and often, too, according to the direction from whence it arrives. We can easily find ourselves overwhelmed by the sheer variety of wind's incarnations, and so must choose only a few examples from an outrageous range of styles.

Toward the tail end of winter, when a few days of unexpected warmth bring whiffs of spring—and a buried store of state-specific memories proper to that season send a few green shoots into one's conscious awareness—the winter will often reassert itself, swooping low at night to chill the walls of your house and repossess the snowy fields. When you step outside in the morning the recently melting surface of the snow has now frozen solid and slick like a pane of glass. And gusting across that glass, a fine mist of crystals speeds past the trunks of trees, some of them tinkling in eddies against the windows as the main current of wind gallops through the fields. Boots slip along the frozen surface as you try to take a few steps, ears and face stung by the icy blast. Nothing in the landscape beckons or reaches out to you, for each bush or branch or telephone pole seems entirely focused on staying in place, every parked car and house holding fast with all its fingers to the ground beneath, each being doing its best to become an inconspicuous part of the ground, a mute lump or appendage of the Earth, affording the wind nothing other than a smooth surface to glide past on its way to wherever. Under the onslaught of the chill wind, each entity subsides into the anonymity of the Earth, and even you, too, find your individuality subsumed into the rigor of standing solid against the icy blasts, as your body makes itself into a smooth stone. Thought is stilled, all interior reflection dissolves, no memory apart from this ancient kinship and solidarity with the density of metal and rock, of heartwood and stone. The outward roar of wind forces one to find the blessed silence of stone at the heart of the mind. Anonymous, implacable, unperturbed—the biting cold of a winter wind returns one to one's unity with the bedrock.

Yet a wind of comparable velocity in the late spring or early summer can have a nearly opposite effect. As when after a long hike one ascends at to a high pass from the eastward slope and peers over into the valley beyond. A moist breeze is riding up the western slope, carrying fresh scents from the forests below, and clouds previously unseen are slowly massing on that side of the range. The wind becomes stronger, more insistent, and you realize that a storm is brewing; it is time to head down and find shelter. Yet something holds you on the pass. As the wind begins to rage, pouring over the crest and rushing down the boulder-strewn slope behind you, it tugs your hair back from your head and fills your cheeks when you open your mouth, whipping your unbuttoned shirt like a kite as an exuberance rises in your muscles. Laughing, crouching and leaping in the wind, facing into it and feeling the first raindrops as you gulp from the charging gusts, imbibing its energy, meeting its wildness with your own as you dance drenched like a grinning fool down the trail—a wild wind can return us to our own vitality more swiftly than any other element. And the needled trees swaying and tossing around us as we descend, jostled by the same wind—are not they, too, caught up in something of the same mood? Not the giddiness, but the exuberant pleasure that lies beneath it, the way the wind challenges us in this season when the sap's already been rising in our veins, testing our flexibility, waking our limbs and our limberness, goading us each into our own animal abandon, our own muscular dance?

There are also the winds of autumn, winds that whirl through the streets tearing the dry, gold-brown leaves from their moorings. Alive with the dank scents of soil and fallen fruit and the composting Earth, the autumn wind teases our nostrils with the sweet scent of smoke as it whooshes past, scattering the humped piles of carefully raked leaves and sending their constituents tumbling across lawns to meet other leaves spiraling down from the branches. Soon the oaks, maples, and beeches stand denuded and exposed, their fractaled complexity silhouetted against the sky. Our own bodies witness this gradual release of leaves, this stripping away of life from the skeletal eloquence of the gray trunks and limbs, and cannot help but feel that the animating life of things is slipping off into the air—that the wind blowing in our ears and moaning in the branches is composed of innumerable spirits leaving their visible bodies behind. We feel enveloped by a moving crowd of unseen essences—sighing, whooshing lives that reveal themselves to us only as fleeting scents, or by a momentary turbulence of dust and spinning leaves. The wind is haunted, alive. Only in this liminal season, before the onset of winter, does the wild psyche of the land assert itself so vividly that even the most reflective and analytic persons find themselves lost, now and then, in the uncanny depths of the sensuous. Their animal senses awaken; the skin itself begins to breathe.

For wind is moodiness personified, altering on a whim, recklessly transgressing the boundaries between places, between beings, between inner and outer worlds. The unruly poltergeist of our collective mental climate, wind, after all, is the ancient and ever-present source of the words "spirit" and "psyche." It is the sacred "ruach" of the ancient Hebrews, the invisible *rushing-spirit* that lends its life to the visible world; it is the Latin anima, the wind-soul that *animates* all breathing beings (all *animals*); it is the Navajo "Nilch'i," the Holy Wind from whence all earthly entities draw their awareness.

Indeed, whenever the native peoples of this continent speak matter-of-factly about "the spirits," we moderns mistakenly assume, in keeping with our own impoverished sense of matter, that they're alluding to a disembodied set of powers entirely outside of the sensuous world. We would come closer to the keen intelligence of our indigenous brothers and sisters, however, if we were to recognize that the spirits they speak of have much in common with the myriad gusts, breezes, and winds that influence the life of any locale—like the particular wind that whooshes along the river at dusk, rustling the cottonwood leaves, or the mist-laden breeze that flows down from the western foothills on certain mornings, and those multiple whirlwinds that swirl and rise the dust on hot summer days, and the gentle breeze that lingers above the night grasses, and the various messenger-winds that bring us knowledge of what the neighbors are cooking this evening. Or even the small but significant gusts that slip in and out of our nostrils as we lie sleeping. We moderns pay little heed to these subtle invisibles, these elementals—indeed we tend not to notice them at all, convinced that a breeze is nothing other than a mindless jostling of molecules. Our breathing bodies know otherwise. But we will keep our bodies out of play; we will keep our thoughts aligned solely with what our complex instruments can measure. Until we have incontrovertible evidence to the contrary, we will assume that matter, itself, is utterly devoid of felt experience.

In this manner we hoard and hold tight to our own awareness—like a frightened whirlwind spinning ever faster, trying to convince itself of its own autonomy, struggling to hold itself aloof from the ocean of air around it.

## Thunderstorm

On a warm afternoon, new leaves creeping out of the just-opened buds, when the apricot trees shamelessly offer their blossoms to a thousand bees, one notices a faint rumble in the air. It dissipates or fades back into the incessant whirring of the bees, until then, sometime later, a similar trembling. The tremor is more felt than heard, a vibration noticed more by our bones and the trunks of trees than by our conscious reflections. Behind the branches of cottonwood, far off to the west, a darkness is growing, massing in the sky, a vague threat on the horizon. Yet now the irregular rumble, once again, more audible now, ominous. The rabbits (who have only just begun appearing near the road and at the edge of the orchard) are sniffing the air, hesitant. And how odd; what's become of all those bees? Now only a few stragglers are moving between the blossoms. Birdlife is more evident, several wingeds calling and swooping between the telephone wires and the trees, expending an unusual amount of energy. Everyone here is now feeling it: the background hush that has come over the land as the clouds thicken into a too early dusk, the rabbits ducking into their digs—a deep hush broken by the alarm call of a bird, and then a bit later by the thudding violence in the near distance. As though the sky was a skin that's been stretched taut. Everyone is finding somewhere safe and hunkering down, tremulous, waiting. And then, quietly, a soft breeze stirs the tips of the grasses, rippling and bending the blades, a kind of pleasure spreading there among the green life, maybe even an eager anticipation very different from the threat vaguely sensed within one's chest and the muscles of other animals.

And then without warning the air splits open, white fire tracing an impossibly erratic path between the sky and the hills opposite, a jagged gash burning itself into one's retinas and turning the entire landscape into a negative afterimage of itself, for an instant, before the shadowed darkness returns. Silence. And then the shattering sound of that splitting, the syncopated cracking, ripping open the world as it explodes in the skull and reverberates off the cliffs. The sound from which all other sounds must come, the Word at the origin of the world. And as the visible world settles back into itself, another bright flame rips haphazard through the gray, and soon the anticipated yet un-prepared-for SHOUT! shatters the air and shudders through the ground underfoot.

Nothing, no creature or stone or flake of paint on the wall of the house, escapes the shattering imperative of the thunderbolt's shout—the way it undoes us and recreates us in a moment. No awake creature is distracted at that moment, no person remains lost in reverie or inward thought; all of us are gathered into the same electric present by the sudden violence of this exchange between the ground and the clouds, the passionate mad tension and static that reverberates through all of us in the valley this afternoon. This rage in the mind.

This passion now rising, it seems, in the branches of the tall Ponderosa pines on the hillside opposite, and soon in the swaying limbs of the closer cottonwoods, and now even the roiling needles of junipers and piñons along the dirt road below the house—some power is moving rapidly across the valley, a tumult of wind in the branches, and the rushing cool sound of...

### Rain

A few drops, at first, on my shoulder and nose, as I hear it begin to pelt the soil of the orchard, and then I am taken up within the cold thicket of drops, soaking my clothes and then the body beneath those clothes, rolling off my nose and dripping off the apricot branches to pool among the grasses, spilling down my arms and gathering in the cuffs of my jeans.

The obvious effect triggered by the rain is release—a steady, dramatic release of tension, like held-back tears finally sliding across our cheeks.

Lightning still flashes through the downpour, and the stuttering of thunder, but all this cascading water drumming on the ground and on my head eases the violence of that darker percussion, drawing my attention back from the splintering tension in the sky to my own cool and shivering surfaces, and the intersecting patterns in the near puddles— returning awareness to the close-at-hand locale. A few minutes earlier, when the lightning seemed to strike nearby, all attention was gripped by the immediate present, yet that present moment was a vast thing, opening onto the whole of the clouded sky, including the whole span and expanse of the valley. A strong rain, however, rapidly shrinks the field of the present down to an intimate neighborhood that extends only a few yards in any direction. For the dense forest of droplets falling all around me is not easily penetrated by my senses. Past and future are utter abstractions, yesterday and tomorrow are far-off fictions; I am gripped in the slanting immediacy of water and mud and skin. I turn my face upward, blinking, trying to follow individual drops as they fall toward me. Difficult. I give up and just open my mouth. The sensuous density of the present moment, and me inside it, drinking the rain.

I head into the house to strip off soaked clothes and towel myself dry. The many-voiced rain sounds steadily on the roof. I stand at the window, staring out. Drops splash against different points on the pane, sliding in scattered droplets down the glass, each droplet picking up others as it descends—every added straggler increasing the velocity of the drop—until they all pool along the bottom.

Even the interior of the house is transformed by the thrumming rain; objects seem more awake and attentive to the things around them—the reclining chairs, tables, and books seem to have shed their distracting ties to the world outside and are now committed citizens of this small but commodious cosmos wholly isolated from the rest of the valley. And the familiar bonds that these objects have with one another, and with me, are all heightened by the sound of the downpour on the sheltering roof and the walls, and the tremble of thunder.

Later, after the rain has dissipated, I open the door onto a different world—a field of glistening, shiny surfaces, of beings quietly turning their inward, protective focus back outward, as creatures poke noses out of burrows, and a thrush swoops down to the edge of a puddle, and then hops in to splash its wings in the wet. Everything glints and gleams, everything radiates out of itself as a thousand scents rise from the soil and the branches and the fungus-ridden trunks, from insect egg-cases and last year's leaves and the moist, matted fur of two squirrels chasing each other along the edge of the roof. A tangle of essences drift and mingle in the mind of this old orchard, each of us inhaling the flavor of everyone else, yielding a mood of openness and energetic ease as the lightened clouds begin to part and the late afternoon sun calls wisps of steam from the grass.

But wait! Are we not simply projecting our own interior moods upon the outer landscape? And so making ourselves, once again, the source and center of the earthly world, the human hub around which the rest of nature revolves?

It is a key question, necessary for keeping us on our toes and turning our attention, always, toward the odd *otherness* of things—holding our thoughts open to the unexpected and sometimes unnerving shock of the real. So are we merely projecting our emotional states upon the surroundings? Well, no—not if our manner of understanding and conceptualizing our various "interior" moods was originally borrowed from the moody, capricious Earth itself. Not, that is, if our conception of anger, and livid rage, has been borrowed, at least in part, from our ancestral, animal experience of thunderstorms and the violence of sudden lightning. Not if our sense of emotional release has been fed not only by the flow of tears but by our experiences of rainfall, or if our concept of mental clarity has long been informed by the visual transparence of the air and the open blue of the sky on those days of surpassingly low humidity. If our sense of inward confusion and muddledness is anciently and inextricably bound up with our outward experience of being enveloped in a fog—if our whole conceptualization of the emotional mood or "feel" of things is unavoidably entwined with metaphors of "atmospheres," "airs," "climates"—then it is hardly projection to notice that it is not only human beings (and human-made spaces) that carry moods: that the living land in which we dwell, and in whose life we participate, has its own feeling tone and style that varies throughout a day or a season.

### The Return of the Repressed

Today, as Gaia shivers into a fever—the planetary climate rapidly warming as oil-drunk civilization burns up millions of years of stored sunlight in the course of a few decades—clearly the felt temper of the atmosphere is shifting, becoming more extreme. As local weather patterns fluctuate and transform in every part of the globe, the excessive moodiness of the medium affects the mental climate in which creatures confront one another, lending its instability to human affairs as well. Our human exchanges—whether between persons or between nations— easily becoming more agitated and turbulent, apt to flare into storms of blame and anger and war as the disquietude in the land translates into a generalized fearfulness among the population, a trepidation, a readiness to take offense or to lash out without clear cause.

Indeed the propensity for random violence becomes more pronounced whenever the sources of stress are unrecognized, whenever a tension is felt whose locus or source remains hidden. And as long as we deny the animate life of the Earth itself—as long as we arrogate all subjectivity to ourselves, forgetting the sentience in the air, and the manifold intelligence in the land—then we'll remain oblivious to what's really unfolding, unable to quell the agitation in ourselves because we're blind to the deeper distress.

For the possibility of a human future, and for our own basic sanity, we need to acknowledge that we're not the sole bearers of meaning in this world, that our species is not the only locus of feeling afoot in the real. To weather the changes now upon us, we must become ever more attentive to the more-than-human field of experience, consulting the creatures and the old local farmers, comparing notes with the neighbors, learning the seasonal cycles of our terrain even as we notice new alterations in those cycles. Listening at once outward and inward, observing the shifts in the animate landscape while tracking the transformations unfolding within us—in this way we weave ourselves back into the fabric of our world.

The violence and disarray of the coming era, its social injustices and its wars, will have their deepest source in systemic stresses already intensifying within the broader body of the biosphere. Yet such systemwide strains cannot be alleviated by scapegoating other persons, or by inflicting violence on other peoples. They can be eased only by strengthening the wild resilience of the Earth, preserving and replenishing whatever we can of the planet's once-exuberant biotic diversity while bringing ourselves (and our communities) into greater alignment with the particular ecologies that we inhabit. Acknowledging that human awareness is sustained by the broader sentience of the Earth; noticing that each bioregion has its own *style* of sentience; observing the manner in which the collective mood of a terrain alters with every change in the weather: such are a few of the ways whereby we can nudge ourselves toward such an alignment.

The era of human arrogance is at an end; the age of consequences is upon us. The presumption that mind was an exclusively human property exemplified the very arrogance that has now brought the current biosphere to the very brink of the abyss. It led us to take the atmosphere entirely for granted, treating what was once known as the most mysterious and sacred dimension of life as a conveniently invisible dumpsite for the toxic by-products of industrial civilization.

The resulting torsions within the planetary climate are at last forcing humankind out of its self-enclosed oblivion—a dynamic spoken of, in psychoanalysis, as "the return of the repressed." Only through the extremity of the weather are we brought to notice the uncanny power and presence of the unseen medium, and so compelled to remember our thorough immersion within the life of this breathing planet. Only thus are we brought to realize that our vaunted human intelligence is as nothing unless it's allied with the round intelligence of the animate Earth.

## Notes

1. See Abram (1991).

2. See Abram (1985).

3. The modern word for the mind, "psyche," originates in the ancient Greek word for wind and breath, much as the word "spirit" derives from the Latin *spiritus*, meaning a breath or a gust of wind. Similarly the Latin word for the soul, *anima*, originates in the older Greek word for the wind, *anemos*. In the ancient world, it would seem, the unseen air was commonly felt to be the very substance of consciousness. Thus the English word "atmosphere," is cognate with the Sanskrit word for the soul, *atman*, through their common origin in the older term *atmos*, which signified both the air and the soul inseparably. The Hebrew word for the spirit, *ruach*, signifies (at one and the same time) the wind, and hence is often translated as "rushing spirit." Such an identification of *air* with *awareness* is found in innumerable indigenous, oral languages. See "The Forgetting and Remembering of the Air," in Abram (1996: 225–60).

4. The theoretical approach of this chapter brings the philosophical tradition of phenomenology—the careful study of direct, sensorial experience—to bear on Gaian ecology. Readers wishing to learn more regarding the phenomenological tradition and its relevance to environmental thought may wish to look at the following books: *The Spell of the Sensuous: Perception and Language in*

*a More-Than-Human World*, by David Abram; *Rethinking Nature: Essays in Environmental Philosophy*, edited by Bruce Foltz and Robert Frodeman; *The Fate of Place: A Philosophical History*, by Edward Casey. *Inhabiting the Earth: Heidegger, Environmental Ethics, and the Metaphysics of Nature*, by Bruce Foltz; *Maurice Merleau-Ponty: Basic Writings*, edited by Thomas Baldwin.

## References

Abram, D. 1985. The Perceptual Implications of Gaia. *The Ecologist* 15 (3): 96–104. Reprinted in A. H. Badiner, ed., *Dharma Gaia: A Harvest of Essays in Buddhism and Ecology*. Berkeley, CA: Parallax Press, 1990.

Abram, D. 1991. The mechanical and the organic: On the impact of metaphor in science. In S. Schneider and P. Boston, eds., *Scientists on Gaia*. Cambridge: MIT Press, pp. 66–74.

Abram, D. 1996. *The Spell of the Sensuous: Perception and Language in a More-Than-Human World*. New York: Pantheon.

Abram, D. 2007. Earth in eclipse. In S. L. Cataldi and W. S. Hamrick, eds., *Merleau-Ponty and Environmental Philosophy*. New York: SUNY Press.

Baldwin, T, ed. 2003. *Maurice Merleau-Ponty: Basic Writings*. New York: Routledge.

Casey, E. 1998. *The Fate of Place: A Philosophical History*. Berkeley: University of California Press.

Foltz, B. 1995. *Inhabiting the Earth: Heidegger, Environmental Ethics, and the Metaphysics of Nature*. Atlantic Highlands, NJ: Humanity Book.

Foltz, B., and R. Frodeman, eds. 2004. *Rethinking Nature: Essays in Environmental Philosophy*. Bloomington: Indiana University Press.

# 14

## Sustainability and an Earth Operating System for Gaia

Tim Foresman

A convergence is being witnessed that combines a growing social awareness of the fragile condition of our Earth's life support systems with the
evolution of spatially enabled technologies (e.g., GIS, RS), which operate
at no cost to the general public from the World Wide Web. This convergence marks a clear signpost for humanity. A new era is being launched
where an Earth Operating System (henceforth EOS) can be designed and
implemented for the benefit of humanity and the natural world. That
this represents a critical juncture for the planet and its living systems is
well documented by the current mass extinction spasm and other catastrophic events for nonhuman and human life, like rapid climate change.
An EOS is proposed as requisite for the continued survival of the Earth's
species and for the betterment of the human condition.

### An Earth Operating System Approach

An EOS is predicated upon the application of a Gaian global system
model for the monitoring and restoration of the planet and its ecosystems. Technologies have evolved from the fields of geographic information system (GIS) and remote sensing (RS) into tangible Digital Earth
implementation systems, as demonstrated by Google, Microsoft, ESRI,
NASA, and other purveyors of 3D geobrowsers. These technologies, in
combination with the activism of communities focused on sustainable
practices, can spearhead the kind of EOS needed to bridge ecological,
economic, and social arenas. Cross-domain management is needed for
our species' future, while orienting to a new, harmonious relationship
between humanity and the Earth's natural systems. This chapter presents
the comprehensive evolutionary paths that create the conditions for an
EOS—a concept defined by Buckminster Fuller half a century ago.

Our universe has been in existence for nearly 14 billion years since the Big Bang, according to the calculations of cosmologists (Hawking and Ellis 1968; Hawking and Penrose 1970). Scholars have been working diligently for millennia to unravel the mysteries of the star's and planet's mathematical choreography and to discover the operating principles for the universe. This cosmic choreography is sufficiently well understood today to allow humans to launch spaceships in our solar system with engineered precision. Regarding the question of "the operating system for the universe," we are intellectually comforted by the advanced discoveries of the laws of physics. The laws of physics govern the universe and can be applied to explain the motions of the planets and the speed of light, as well as gravity and how televisions work. Whether these laws were created by deistic design, or simply are, is fodder for a never-ending dialogue between the faithful and the secular; even so, the physical laws of the universe as formulated and refined in the twentieth century are egalitarian and work for everyone—sinner and saved. Still, much remains to be discovered in this universe and physicists are increasingly allowing for metaphysical phenomena to coexist within their mathematical maize as they seek ultimate unification laws for explaining the cosmos (Radin 1997).

As has been well articulated in the Gaian literature, the Earth began 4.5 billion years old from a molten and sterile state that finally cooled enough to allow (in less than a billion years) for the incubation of complex molecular compounds of exotic chemistry, fostering basic life forms in shallow seas (Lovelock 1965, 1969, 1988; Margulis and Lovelock 1974). The laws of evolution, put forth by Charles Darwin in his *On the Origin of Species*, have acted for millennia to keep the life forms of our planet unfolding toward ever-increased complexity and diversity, bound by constraints of the physical world, creating wondrous ensembles of life (Darwin 1859; Dennett 1995).

If we are to ask "what is the operating system for the planet Earth," it might be reasonable to regard the laws of evolution as representing the Earth's operating system, albeit in full compliance with the universe's own operating system. Another view regarding an EOS, however, is raised for the reader to consider: that which can be studied with the aid of computing and technology. The mechanistic and ecological system components along with their feedback loops can be understood, measured, and perhaps modeled. These Gaian components include atmospheric constituents and dynamics, the climate system, water distribution (as well as its abundance and quality), the oceans and their life forms, global biodiversity, and other parameters that can be assessed both quantitatively and qualitatively. If we wish to advance a Gaian, whole-Earth perspective into the mainstream of human operations for international collaboration and cooperation, then the introduction and application of an EOS, as a *lingua franca*, might prove to be beneficial.

A conceptual path for the foundation of these ideas can be traced to the visionary writings of R. Buckminster Fuller. His seminal 1968 book, *Operating Manual for Spaceship Earth*, explores the metaphor of Earth as a self-contained life-support system that travels through the universe bounded by requirements similar to those for a human-occupied spacecraft. These requirements incorporate a raft of basic life-support components, such as an energy source, basic chemical elements and molecules, and the requirement to recycle and regenerate within a closed system containing all of these constituent parts. His contribution was timely, concurrent with the birth of the space age, in creating a metaphor that would resonate with an exploding human population clearly beginning to overtax the Earth's carrying capacity. Fuller's contribution was all the more poignant as the Earth's denizens became mesmerized by the landmark Apollo photo of our unique and lonely blue planet set against the immensity of the universe.

While Fuller may be remembered for creating the arresting metaphor of Earth as "spaceship," what he offered was a better understanding of how the living components were to remain functional for enduring periods for human survival within a healthy biosphere. It was here that Fuller called for cybernetics to enable humans to gather, store, and manipulate vast quantities of data about the planet and societal commerce; modern cybernetic assistants—computers—would also reveal the consequences were we to remain unable to focus our collective attention on issues of survival and sustainability (Fuller 1981). Indeed Fuller foresaw and discussed the creation of a "geoscope" decades before operational Digital Earths became computer realities. He understood the need for technology and computers to assist humans in gaining knowledge about the planet, as well as command over our unsustainable economic and consumer patterns.

Gaia's ecological systems and services are the life support for all known species, and they are therefore critical parameters that require our most accurate understanding and monitoring. An EOS should therefore be designed to facilitate the capacity to assemble, integrate, and report information about Gaia life-support parameters. Subsequent to Fuller's books, the Club of Rome (in the landmark work *Limits to*

*Growth*) produced a series of startling projections for some of these critical Gaia parameters regarding our increasingly overpopulated and globalized planet. In retrospect, most of their so-called alarmist projections proved to be accurate, with population levels, for example, increasing precisely as had been calculated using the best growth models of the era (Meadows et al. 1972).

To our best knowledge all systems on Earth continue to exhibit a disheartening decline while human population and resource consumption grow. Data for this trend have been painstakingly assembled through the premier global assessment series produced by the United Nations Environment Programme (UNEP). An international community of scientists has focused on the comprehensive and objective measuring the planet's environmental conditions and trends, derived through a consensus process and reported in the *Global Environmental Outlook* (GEO) series (UNEP 2002, 2007). No place on Earth is unaffected by human activities including remnant wilderness and protected areas. Atmospheric chemicals are polluting even these semi-isolated natural pockets, which are also showing signs of being adversely affected by climate change. There are indeed no safe havens from the system-wide perturbations of Gaia.

Concomitant with dangerous climatic dynamics is the disturbing news of the human-driven mass extinction underway (Wilson 2002). Catastrophic species losses include extinctions of mammals, birds, reptiles, amphibians, and fish, as well invertebrates, plants, and microorganisms. Large-scale fisheries collapse is predicted by midcentury, with disastrous consequences both for marine ecosystems and for the one billion people who depend on fish protein (Murawski et al. 2007). Soil degradation moreover is compounding the agricultural communities' ability to feed a world population of 6.7 billion (and growing). Water for consumption is seriously lacking for approximately one billion people whose lives can best be described as barely tenable under severe hardships; and the water crisis is expected to deepen. Air pollution continues to create adverse ecological and societal impacts, chronically affecting the health of our young and most vulnerable citizens. It is a sober conclusion of broad consensus that our only hope lies in people joining in "urgent and cooperative action" to address the extraordinary challenges we face (Millennium Ecosystem Assessment 2005; Gore 1992, 2006; Laszlo 2001; Strong 2000; SEC 2007). In the immediate future, humans will witness an unprecedented transition for the biosphere and our life within it.

From the perspective of humanity and many current life forms and ecosystems, prognosis for Gaia, though dire, need not be hopeless. We have an array of powerful technologies that will allow for the comprehensive monitoring of the planet's living and nonliving systems in real time. While technology alone cannot furnish the full solution—but must be grounded and contextualized within a new biospheric ethic—the prowess and promise of technological innovations do give hope of a foothold. NASA, for example, launched the "Mission to Planet Earth" initiative toward the end of the last century in which it raised the clarion call for an international EOS. This concept, which has since morphed into the Global Earth Observation System of Systems (GEOSS 2005), was conceived as a connected constellation of orbiting satellites with millions of *in situ* ground sampling stations to map, measure, and model the Earth's systems for atmosphere, oceans, and land.

Earth observation systems (e.g., GEOSS) are in harmony with Fuller's visionary prescriptions that these systems could be used to reveal invisible processes and interactions among the Earth's biotic and abiotic systems. What is needed, however, is the capacity to integrate and broadcast the data and information about the Earth systems to all of Earth's citizens. GEOSS and its deliberating body of international representatives may take decades before reaching even their primary bureaucratic goals for interoperability among the varying constituent nations' monitoring systems. There are, however, alternatives that portend short cuts toward improving the interoperability and data exchange needed to understand and care for the planet's systems. NASA's Digital Earth initiative, launched in 1998, provides a refreshing model for both technological infrastructure and enhanced human collaboration, on a broad scale, that could be viewed as requisite conditions for an EOS.

## Digital Earth Foundations

Vice President Al Gore, in a 1998 speech addressed to a crowded auditorium in Los Angeles, portrayed a future when a girl could sit before a virtual or three-dimensional representation of the Earth, and receive information in response to inquiries about the planet, living systems, and human ecologies. This Digital Earth vision launched a movement in the US government, with a metaphor that scientists and nonscientists alike could comprehend and use to accelerate the development of visual-information technologies (Gore 1998). This vision was able to bring together heretofore disconnected groups whose common ground resided in the study of Earth—its dynamics, resources, and ecosystems. The vision would also allow disenfranchised groups access to helping society define how to direct our actions toward a sustainable future. Gore's position as vice president of the United States added gravitas to the Digital Earth initiative, led by NASA in collaboration with other US agencies.

International enthusiasts began constructing major components of the Digital Earth vision yielding the present set of early prototypes to support an EOS. This movement has also been fueled by the tremendous quantities of satellite and remote sensing data, as well as by the efforts of collaborative data resource sharing among digital frameworks like Global Map and the Global Spatial Data Infrastructure (GSDI 2008; ISCGM 2008). Combined with city- and national-level GIS data, this Digital Earth design and framework enables a host of applications across a wide range of human and physical scales. Partnerships that include business, government, NGOs, and universities are developing to capture Digital Earth's promising technological solutions and to direct focus on the increasing calamities facing both developing and developed nations from anthropogenic pressures.

Operational, web-enabled 3D geobrowsers and their applications were comprehensively premiered at the Fifth International Symposium on Digital Earth (Foresman 2008a). These geobrowsers represent the first wave of novel and ubiquitous user-interface tools for harnessing the Digital Earth vision for an Earth Operating System. Many technological partnerships exist in this field, including NASA Worldwind, Microsoft Virtual Earth, ESRI ArcGIS Explorer, GeoFusion, and Google Earth. These achievements support the optimistic proposition that the elements from the Digital Earth community can and will provide the operational components for use in an EOS.

## Google Earth as EOS Prototype

Google Earth is the most widely recognized player in the virtual Earth world, and those familiar with its platform functions can easily visualize further evolutions of an EOS for Gaia. Google Earth—originally Keyhole Technologies—focused on creating a tessellation engine (a virtual globe that data and images can be electronically pasted onto) that would be accessible via the internet. Environmental and humanitarian applications were not components of Keyhole's original mission until UNEP developed a global environmental assessment software application in 2000. Google purchased Keyhole in 2004 and launched Google Earth in 2005 spawning tremendous interest in connecting information searches with geo-locations. These geo-locations have been visually intuitive and appealing to masses of nontechnical public users on the internet (Foresman 2008b).

An avalanche of community-initiated applications has emerged to provide remarkable reporting on leading social and environmental issues. Stunning and unexpected developments include the documentation of genocide in Darfur, pernicious forest clear-cutting in Burma, and tragic mountain-top removal for coal in Appalachia (Foresman 2007). The successful strategy of Google Earth lies in the accessibility and simplicity of tools that enables users to interface with KML (keyhole markup language) to display their information and satellite images from the internet overlaid on a 3D virtual globe. This startling interface between public technology and current events demonstrates the powerful applications that virtual 3D Gaia can have both in social-justice and ecological arenas.

There remains much to be done, however, in attempting to align the major organizations that are competing for market recognition and leadership to arrive at the conditions necessary for a shared EOS. Standards exist for interoperability that include quality control metadata and open-source protocols for data exchange, but these have yet to be universally adopted thereby postponing the higher performance and achievement capabilities of web-based 3D geobrowsers. The international information and communications framework that has been building over the last two decades through collaborative efforts (e.g., the Spatial Data Infrastructure community) has not reached maturity (FGDC 2008). This framework will also need additional domain experts for facilitating the information and data flow associated with each and every location on the planet.

Collaborative science communities are also increasingly seeking wiki-based tools to enable expansion of semi peer-reviewed information catalogs. Although counterintuitive to many scientists and librarians, the wiki-tools have proved to be as reliable, or more accurate, than conventional approaches to encyclopedic knowledge repositories, and are also temporally more appropriate (Earth Portal 2008; Giles 2005; Hawken 2007). Most information and understanding necessary for sustainability, and conducive for the continued existence of all Gaia's life forms, are not restricted to the scientific community (LOHAS 2008). Indeed more connected and commonsensical knowledge, especially that which is grounded in local communities, can be supported by wiki-tools contributing interconnections and networking between people in an EOS framework. The collaborative interfacing of active citizens through a Digital Earth constitutes a key technology for local communities to move forward (and perhaps even to survive). One of our greatest challenges is how to transform society thereby enabling create grassroots implementations of what an EOS might reveal about the state of human and nonhuman ecologies and their harmonious integration. Mike Carr offers some insight into societal reorganization at bioregional levels (2001).

There are encouraging signs that more and more people around the world are asking questions regarding their individual and collective consumer behaviors. There is increasing demand to know about the origins and conditions of production of consumer goods such as wood products, food, clothing, toys, and water. An EOS could help with "truth in labeling" and could even help stimulate needed shifts in consumer behaviors. New social outlets for peace are being investigated by groups of women attempting to map episodes of violence, on a village by village case, using EOS-type technology (NWGPS 2005). The importance of such initiatives lies also in the potential of providing ecosystem information and feedback directly to women leaders who are attempting to affect social and environmental changes in their communities. As noted by Nobel Laureate Wangari Maathai, self-governance will not succeed if ecological conditions are ignored, and ecological restoration programs will fail if sustainable, democratic governance is not included in the equation (Maathai 1985).

Forward progress for an EOS is urgently required to translate the cybernetic underpinnings of Gaia theory into practical, day-to-day tools. Access to free, web-based Gaian informational systems is rapidly evolving, as evidenced by technical evolutions along the Digital Earth visionary path. Attention should next be focused on stepping up the implementation of technology-based systems for increased citizen access. A grassroots tide of deepening awareness and knowledge may effectively counter the seemingly irrepressible onslaught of biodepletion, resource exploitation, and climate change.

## Conclusion

Given today's litany of global challenges, solutions that enable rapid and collaborative action by the majority of the planet's citizens are needed. An EOS for Gaia could provide the means for collective communication and action. The most promising expectation within a global setting of competing social, governance, ecological, and economic demands is the nonlinear growth of web-based social networks; such an acceleration seems plausible given society's phenomenal adoption of web-based networking and tools. If globally distributed social networks—working with NGOs, industry, universities, and governments—can be attuned directly to the realities of Gaia's systems through EOS-based knowledge and insights, then there is greater hope of directing human actions toward sustainability.

Paul Hawken's optimism is his latest work *Blessed Unrest* stems from finding that the largest social movement in history is occurring in our time: this movement is concerned with redressing both environmental and social-justice issues from local to global levels. "If you meet the people in this unnamed movement and aren't optimistic," Hawken submits, "you haven't got a heart" (2007). The fruition of this optimism will require unprecedented collaborative work, utilizing an emerging EOS, to effect the profound changes required to preserve Gaia's natural systems in harmony with human worlds.

# References

Carr, M. 2001. *Bioregionalism and Civil Society: Democratic Challenges to Corporate Globalism*. Vancouver: UBC Press.

Darwin, C. [1859] 1964. *On the Origin of Species*. Cambridge: Harvard University Press.

Dennett, D. C. 1994. *Darwin's Dangerous Idea: Evolution and the Meanings of Life*. New York: Simon and Schuster.

Earth Portal. 2008. www.earthportal.org.

FGDC. 2008. Federal Geoinformation Data Committee. www.fgdc.gov.

Foresman, T. W. 2007. Digital Earth as an operating system for a troubled planet. *Imaging Notes* 22 (1): 14–15.

Foresman, T. W. 2008a. Fifth International Symposium on Digital Earth. *International Journal for Digital Earth* 1 (1): 168–70.

Foresman, T. W. 2008b. Evolution and implementation of the Digital Earth vision, technology, and society. *International Journal for Digital Earth* 1 (1): 4–16.

Fuller, R. B. 1968. *Operating Manual for Spaceship Earth*. Carbondale, IL: Southern Illinois University Press.

Fuller, R. B. 1981. *Critical Path*. New York: St. Martin's Press.

GEOSS. 2005. *Strategic Plan for the U.S. Integrated Earth Observation System.* Washington, DC: NSTC Committee on Environment and Natural Resources.

Giles, J. 2005. Special Report Internet encyclopaedias go head to head. *Nature* 438: 900–901.

Gore, A. 1992. *Earth in the Balance: Ecology and the Human Spirit.* Boston: Houghton Mifflin.

Gore, A. 2006. *An Inconvenient Truth.* Emmaus, PA: Rodale.

Gore, A. 1998. www.isde5.org.

GSDI. 2008. *Global Spatial Data Infrastructure.* www.gsdi.org.

Hawken, P. 2007. *Blessed Unrest: How the Largest Movement in the World Came into Being and Why No One Saw It Coming.* New York: Viking, p. 4.

Hawking, S. W., and G. F. R. Ellis 1968. The cosmic black-body radiation and the existence of singularities in our universe. *Astrophysical Journal* 152: 25–36.

Hawking, S. W., and R. Penrose. 1970. The singularities of gravitational collapse and cosmology. *Proceedings of the Royal Society of London* A314: 529–48.

ISCGM. 2008. Interagency Steering Committee for Global Map. www.iscgm.org.

Laszlo, E. 2001. *Macroshift: Navigating the Transformation to a Sustainable World.* San Francisco: Berrett-Koehler.

Lovelock, J. E. 1965. A physical basis for life detection experiments. *Nature* 207: 568–70.

Lovelock, J. E. 1969. Planetary atmospheres: compositional and other changes associated with the presence of life. In O. L. Tiffany and E. Zaitzeff, eds., *Advances in the Astronautical Sciences.* Tarzana, CA: AAS Publications Office, pp. 179–93.

Lovelock, J. E. 1988. *The Ages of Gaia.* New York: Norton.

LOHAS. 2008. *Lifestyles of Health and Sustainability.* www.lohas.com.

Maathai, W. 1985. *The Green Belt Movement: Sharing the Approach and the Experience.* New York: Lantern Books.

Margulis, L., and J. E. Lovelock. 1974. Biological modulation of the Earth's atmosphere. *Icarus* 21: 471–89.

Meadows, D. H., D. L. Meadows, J. Randers, and W. W. Behrens. 1972. *The Limits to Growth.* New York: Universe Books.

Millennium Ecosystem Assessment. 2005. www.millenniumecosystemassessment.org.

Murawski, R. Methot, G. Tromble, R. W. Hilborn, J. C. Briggs, B. Worm, E. B. Barbier, N. Beaumont, J. E. Duffy, C. Folke, B. S. Halpern, J. B. C. Jackson, H. K. Lotze, F. Micheli, S. R. Palumbi, E. Sala, K. A. Selkoe, J. J. Stachowicz, and R. Watson. 2007. Biodiversity loss in the ocean: how bad is it? *Science* 316: 1281–84.

NWGPS. 2005. *From Local to Global: Making Peace Work for Women.* New York: NGO Working Group on Women, Peace, and Security.

Radin, D. 1997. *The Conscious Universe*. New York: HarperCollins.

Scientific Expert Group on Climate Change (SEC). 2007. *Confronting Climate Change: Avoiding the Unmanageable and Managing the Unavoidable*. R. M. Brerbaum, J. P. Holdren, M. C. MacCracken, R. H. Moss, and P. H. Raven, eds. Report prepared for the United Nations Commission on Sustainable Development. Sigma Xi, Research Park, NC, and the United Nations Foundation, Washington, DC.

Strong, M. 2000. *Where on Earth are We Going?* New York: Knopf.

UNEP. 2002. *Global Environmental Outlook 3*. United Nations Environment Programme, Nairobi.

UNEP. 2007. *Global Environmental Outlook 4*. United Nations Environment Programme, Nairobi.

Wilson, E. O. 2002. *The Future of Life*. New York: Vintage Books.

# 15

## The Gaian Generation: A New Approach to Environmental Learning

Mitchell Thomashow

How would schooling change if it were completely overhauled so as to educate students to observe, assess, and interpret environmental change? What if our most prominent educators and scientists developed an approach to K–16 schooling in which an understanding of the biosphere—a Gaian approach—became the foundation of an entire curriculum? How might we train a "Gaian generation" of environmental learners? This chapter is a speculative attempt to answer that question. We are on the verge of a new twenty-first century environmental science, and we urgently need cohorts of learners who can apply this science to the daunting task of planetary well-being.

Our challenge is compounded by the prevailing absence of natural history knowledge and awareness. Few children spend time outdoors (Louv 2006).[1] Fewer and fewer children can identity the local flora and fauna of their neighborhoods and communities. And yet, with just basic computer skills, they have access to a global network of environmental information and tools. How do we revitalize an interest in the natural world, supplement it with the vast information repository that's available, and educate a new generation of environmental learners?

### Reformulating Environmental Education

"Pattern-based" environmental learning must become the conceptual foundation of an integrated environmental change science curriculum. This can be accomplished, in part, by linking a hands-on, empirically oriented, observational approach to natural history (visceral learning), with a broader conceptual, computer-enhanced, pattern-based approach to environmental science (virtual learning). Make no mistake. Gaian learning starts with intimate awareness of local natural history. Direct observation of the natural world is the curricular substrate for understanding the biosphere. But such learning also requires an awareness of spatial and temporal variation. With the power of laptop computers, interactive databases, and the scaling tools that both facilities enhance, a pattern-based approach to environmental learning is at our fingertips.

I am urging a reformulation of K–16 science, an approach that is substantively informed by but also linked to new conceptual frameworks. What are the developmental structures, the cognitive orientations, and the perceptual foundations that form the basis of this reformulation? In this chapter, I propose exactly such a reformulation, informed by state-of-the-art global change science, culminating with concrete suggestions for educational institutions.

## The Mandate (the IGBP Challenge)

The International Geosphere-Biosphere Programme (IGBP) is an interdisciplinary consortium of research scientists who are primarily concerned with the Earth system challenges posed by global environmental change. Its research agenda "comprises a suite of research projects focused on the major Earth system components (land, ocean, and atmosphere), the interfaces between them (land-ocean, land-atmosphere, and ocean-atmosphere) and systemwide integration (Earth system modeling and paleoenvironmental studies)."[2]

IGBP publishes a series of comprehensive environmental change science anthologies (*The IGBP Series*) representing the epitome of peer-reviewed, international, interdisciplinary, innovative, approaches to a holistic, biospheric assessment of the Earth system.[3] Anyone interested in developing a deep understanding of the complexities of environmental change science should be familiar with these volumes. The seminal work *Global Change and the Earth System* (2004) provides both a comprehensive assessment of the various stresses and pressures on the Earth system and a compelling epistemological approach for researching, interpreting, and communicating concepts of environmental change. The final chapter, "Towards Earth System Science and Global Sustainability," offers an "Earth system science toolkit."

The guiding premise of the IGBP approach is that an "integrative Earth system science is beginning to unfold" as "observations of Earth from the surface and from space are yielding new insights almost daily." They suggest that a conceptual reorientation is necessary and the "biggest challenge" facing the scientific and educational communities "is to develop a substantive science of integration" (Steffen et al. 2004: 264). The IGBP mandate trumpets a challenge to reorient environmental science education to provide students with the conceptual tools for interpreting, assessing, and comprehending global environmental change.

I will describe this challenge in some detail as it offers an authoritative, compelling, and ultimately urgent case for such a reorientation. The IGBP mandate provides a biospheric perspective on environmental change science, with an emphasis on both the analytical and cognitive orientations that such a science demands. Its additional emphasis on sustainable solutions links theory and practice, providing a tangible reality to environmental change. Human life, ecosystem integrity, and planetary health will be profoundly impacted by Earth system changes. Hence urgency, if not a moral imperative, is the foundation for this mandate.

In 2001 the Global Analysis, Integration and Modeling Task Force (GAIM), a subcommittee within the IGBP, "developed a set of overarching questions as a challenge to the scientific community concerned with global change." These questions were organized into four categories: analytical, operational, normative, and strategic. The analytical questions are of particular interest for environmental science education (ibid.).

1. What are the vital organs of the ecosphere in view of operation and evolution?
2. What are the major dynamical patterns, teleconnections, and feedback loops in the planetary machinery?
3. What are the critical elements (thresholds, bottlenecks, switches) in the Earth system?
4. What are the characteristic regimes and time-scales of natural planetary variability?
5. What are the anthropogenic disturbance regimes and teleperturbations that matter at the Earth system level?
6. Which are the vital ecosphere organs and critical planetary elements that can actually be transformed by human action?
7. Which are the most vulnerable regions under global change?
8. How are abrupt and extreme events processed through nature-society interactions?

Further the IGBP mandate poses a series of conceptual challenges, dictated by the characteristics of a complex, multilayered template of interconnected biospheric systems. For example, a student of environmental change science must be able to cope with complexity and irregularity. "Most environmental systems are characterized by a multitude of nonlinear internal interactions and external forcings" (ibid.). How do you learn to interpret nonlinear Earth system behaviors? How do you recognize thresholds and irreversible changes? How do you accommodate for indeterminacy or intrinsic uncertainty? How do you recognize the characteristics of emergent properties and complex systems? Finally, and at the core of the toolkit, is an understanding of scaling effects, recognizing the interactions and distinctions between local, intermediate, and global spatial scales, as well as interpreting vastly different temporal relationships. The IGBP mandate describes these as the "visionary tools" that are a prerequisite for global change research.

## The Earth System Science Toolkit

These conceptual challenges are the cognitive foundation for an "Earth System Science Toolkit...an interlinked suite of probes and processors that sense and interpret Earth System behavior in a holistic way" (ibid.). This suite includes paleoscience, contemporary observation and monitoring, Earth system experimentation, global networks, and the simulation of Earth system dynamics.

Although rapid environmental change presents complicated "no-analogue states," (viz., that the "Earth System has recently moved well outside the range of the natural variability exhibited over the last half million years"), the use of "multi-proxy" approaches remains crucial. Paleoscience emphasizes the recovery of "key archives of past change" (ibid.). Those archives include mountain glaciers, coral reefs, tree ring records, biological species assemblages in lakes, boreal peat lands, coastal environments, coastal tropical wetlands, or any ecosystem in rapid transition where data gathering from the more recent past provides an historical context for assessing rapid environmental change. The collection, interpretation, and assessment of these data must become a foundation for environmental science teaching.

In the last few decades we have seen a proliferation of Earth system data, enabled by extraordinary advances in computer technology, observation of the Earth from space, and sophisticated monitoring techniques. Through global computer networking and the relative accessibility of the Internet, much of these data are publicly available and accessible. This global change information base should be effectively organized so that educators can use it as the basis for teaching environmental science.

Rapid environmental change results in dramatic Earth system experimentation—altered biogeochemistry of the oceans, the introduction of alien species, the removal of endemic species—processes that reflect a contemporary, ubiquitous, perceptual challenge. Any student of environmental science can observe simulations of future environmental conditions on Earth by studying "the structure and functioning of ecosystems under new combinations of atmosphere and climate" (ibid.).

The depth, richness, and complexity of these data require a global network of thousands of trained, dedicated observers, who use similar protocols, and who have access to this shared data. "Planetary patterns emerge more clearly when small-scale or site-specific measurements and process studies are carried out in a consistent and comparative way across the globe" (ibid.). Emerging global computer networks facilitate the exchange and accessibility of these data. Such a global initiative should be linked to a similarly comprehensive network of schools and other educational institutions.

The portability and power of computer technology also supports increasingly instructive and dynamic "virtual" simulations, scenarios, and experiments. Although highly technical knowledge is required, for example, to "simulate mathematically the physical dynamics of the atmosphere and the oceans and their coupling" (ibid.), or to incorporate the dynamics of major biogeochemical cycles, more simplified versions of these models serve to enhance a student's understanding of Earth system processes. Why not provide school systems, teachers, and students with the software and training to explore such simulations in environmental science classrooms?

For these purposes I will present hands-on, educational approaches that integrate the IGBP mandate as the basis for environmental science education. To create a resilient and comprehensive understanding of biospheric processes, environmental science must emphasize the interpretive dimensions of the eight analytical questions as suggested by the GAIM task force. What are the conceptual, developmental, and perceptual challenges intrinsic to their investigation? This is the educational essence of the IGBP mandate. What specific challenges do they create for environmental learning?

## Pattern-Based Environmental Learning

How can we train an entire generation of students and teachers to reorient their approach to learning so as to enhance their understanding of the biosphere? This is both a perceptual and substantive challenge. Learning about biospheric processes requires a perceptual reorientation, an educational approach that stresses pattern-based learning. The task for the science educator is to develop a conceptual curricular sequence that helps students perceive, recognize, classify, detect, and interpret biospheric patterns. At the core of this approach is an emphasis on scale, an understanding of how to interpret spatial and temporal variability, linked to the dynamics of biospheric processes and local ecological observations.

Consider some of the dynamic biospheric processes that are crucial to understanding global environmental change: biogeochemical cycles, watersheds and fluvial geomorphology, biogeographical change (e.g., species migrations, radiations, and convergences), plate tectonics, evolutionary ecology, and climate change. What if these concepts became the basis of science teaching as soon as a child starts school? You can teach a first grader to follow the hydrological cycle, to observe the flow of water in a river, to observe phenological changes, to understand animal and plant migration. You can teach an elementary school child about plate tectonics, climate change, seed dispersal and pollination, or atmospheric and oceanic circulations.

I believe that the source of this learning is a pattern-based orientation. Once you understand the basic earth/land/water movements of a biogeochemical cycle and the various teleconnections between these mediums, you have perceptual awareness of a fundamental biospheric process. Depending on the grade level and learning sequence of the curriculum, the substantive depth of investigation may be enhanced. Each year, K–16, a student can study the carbon cycle, with additional layers of complexity as the necessary mathematics, modeling, or mechanics become enhanced. The curricular substrate is the ability to interpret the patterns that are intrinsic to biogeochemical cycles as linked to a growing understanding of scale and connectivity.

Variable scalar hierarchy is an important conceptual tool for biospheric perception. The observer learns that causation depends on context. Depending on the scale of your observation, you learn to link different phenomena, and you understand that the dynamic changes inherent in any landscape are a function of spatial and temporal boundaries. There is a pattern language that transcends scale. The emerging science of landscape ecology, for example, works with a taxonomic lexicon that implies such a structured pattern language: corridors, gaps, mosaics, borders, and boundaries. Observing such structures through tangible, hands-on, research projects provides students and teachers with the opportunity to explore these patterns.

A deeper exploration of biospheric patterns and processes (as in the case of landscape ecology) yields mathematical and linguistic learning opportunities that further deepen the curricular sequence. Should this change how we teach math and language? Would math instruction be more meaningful if it was coordinated with the observation of biospheric patterns? Can such coordination be linked to the earliest years of schooling?

Pattern-based environmental learning is the conceptual foundation for a biospheric curriculum. This approach is necessarily both visceral and virtual. It must proceed, on the one hand, with hands-on, outdoors-based, field observations, taking advantage of the perceptual gifts of the five senses. There is no better educational approach for biospheric learning, than intimate, empirical observations of natural history. However, pattern-based learning also requires the ability to explore and practice the manipulation of data by experimenting with scale. Through the use of computers and other forms of instrumentation this manipulation can occur through magnification and miniaturization. Science teaching has some remarkable perceptual tools that are now widely available. How might they further enhance biospheric perception? First, let us look at the visceral approach and why intimate awareness of local natural history is a prerequisite for pattern-based environmental learning.

**The Visceral Approach: Biospheric Natural History**

Richard Louv's 2006 book *Last Child in the Woods: Saving Our Children From Nature-Deficit Disorder*, suggested that an entire generation of North American youth no longer play outside. This became a rallying cry for dozens of environmental organizations, culminating in the sponsorship of federal legislation entitled the No Child Left Inside Act of 2007, an effort to restore and revitalize environmental education funding for American public schools. The explicit assumption of such legislation is that less time outdoors results in declining awareness of and interest in ecological issues and knowledge. Implicitly, it assumes that the dominance of computers, television, video games, and other electronic entertainment, leads to inactivity, a decline in physical fitness, and less curiosity and interest in the natural world.

I am not sure there is enough evidence to warrant a clinical psychological term such as "nature-deficit disorder," but Louv's essential point is well taken. One can presume a declining awareness of natural history, and such a decline can only be reversed with a dedicated effort on the part of schools, communities, and families to promote outdoor learning. Louv suggests that outdoor play is crucial to the healthy psychological development of children. Environmental educators insist that outdoor play is a prerequisite to promoting an ecological understanding of the natural world.

What is the relationship between outdoor play and biospheric natural history? Intimate awareness of local natural history is the educational foundation for interpreting biospheric patterns (Thomashow 2002). There are "exemplary biospheric naturalists," scientists whose lifework is to study the ecological, evolutionary, and geological dimensions of Earth system science, and whose insights are grounded with their natural history skills. Lynn Margulis and Tyler Volk (see their work elsewhere in this volume), derive their Gaia-based interpretations from a combination of lab-based studies (using sophisticated instrumentation) and avid field observations. Margulis, a remarkable science educator, as well as a great theorist, has written a series of outstanding "five kingdom" field guides that stress how immersion in field-based observations yields rich insights into environmental evolution (see Margulis and Schwartz 1998). Volk's work emphasizes field-based observations of biogeochemical cycles, linked to observing the interfaces between oceanic, atmospheric, and terrestrial milieus (1998).

Charles Darwin, surely an exemplary biospheric naturalist, is a particularly interesting educational case study. How do Darwin's impeccable field observations, detailed analytical investigations, and insatiable curiosity lead to his expansive theoretical view? *The Voyage of the Beagle* is the ultimate biospheric field trip, a circumnavigational data-collecting journey, which enabled Darwin to juxtapose data from different habitats, link ornithological and geological observations, and speculate on both spatial and temporal variation. Yet some of the most compelling reading in *Voyage of the Beagle* is Darwin's Galapagos material, specifically, his comprehensive experimental observations of *Amblyrhynchus*, a "remarkable genus of lizards." Here Darwin demonstrates his extraordinary capacity for asking profound ecological questions. His deeply interpretive, sharply analytical questioning process throughout *The Voyage* depicts an attention to detail, ultimately linked to broader patterns. The source of Darwin's inspiration and perceptual awareness originates in his outdoor, field-based investigations, the basis for his investigative protocol. The integration of hands-on field exploration with global travel is the milieu of nineteenth-century natural history, and serves as the origins of evolution, ecology, geology, and ultimately Earth system science.

Visceral approaches to natural history provide an intimate awareness of species and habitats. The outdoor experience provides a dynamic learning milieu and an inspirational and motivational context. There is sufficient narrative evidence to suggest that outdoor, immersive, field-based studies are crucial to developing the observational capacity that leads to biospheric awareness. Only extraordinary individuals have the motivation and perseverance to pursue such learning on their own. Like all forms of learning, this approach requires supervision, structure, mentoring, and a learning community of like-minded collaborators. What are the implications for environmental learning in schools? How is this approach incorporated into a unified environmental change science curriculum?

Consider the curricular potential of phenology (the scientific study of periodic biological phenomena, such as flowering, breeding, and migration, in relation to climatic conditions). Phenology is essentially the study of the changing of the seasons. People interested in phenology might study plant budding and floral blooms, spring and fall migration of birds or butterflies, and the appearance of insects. Of particular interest, phenological observations can be tracked locally, compared to other data on an annual basis, and then compared between places. You can study changing climatic circumstances, the life cycles of specific plants and animals, and other indicators of biological and climatic change.

In a 2001 article in *Science*, Josep Penuelas and Iolanda Filella report that although phenological changes differ from species to species, there are geographically diverse, substantial climate-warming induced changes in a variety of habitats. The report cites several dozen studies in peer-reviewed scientific publications indicating short-term phenological change is a global phenomenon, linked to climate change. They conclude "as in many areas of environmental science, the key requirement is long-term data sets."

Today, thousands of people—professionals and volunteers—record phenological changes all over the world, as do international and national phenological monitoring networks such as Global Learning to Benefit the Environment (GLOBE) or the European Phenology Network. Together with remote sensing, atmospheric, and ecological studies, these data will help to answer the many questions raised by the recently reported climate effects on phenology: What are the limits of the lengthening of the plant growth season and the consequent greening of our planet? Will the (less seasonal) tropical ecosystems be less affected than boreal, temperate, and Mediterranean ecosystems? How will different aquatic ecosystems respond? How will responses to temperature and other drivers of global change interact to affect phenology and the distribution of organisms? How will changes in synchronization between species affect population dynamics both in terrestrial and aquatic communities? Will appropriate phenological cues evolve at different trophic levels? (Penuelas and Filella 2001)

Answers to these questions all require field-based observations, locally gathered data sets, and scores of professional and volunteer observers. What an ideal learning opportunity for science classrooms. Students and teachers can track the weather, keeping daily logs of moisture, sunshine, cloud patterns, and the accompanying landscape changes. These on-the-ground observations can be linked to satellite photos and other global climate patterns. Gardening serves as a fine introduction to both local natural history and global climate patterns, or as an introduction to plant domestication, evolutionary ecology, and coevolution. Watershed studies teach the movement of water in a landscape, hydrological cycles, and basic geomorphology.

The outdoor field experience is crucial in serving as the foundation for pattern-based environmental learning. The visceral, hands-on experience—integrating sensory observations with empirical data collecting—provides an enduring, whole body/mind perceptual approach to learning about the biosphere. It serves as the template for more abstract learning, and deeper explorations of the scaling phenomena that is fundamental to understanding biospheric patterns.

## A Virtual Approach: Exploring a Biospheric Pattern Language

As much as environmental educators rue the great numbers of children left inside, there is another side to the increasingly screen-filled hours of childhood. Video games, internet-based communications, cell phones, digital photography, and digital recording programs have profound conceptual impact on their users. I take a McLuhanesque view—the use of these technologies promotes specific, pattern-based conceptual practice

(McLuhan 1964). Concepts of connectivity (networking), scaling (magnification, miniaturization), and complex systems (emergent properties, nonlinearity) are all intrinsic to the use of computers and the Internet.

A basic word processing program teaches its users how to instantaneously change the size of text, rearrange text on a page, organize notes and information, create layers of text within text, and how to share text with other users. Any basic digital photography program provides its users with remarkable scaling tools—changing the size and detail of pictures, rearranging them, linking them to music, turning them into slide shows. A power point presentation (when skillfully arranged) can be a magnificent exercise in juxtaposing scale.

Consider a highly popular computer game like *The Sims*. In this simulation you observe and manipulate a community of individuals who interact differently depending on how you program them. Their social interactions are a lesson in emergent properties. Entirely unanticipated situations can occur. Based on the variables that contribute to this emerging sociology, you can change the social settings and characters accordingly. *The Sims* is a "simplified simulation" of complex systems. Any computer user can freely download Google Earth, which gives you the ability to instantaneously find your neighborhood, zoom out to a spinning globe, and then come back again. This is an extraordinary, hands-on experiment in scaling, a global atlas of unprecedented conceptual power.

We have raised an entire generation of computer-oriented, screen-based learners who already have many of the conceptual skills (scaling and networking) that are a prerequisite for biospheric perception. Indeed, in ways that we cannot even imagine, perhaps we are on the verge of a true Gaian generation of educational opportunity. What if you take all the conceptual skills that are so easily learned with the use of computers and the Internet and apply them to pattern-based environmental learning? Exploring the spatial and temporal dimensions of biospheric processes requires scaling and connectivity tools. These are at the fingertips of anyone who has access to the Internet and a computer.

I am suggesting that the scaling, networking, and complex systems skills that are intrinsic to the IGBP mandate are already being taught by virtue of computer technology and the Internet. Our challenge is to apply those skills with environmental change science in mind. This can only be done through organized curricular approaches, integrated in formal and non-formal educational settings. Imagine if the power of Google Earth became the foundation for a K–16 environmental change science curriculum. Surely elementary school children raised on computers and video games would be comfortable with Google Earth software, as they already have the conceptual ability to intuitively navigate the software with minimal supervision. But what exactly do we ask them to do with Google Earth?

At this moment I am looking out the window of my small cottage in rural, central Maine, watching a dynamic shower pass through the landscape. It's mid-September and the wetland maples have already turned to shades of red and orange. The strong winds accompanying the shower are sending the first wave of brown leaves to the ground. I fully expect a wave of migrating warblers to arrive on tomorrow's northwest wind. My gaze shifts from the window to the laptop. I visit an appropriate website so I can trace the storm on a weather map. I notice the heavy showers from this morning over down-east Maine. I see that the current shower is part of a thin band of rain, and that the heaviest rain has passed. I zoom out on the map and notice there is one more band of showers in New Hampshire, still a few more in New York State, and dry air will soon follow.

But I am not satisfied. I wish that from this same Internet mapping location I could view a wide-ranging series of maps to challenge my ecological curiosity. I would like to view a biogeographical portrait of the changing leaf patterns, or a map of bird, insect, and bat migrations. I imagine collecting daily ecological or meteorological data and inputting them on these maps. I would like to know about other folks who have similar interests and communicate directly with them about what they're seeing.

All these requests are feasible. They are technologically available, inexpensively provided, easy to use, and absolutely pertinent to the ecological portrait of the planet. How can the use of the Internet and computers and all of the conceptual skills they embody be integrated with hands-on field observations? And how might this integration serve as the basis for a comprehensive environmental change science curriculum?

### The Cognitive Perceptual Challenge: An Integrated Framework for Teaching Environmental Change

How exciting it would be to organize a conference for an internationally statured group of cognitive theorists, anthropologists, educational researchers, environmental change scientists, classroom teachers, and experts in traditional ecological knowledge who would be convened to organize a K–16 environmental change science curriculum that is developmentally appropriate. Is there an exemplary sequence of instruction and an effective layering of teaching methodologies that coordinates multiple intelligences, childhood and adolescent development, and cognitive development so as to optimize learning about environmental change?

Pending the research agenda necessitated by such a charge, I offer some tentative suggestions, influenced by reading dozens of autobiographical and biographical accounts of "exemplary biospheric naturalists," as well as observing dozens more undergraduate and graduate environmental studies students. These suggestions are merely an example of paths that may facilitate pattern-based environmental learning, based on relative "success stories," that is, individuals, who have always been attracted to studying ecological and biospheric phenomena. My assumption is that the single greatest conceptual challenge in perceiving environmental change is the difficulty in interpreting spatial and temporal relationships. The challenge then is how to develop the ability to observe what is close at hand (intimate awareness of local natural history) and link those observations to biospheric phenomena. How do educators sequence such learning?

Exemplary biospheric naturalists understand how to juxtapose scale, see multiple spatial and temporal dimensions in a landscape, and move conceptually through ecological space and geological time. I suggest there are three interconnected learning approaches that form the basis of this awareness—field-based natural history, interpretive questioning, and an ability to observe patterns at different scalar levels. These are coordinated dimensions of learning, appropriate at all age levels, but with increasing degrees of sophistication. With greater depth of knowledge, more refined perceptual awareness, and greater sophistication of expression, the learner is increasingly capable of discovering and understanding the patterns of environmental change.

For a child or adolescent, field-based natural history often is organized around a natural history collection of some kind, often informed by either a standard (keys and taxonomies) or an improvisational classification scheme.[4] The child typically plays with this collection, using it as the basis for understanding order and structure. Young naturalists gather these collections by immersing themselves in whatever local habitats are available, often experiencing sensory exploration of the outdoors. These collections are further enhanced with note-taking, visual illustration, or other forms of coding and explanation.

I suggest that natural history collections should be a priority for an integrated environmental change learning sequence. Such collections can take the form of photographs, note-taking, mapping, other forms of visual illustration, as well as a "leave no trace" approach to handling natural artifacts. However, what's most important is how these collections become the basis for interpretive classification schemes. It's not enough to collect things and sort them. The purpose of the collection is to heighten your observational awareness—to know what's common and what's rare, to know what can be found here and what can be found there, to observe associations, characteristics, and correspondences.

By an interpretive classification scheme, I refer to a method for asking and answering questions about environmental change. Why do so many birds migrate from the North to South and back again? Why have invasive species become so dominant in this landscape? Why are there more (or less) Monarch butterflies in the garden this year? How much carbon is there in this forest? How much carbon is there in the atmosphere? When is there too much carbon in the atmosphere?

You cannot ask questions such as these unless you first know what you are observing. Collection, identification, and classification are meaningless without interpretation, causation, and sequence. Taken together, collection and interpretation lead to observations of scale. The essence of good interpretive questions is the juxtaposition of time and space. How did events over there influence what is happening here? How did events from the past set up the circumstances of the present?

Ultimately, to satisfy the learning requirements of the IGBP mandate, an educational curriculum should aspire to cultivate "pattern-based environmental learning." There are patterns that transcend scale, that emerge in a variety of landscapes and milieus, that link atmospheric, oceanic, terrestrial, and organismic phenomena, and that show the relationship between spatial and temporal variation. The purpose of environmental change science is to detect, interpret, and assess these patterns, and use them as a basis for public policy.

This is the essence of the cognitive perceptual challenge: how to derive a curriculum and a teaching methodology that allows the observer to detect such patterns. My educational hypothesis is that such pattern-recognition is the conceptual foundation for understanding how to cope with complexity and irregularity—the core of the Earth system science toolkit as proposed by the IGBP. Understanding nonlinearity, thresholds, irreversible changes, indeterminacy, complexity, emergent properties, and scaling effects requires an environmental change pattern language.

Landscape ecology provides an approach that illustrates concepts of ecological spatial variation (mosaics, gaps, boundaries, corridors, patches, edges, fragments, etc.). How might we elaborate such a pattern language as a template for teaching environmental change science? What are the patterns of connectivity (networks, nodes, and link)? What are the patterns of oceanic and atmospheric circulations (wave, rhythm, flow, fluidity, and fluctuation)? Is there a language to discern various rates of change? What is the relationship between a trend and a discontinuity?

In teaching how to observe environmental change, concepts such as waves, thresholds, and cycles are crucial, and with supervised curricular attention, can be taught throughout the K–16 learning sequence. Waves appear ubiquitously as visual and acoustic representations of rates of change. They reflect frequency, longevity, and periodicity. They can be evaluated mathematically as ratios and rates. A wave is a tangible manifestation of environmental change, observed both virtually and viscerally.

Waves can be used to teach about thresholds. A threshold describes a point, level, sequence, event, or flow that causes a dramatic shift in condition. When is a threshold reached? How do you know? At what point does it cause an irreversible condition? Can thresholds be predicted? Is a threshold a discontinuity in a cycle?

A cycle is a continuous and predictable series of relationships within a system, in which the flow and exchange of materials, ideas, or events, move according to repeatable, yet variable patterns. Of particular interest is the relationship between cycles, which may form another system of cycles or have nonlinear emergent properties. School children can observe cycles, and yet it is the depth and complexity of cycles that is so crucial to understanding environmental change.

An integrated cognitive framework for teaching environmental change is an epistemological challenge. It requires a reconsideration of how science is taught, how it is linked to mathematics, language, and the arts and how it serves to empower students to assess and propose solutions for problems of planetary significance. It starts with emphasizing how important it is to promote ecological awareness and observe natural history. It is deepened and enriched with the use of computers and the Internet and the implicit scaling conceptualizations embedded in their use. It is coordinated with substantive curriculum about the earth system. It is applied by changing the meaning and purpose of schooling.

## Schools to Teach Environmental Change: A "Gaian Generation"

I propose developing an international network of high schools organized around teaching to the IGBP mandate, designed to train a new "Gaian" generation of environmental change science researchers. Let's design these schools as educational laboratories for teaching environmental change science. Let's organize them so that the schools become nodes in a research network, each becoming a center for long-term environmental change research, with teachers supervising students through community-based projects, linked to partner schools in an international network. These schools will share both their teaching approaches and results, while compiling databases of biospheric observation. Let's organize art and music instruction, literature and philosophy, social studies and psychology, around environmental change.

As a starting point, consider a field-based approach (linking the visceral and virtual), as informed by the IGBP "Earth system science tool kit." For example, using *paleoscience* as the foundation for hands-on field natural history, provide students with the interpretive skills to reconstruct past environments at a variety of spatial and temporal scales and at different organismal levels. Every habitat has a uniquely interesting history. Teach students to reconstruct a habitat using a sequence of time scales, starting with the immediate past to a historical time frame, to a Pleistocene approach and then finally a geological time scale. What creatures walked this place ten million years ago? Were there mountains here or was this place covered with ocean? And then have the students envision what the place will look like in the future (ten years, one hundred years, one thousand years).

The IGBP mandate stresses *contemporary observation and monitoring*. Teach the students how to understand, develop, and assess indicators of ecosystem health, and to apply those indicators to human well-being. Let the school become the center for assessing ecosystem health. Equip the school with laboratory capabilities to become a regional monitoring center for ecosystem health. Publish those observations on a school website, in local newspapers, as public demonstrations of the vitality and usefulness of such learning.

Emphasize ecological monitoring of the school itself. How much energy does it use? Where does its food come from? How much carbon does the school emit? To what extent is the school a living laboratory for sustainability initiatives? How are those initiatives linked to a broader conception of global environmental change?

Let this monitoring become the basis for integrated regional studies. What environmental issues does your community face? How can the school collect data to better inform public decisions about those issues? What role can the students and their teachers play in informing the public about local environmental issues? How might these regional studies involve local politicians and businesses? Let's elevate our high schools by making them centers for community deliberations about urgent environmental issues.

These regional studies can be the basis of international partnerships and learning affiliations. The IGBP mandate recommends *global networks* for sharing research data. Schools can have both "sister" schools in diverse regions and affiliations with relevant NGOs, especially those that are nodes in long-term environmental change networks. Students can spend a year at their partner schools. They can be sponsored by science education facilities (museums of natural history) or service organizations (Rotary International). They learn to see their work as international in scope and importance.

Finally, the IGBP mandate recommends *Earth system experimentation and simulating Earth system dynamics*. Depending on the scale, one can design "what if" scenario-based curriculum. What will happen to a given place given several different climate change scenarios? How will the habitat change if a particular invasive species travels here? How are these local changes linked to more complex, biome-scale variables?

This is an excellent milieu for using innovative computer software. Some years ago, Electronic Arts released two outstanding computer games, *Sim Earth* and *Sim Life*, modeled after their commercially successful *Sim City*, and then followed by the remarkably successful *The Sims*. Unfortunately, *Sim Earth* and *Sim Life* lacked that same commercial success. However, they were remarkable simulations about Earth system experimentation. *Sim Earth*, designed with Gaian principles in mind, allowed the user to explore a range of atmospheric, oceanic, and biological variables. *Sim Life* allowed you to tinker with ecosystems at the community and genetic level. What if Electronic Arts and other computer game designers were commissioned by the National Science Foundation to develop a new generation of these simulations, linked to an international network of environmental change pedagogy? Might the NSF partner with the IGBP in developing such software for use in schools, in combination with a comprehensive approach to pattern-based environmental learning?

These suggestions, by way of example, are merely a few of the possibilities that are within the reach of imaginative educators and scientists. They can be applied in diverse educational environments, anywhere on the K–16 learning spectrum, modified accordingly. None of them are beyond the educational capacities or the international learning infrastructure of twenty-first century schools, colleges, and universities. But they do require a mobilization of resources in service of environmental learning. And they require an urgency of purpose, a common awareness that the future of the planet is at stake.

We live at a time when extraordinary learning resources are available for schools everywhere. We are on the threshold of a deeper planetary awareness, an emerging understanding of biospheric dynamics, a comprehensive "science of integration." But none of this will occur without challenging the status quo of science education. We should be planning schools so as to train a Gaian generation of learners, students who see the biosphere in every habitat and organism, who are equipped to interpret environmental change, who are keen to observe the natural world, and who know that their very survival may depend on it.

## Notes

1. Louv's book triggered a national movement in environmental education, culminating with proposed national No Child Left Inside legislation. See the following website for more information. http://www.naaee.org/ee-advocacy.

2. See page 3 of the IGBP Brochure.

3. The book series is described at http://www.igbp.kva.se/page.php?pid=230.

4. For an interesting anthropological approach to collections, natural history, and the organization of ecological knowledge, see Atran (1993).

## References

Atran, S. 1993. *The Cognitive Foundations of Natural History: Towards an Anthropology of Science.* New York: Cambridge University Press.

Darwin, C. 1989. *The Voyage of the Beagle.* New York: Penguin.

IGBP Brochure. http://www.igbp.kva.se/page.php?pid=363.

Louv, R. 2006. *Last Child in the Woods: Saving Our Children From Nature Deficit Disorder.* Chapel Hill: Algonquin Books.

Margulis, L., and K. Schwartz. 1998. *Five Kingdoms. An Illustrated Guide to the Phyla of Life on Earth.* New York: Freeman.

McLuhan, M. 1964. *Understanding Media: The Extensions of Man.* New York: McGraw-Hill.

Penuelas, J., and I. Filella. 2001. Responses to a warming world. *Science* 294 (October 26): 793–95. doi: 10.1126/science.1066860.

Steffen, W., A. Sanderson, J. Jager, P. D. Tyson, B. Moore III, P. A. Matson, K. Richardson, F. Oldfield, H. J. Schellnhuber, B. L. Turner II, and R. J. Wasson. 2004. *Global Change and the Earth System.* Berlin: Springer.

Thomashow, M. 2002. *Bringing the Biosphere Home: Learning to Perceive Global Environmental Change.* Cambridge: MIT Press.

Volk, T. 1998. *Gaia's Body: Toward a Physiology of Earth.* New York: Copernicus.

# 16

# Gaia Theory: Model and Metaphor for the Twenty-first Century

Martin Ogle

James Lovelock's "Gaia theory" is the most recent and complete rendition of the scientific view of Earth as a living system. Lovelock (2000: 11) has described Gaia as "the Earth seen as a single physiological system, an entity that is alive at least to the extent that, like other living organisms, its chemistry and temperature are self-regulated at a state favourable for its inhabitants." In *Scientists Debate Gaia* (2004: 3) he also characterizes Gaia as "a new way of organizing the facts about life on Earth, not just a hypothesis to be tested." Gaia Theory is both of these descriptors and more.

Earlier scientists presaged the idea of Earth as a living system. For example, Russian scientist Vladimir Vernadsky (1998, translated from 1926) discussed how processes at the level of the organism were reflective of processes in the biosphere, writing that "there is a close link between breathing and the gaseous exchange of the planet." In *Elements of Mathematical Biology* (1956: 16, originally published in 1924), ecologist Alfred Lotka wrote that "it is not so much the organism or the species that evolves, but the entire system, species and environment. The two are inseparable." Aldo Leopold, in a pioneering 1923 article (published in 1979: 140) entitled "Some Fundamentals of Conservation in the Southwest," wrote of "the indivisibility of the earth—its soil, mountains, rivers, forests, climate, plants, and animals," urging us to "respect it collectively not only as a useful servant but as a living being."

## The Impact of Lovelock's Gaia

Forerunners notwithstanding, it was Lovelock who created a compelling and enduring research program into how our planet operates in ways *analogous* to a self-regulating organism. Heeding the advice of novelist

William Golding, Lovelock named his idea "Gaia" to reflect the fact that contemporary science is rediscovering early cultural views of the Earth as living being. The living planet metaphor has engendered both interest and controversy, raising challenging questions for science: in what sense is the Earth alive, how does the Earth work, and how do humans fit into this reality? Metaphor should not intrude unnecessarily into the exactitude of given scientific tests and processes nor should scientists subject metaphor to a rigorous peer-review process: both science and metaphor of the Earth as a living system *can* enrich and expand each other. This chapter submits that "pure science" (often referred to as Earth system science) and the "pure metaphor" (that of the Greek goddess of the Earth, but extending to the widely held indigenous view of the Earth as a living entity) are of great benefit to our contemporary world. Moreover the synergy between the science and the metaphor can add to our overall understanding of the planet without compromising the integrity of either.

In accepting the Philadelphia Liberty Medal at Independence Hall in 1994, Vaclav Havel alluded to this expansion of the scientific mind. He pointed to the Gaia hypothesis as a reason for his optimism about the future, referring to it as "postmodern—a science that in a certain sense allows it to transcend its own limits." The image of a mythical living Earth, of which humans are a part, has already prompted valuable questions that scientists might not have dreamed about asking just a few decades ago; such a challenging enterprise will likely continue. Lovelock's preface to his book, *Gaia: The Practical Science of Planetary Medicine* (2000), also contains an excellent overview of this and similar sentiments on science and metaphor.

Today, we possess unparalleled knowledge and technology for solving discrete problems and challenges. Whether it provides cures for diseases, the use of satellites for communication, or developments in space travel and biotechnology, modern science is unsurpassed in its ability to make things work. In the midst of this awesome power, however, humanity suffers some of the most entrenched and large-scale problems ever known; consequently, we are perched precariously on the edge of massive disruptions in energy availability, climate change, and food production. There are many reasons for this paradoxical juxtaposition of dire problems in the presence of unsurpassed knowledge: solutions to narrowly defined problems often cause new sets of problems; a perceived entrenchment of a cultural divide between "pure science" and "faith and values" has clouded effective understanding; and underlying assumptions in technological developments often remain unquestioned, skewing progress down harmful paths instead of ecologically and socially healthy ones.

Many of our intractable problems stem from an imbalance between human activity (especially resource use) and the integrity of living systems. We stand in great need of an interdisciplinary understanding of how Earth systems work and how human systems can fit harmoniously; we also need a holistic context that will allow us to perceive and solve large-scale problems. Gaia theory offers both the knowledge and perspective required. In this chapter, I discuss both the science and metaphor of Gaia, including examples of their synergy, and then sketch the implications of Gaia theory for energy policy, global warming, and agriculture.

## Gaia Theory Offers Interdisciplinary Context

Ecology, the science often referenced vis-à-vis environmental issues, is defined as the study of interrelationships among organisms and their environment. Gaia theory can then be viewed as the fullest expression of ecology available to us today. It provides a context in which the largest possible scope of interrelationships (including those involving human beings) can be examined because it views the surface of the planet as one living system. This was definitely *not* the science taught to most of us in high school and college—where we received an image of the Earth conveniently orbiting the sun at just the right distance so as to neither burn nor freeze.

One of the key examples of how Gaia theory transcends traditional biology and geology is the postulate that the Earth has reacted as a single, living system in response to the sun's becoming hotter during the past 3.8 billion years. Pre-Gaian views of life on Earth reflected in biology and geology textbooks published over the past few decades explained the Earth's atmosphere, for instance, by way of what Joseph (1990) called the "greenhouse metaphor." This rather mechanical view examined the effects of human-made greenhouse gasses on the temperature of the Earth, but did not describe a dynamic system of feedback mechanisms. According to Joseph, a "membrane metaphor" suggested by Gaia theory is a more apt description; it views the Earth's atmosphere as more analogous to the semi-permeable cell membrane than to the glass panes of a standard greenhouse. Physician and essayist Lewis Thomas (1974: 171) celebrated this idea in his book, *The Lives of a Cell*, in which he wrote: "Viewed from the distance of the moon, the astonishing thing about the earth, catching the breath, is that it is alive.... Aloft, floating free beneath the moist, gleaming membrane of bright blue sky is the earth, the only exuberant thing in this part of the cosmos." The "membrane metaphor" represents a paradigm shift of scientific inquiry.

Pre-Gaian textbook science often did not address the fact that the sun's luminosity has increased at least 25 percent during life's tenure on Earth. Armed with a "greenhouse metaphor," it would have had to conclude that the Earth's temperature would have been expected to rise to levels impossible for life as we know it (Lovelock 1991, 2000). On the contrary, our planet has experienced a temperature regime much *cooler* than would be expected by its distance from the sun—around 40°C cooler, depending on the calculation (Lovelock 1979; Volk 1998; Schwartzman and Volk 2004). And, although there have been variations in temperature over time, the overall trend has been remarkably stable— a stability largely, if not mostly, attributable to living processes (Harding 2006). According to Gaia theory, it is the living system, consisting of tightly coupled organic and inorganic components, that has exerted this moderating influence on climate and other features of the Earth.

The maintenance of somewhat stable surface temperatures by the Gaian system, even in the face of increasing solar luminosity, may be regarded as *roughly* analogous to our own bodies that maintain a core temperature even as external temperatures change. The Earth's living system maintains conditions that are quite different than what would be expected through chemistry and physics alone. Among other factors, the living system heavily influences cloud formation, levels of carbon dioxide and other gasses in the atmosphere, and the color (and thus albedo) of the Earth's surface (Lovelock 2000, 2004). While Lovelock and some of his colleagues (e.g., Lenton 2004; Harding 2006) have characterized these moderating influences of life as "self-regulation," others, like Tyler Volk and David Schwartzman, have asserted that the idea of "self-regulation" is misleading since there are no system set points in terms of atmospheric gas composition, temperature regimes, or other factors (Volk 2003; Schwartzman and Volk 2004). At least part of the disagreement over self-regulation stems from the term's connotations of purpose. Some Gaian thinkers have tried to steer away from teleological implications by eliminating the concept of self-regulation altogether. Regardless of whether self-regulation is regarded as acceptable terminology, Gaian theorists certainly converge on the premise that life powerfully shapes surface conditions.

There is virtually no evidence of self-regulation around *un*varying set points, nor evidence that self-regulation is equally strong for all factors. However, Lovelock (2000: 141) has used the term "homeorrhesis" to describe the dynamic stability of, for example, temperature, oxygen, and ocean salinity around *shifting* balance points over vast periods of time. "Gaia's history," he noted, is "characterized by homeorrhesis with periods of constancy punctuated by shifts to new, different states of constancy. With some variables, such as temperature, the changes are small...with others, such as gaseous abundance, the levels of homeostasis have progressively changed in steps." Oxygen, for instance, which was present in only trace amounts at the beginning of life, rose rapidly with the advent of photosynthesis and was stable over vast periods of time before increasing to new plateaus and staying relatively constant again for long geological periods (Lovelock 2000; Lenton 2004; Volk 1998).

Research by Lovelock and colleagues has shown that the Gaian system may moderate not only oxygen but other atmospheric gases, including methane, carbon dioxide, hydrogen sulfide, among many more. Some gas levels do indeed stay within narrow limits over significant periods of time. Oxygen has hovered around 21 percent in the atmosphere for at least the millions of years that large vertebrates (that require such levels) have been on Earth (Lovelock 2000, 2004; Volk 1998; Lenton 2004). Oxygen is consumed in great gulps in fires and in the oxidation of elements from the Earth's interior; it is exchanged in photosynthesis and respiration and is otherwise being pulled out of and added to the atmosphere. Given oxygen's great reactivity in both organic and inorganic processes, just the fact that oxygen levels have consistently remained very close to 21 percent over even a few million years can be regarded as a remarkable testament to the self-regulative tendency of the Gaian system.

Harding (2006) wrote extensively about how life heavily influences ocean alkalinity and salinity, temperature, and other environmental factors, all of which show remarkable stability. For instance, global cycles of calcium, phosphates, and sulfur are moderated by the activity of microscopic algae called coccolithophores. Through their metabolism and adaptations for maintaining salt balance in their own bodies, these tiny organisms release gases that influence cloud formation, form skeletons that are part of limestone deposition, and otherwise exert significant influence on the global system. Organic and inorganic processes form a seamless continuum in the new understanding of the living Earth articulated in Gaia theory.

Gaian-oriented research also provides a context within which we can account for all aspects of human biology as part of this seamless continuum. In a traditional ecological study of a pond, we would not arbitrarily decide to leave out the biggest fish or its behavior, for to do so would obviously constitute an incomplete study. Until recently, however, ecological studies have made scant reference to human activity and, even today, often leave out our behavior. In relegating determinants of our behavior (belief systems, metaphors, symbol formation, etc.) to completely separate fields such as philosophy and religion, we severely limit our understanding of not only our *relationship* to the Earth's living system but, indeed, of the living system itself.

## The Metaphor of a Living Planet

When we conceive of human emotion as part of our biology, and thus as a part of the Gaian system, we can discern the value of metaphor and myth more clearly. Metaphor and myth may actually be biological adaptations unique to us as creatures with high levels of self-awareness and awareness of time. They are important parts of our behavior to understand and tap into as we move into an uncertain future. Our emotional connection with Gaia is profoundly affected by symbolism, stories, and myths, as it is by reasoning and scientific observation. Just as we need to be guided by compelling and accurate science, we also need to be moved emotionally by compelling and accurate metaphors.

Physicist Freeman Dyson placed great importance on human emotions, seeing them as integral to our relationship to the Earth. In *From Eros to Gaia* (1988: 343), he maintained that "the central complexity of human nature lies in our emotions, not in our intelligence. Intellectual skills are means to an end. Emotions determine what our ends shall be." Dyson recognized how the human brain's hardwiring is integrally linked to the prospects for a healthy relationship with the living system of which we are a part. He regarded "one hopeful sign of sanity in modern society" to be "the popularity of the idea of Gaia, invented by James Lovelock to personify our living planet. As humanity moves into the future and takes control of its evolution," he added, "our first priority must be to preserve our emotional bond to Gaia" (Dyson 1988: 345).

Joseph Campbell (1972), one of the world's foremost authorities on mythology, described mythology as "coeval with mankind," noting that myths are present in every culture, past and present, and exist because of the evolution of an intense awareness of self and of one's own immi-nent death. In an interview with journalist Bill Moyers shortly before his death in 1987, Campbell shed light on the importance of incorporating the mythology of a living Earth into our society. In response to a ques-tion of whether new myths would come from "the Gaia principle," Campbell responded that "myths come from the realizations of some kind that have then to find expression in symbolic form. And the only myth that is going to be worth thinking about in the immediate future is one that is talking about the planet, not the city, not these people, but the planet and everybody on it. That's my main thought for what the future myth is going to be. And until that gets going, you don't have anything" (see Flowers 1988: 32).

Even the most practical of human endeavors make use of symbols and metaphor to create modern mythologies. NASA purposefully selected names like Mercury, Gemini, and Apollo for its missions. In 1960 Abe Silverstein, director of Space Flight Development, proposed that NASA's manned trip to the moon be named Apollo. After consulting a book of mythology one evening, he concluded that the image of "Apollo riding his chariot across the Sun was appropriate to the grand scale of the proposed program" (NASA).

Do the metaphor and symbolism of Gaia matter? Consider the words of Tim Flannery, author of *The Weather Makers* (2006: 17), a superb work on global warming. "Does it really matter whether Gaia exists or not?" he asked. "I think it does," he continued, "for it influences the very way we see our place in nature. Someone who believes in Gaia sees every-thing on Earth as being intimately connected to everything else, just as organs in a body.... As a result a Gaian worldview predisposes its adher-ent to sustainable ways of living." This is not to imply that believing in the metaphor or the science of Gaia necessarily predisposes *all* adherents or predisposes them *perfectly*. In fact many find Lovelock's own prescrip-tion of nuclear power to be contrary to sustainability. To be fair, however, Lovelock has long pointed to the "Three C's" (cars, chainsaws, and cattle) as the biggest impacts on the planet, a sentiment with which many concur who also take issue with his stance on nuclear energy. The point is that the science and metaphor of the Earth as a living system compel new views on how humans fit with that whole and, at least on balance, drive the search for self-preservation (read "sustainable") activities.

The metaphor of Gaia helps us to see beyond the blinders often set up by reductionist science. It allows us to intuit a living planet of stunning beauty, vibrancy, and mystery where before we had seen a rock on which organisms lived at the mercy of physical and chemical circumstances—including a precise distance from the sun. Just as other metaphors help us grasp large or complicated ideas, Gaia allows us to empathize with a complex living system in ways that we are just beginning to understand. The metaphor of Gaia enables a cohesive inquiry into the nature of the living system while still debating whether regulation, self-regulation, homeostasis, homeorrhesis, or other terms are the most accurate and descriptive. And, in a very real sense, the metaphor of Gaia is a window through which we can connect with those before us who sensed the existence of a living Earth. No matter how far we may have come with our science, we are beginning to rediscover knowledge that our ancestors might have accessed in different ways.

The celebrated image of the Earth from outer space immediately calls forth a sense of limits. The world, which may appear to be infinite from the vantage point of being *on* the planet, is suddenly perceived as finite. After astronauts saw, photographed, and described the image of the Earth from space, terms like "thin film of life" and "tiny blue ball" have become more common in our communications, reinforcing the sense of limits. Gaian science also sheds light on limits and offers powerful lessons and insights for human endeavors.

### Lessons from Gaia for Human Systems

We now understand key aspects of evolutionary change that have allowed life to persist in the face of various challenges ranging from ever-increasing solar luminosity to a simple exhaustion of food resources. Early in Earth's history living things consumed the "primordial soup"—high-energy molecules thought to have been spontaneously formed due to the interaction of light/UV radiation with molecules in water. Some microorganisms consumed these molecules; some consumed other microorganisms. As these organisms multiplied, however, they could have come to a grinding halt when all available high-energy molecules in the form of the primordial soup and other organisms were digested—broken down into simpler, low-energy molecules.

Lynn Margulis and Dorion Sagan (1997), as well as Elisabet Sahtouris (1989), described developments at this point in the story of evolution that allowed life to transcend this dilemma. First, bacteria evolved the ability to photosynthesize—to use sunlight to re-energize the low-energy molecules around them and turn them into food and useful energy. Purple photosynthesizing bacteria were the first to do this, using carbon dioxide and hydrogen or hydrogen sulfide as the raw materials for their bodies and energy for their activities. Subsequently, blue-green bacteria developed a more productive form of photosynthesis that used water in place of less common hydrogen molecules.

Second, when some larger organisms ingested blue-green bacteria, instead of breaking them down for food, which would result in more low-energy molecules, they evolved permanent interactive, physical partnerships with them. The chloroplasts (the solar energy-using packets) of plants all around us are the evolutionary remnants of free-living, photosynthetic bacteria that formed seamless symbiotic ventures with other organisms (Margulis 1998; Margulis and Sagan 2002). The endosymbiosis theory of cell evolution was elaborated by Margulis, inspiring her to endorse Lovelock's Gaia, because the Gaian system pointed toward a larger symbiotic unity—symbiosis writ large, or perceived from the perspective of outer space.

Such realizations of how evolution works highlight at least two lessons applicable to human systems: the importance of emulating photosynthesis through the use of daily incoming solar radiation as the basis for our energy consumption, and the need to envision symbiotic systems for energy production and use. The first lesson is a study in limits. Energy use is perhaps the most important place to start for it drives and limits the growth of human systems. Fossil fuels and nuclear energy have been harnessed in huge amounts historically, resulting in unsustainable impacts on the planet, including and extending well beyond the impacts of their extraction and pollution. It is now evident that supplies of some of these fuels are becoming limiting factors for growth, resulting in a feverish search for alternative energy sources such as renewables and a revived (and expanded) nuclear industry. Further there is debate on just how much energy could be supplied practically by renewable resources. Citing "limits to growth," McCluney asserted in a report for the Florida Solar Energy Center (2003: 12) that "it is clear that attempts to solarize the world economy are fated to run into serious obstacles unless population and per capita consumption are drastically reduced." Contrary to this assessment, reports in 2007 for the American Council on Renewable Energy (ACORE), the American Solar Energy Society, and the Institute for Energy and Environment offered much more optimistic views that renewable energy can partially-to-mostly offset fossil fuels and nuclear energy within two or three decades. Limits and challenges are noted, however, within these reports. For instance, the ACORE report qualified its predictions based on "right policies and conditions." Other practical limits to renewable energy, such as storage and transmission capability, have also been widely discussed.

Regardless of the feasibility, capability, or pollution levels of future energy sources, however, a Gaian view of Earth as a living system reminds us that there are other limits as well, even if we successfully harness renewable energy supplies. In his Pulitzer prize-winning book, *Collapse: How Societies Choose to Fail or Succeed* (2005), Jared Diamond listed 12 major environmental problems that confront modern society including loss of natural habitat, loss of topsoil, water shortages, and others. He noted (p. 498) that "our world society is presently on a non-sustainable course, and any of our 12 problems of non-sustainability ...would suffice to limit our lifestyle within the next several decades. They are like time bombs with fuses of less than 50 years." Limiting our energy use, no matter from what source, strikes some as being a recipe for miserable human existence—the proverbial "freezing in the dark." But this need not be the case.

Within the limitations imposed by the use of renewable energy, efficiency, and conservation will allow for comfortable and fulfilling human life. Efficiency and conservation are not the same, as can be illustrated in the construction and operation of a house. Orienting a house to the south, building a sunroom, installing insulation and heat-storing materials, and buying appliances and light bulbs that need less energy make a house efficient. Conservation, however, is largely about human *behavior*—making value-based decisions such as paying more for local materials or those with low-embodied energy, taking the time to operate windows and house fans in the proper manner, or heating and cooling to moderate levels only when necessary. Both conservation and efficiency become more feasible when we adopt the attitude that we are in a symbiotic relationship with the rest of the living system. This applies at the level of personal behavior and decisions; but to take hold and make a difference, conservation must become a shared ethic at the levels of culture and society. For human culture to be sustainable, we must find ways to conduct our affairs using just a fraction of the energy we use now—what Lovelock (2006) has called a "sustainable retreat."

In the final analysis our ability to reduce energy consumption purposefully will be determined by our stories, myths, and symbols and by whether they imbue in us a sense of limits. As Dyson (1988) and Flannery (2006) suggested, the metaphor of Gaia may be the best metaphor to inspire this change.

## Global Climate Change

Over the eons living processes have incorporated carbon dioxide—the gaseous form that carbon takes in the atmosphere—into solid rock such as limestone ($CaCO_3$) and coal (largely carbon) and into other fossil fuels. With large amounts of carbon thus buried and sequestered away from reacting with oxygen, carbon dioxide levels began to decrease rapidly, from perhaps 95 percent of atmospheric gas when life began to the 0.03 percent it is today. Carbon dioxide is an effective greenhouse gas that traps heat in the atmosphere and slows its escape to space. Although the sun is about one-third brighter now than it was when life began, the thinning blanket of carbon dioxide (along with many other mechanisms) has resulted in surface temperatures that are much cooler than they would otherwise be. Life as we know it is dependent on this temperature regime. Viewing Earth as a living system allows us to see that this phenomenon of carbon-sinks is analogous to the healthy state of an individual organism. For instance, in healthy human beings, calcium resides largely in the bones and stays below certain levels in the bloodstream. The disruption of this balance causes the disease osteoporosis. Analogously, when huge amounts of carbon are released from solid carbon-sinks, an imbalance in the entire system occurs. Although the Gaian system will adjust to a new equilibrium, many organisms (including human beings) that are dependent on current conditions may not fare well (Tickell 2004). Keeping carbon in its buried or otherwise sequestered form is healthy for human beings.

Lovelock favors nuclear energy as a short-term energy solution because he considers it the only way to prevent catastrophic global warming. In his more recent book, *The Revenge of Gaia* (2006), he argues that levels of energy (and other resource) consumption *will* have to be radically lowered in the near future, no matter what energy path we follow. He describes a future in which travel occurs in sailboats and food is synthesized so that farmlands can be returned to their natural and semi-wild conditions. Many might take issue with this view of the future. Indeed the dominant economic model of modern Western society (one that shows constant growth) compels visions of a much more resource-intensive world; with just a 3 percent economic growth (considered a modestly healthy rate), however, the economy would double in 23 years! Even allowing for economic growth not tied entirely to resource use, such growth will stress the Earth's living systems tremendously in ways not conducive to the well-being of humanity and countless nonhuman species.

There is no doubt that carbon dioxide emissions need to be curbed drastically, but if we rely solely on a transition from fossil fuel to nuclear energy to affect this change, will we be able to wean ourselves from the need for constant growth? With the massive power of nuclear energy at our fingertips, what is to prevent us from immediately bumping up against *other* critical limits, especially given that we are already at tipping points on many of them? What will prevent us from converting more forest and marsh to farmland, more living material into just so many consumer goods, or otherwise impoverishing living systems? The answer is *nothing*—unless retreat from this runaway growth becomes our stated and serious goal. We must develop a conservation ethic that flows from our scientific understanding of the Earth as a living system.

The view of the Earth as a living system speaks directly and powerfully to conservation because of the central realization of limits that it spawns. Conservation—an actual reduction of energy and resource consumption—should be at the forefront in any serious solution for global warming and in our attempts to live sustainably *as part of* the Earth. I believe Gaia theory points us in the direction of renewable energy for most of our energy needs. Whether we make this transition quickly or whether we stay reliant on today's predominant energy sources indefinitely, our main goal must be to use *far less* energy overall. If we do not effect this change now with relatively little pain, we may be forced to do so soon enough but in an uncomfortable manner.

### Agriculture

Agriculture is perhaps the most important relationship between human beings and the Earth as a whole wherein the transition to food production based on inherent limits of the living system may be our biggest challenge. Lovelock (2000) has argued that "by far the greatest damage we do to the Earth, and thus by far the greatest threat to our own survival, comes from agriculture." In fact one reason that Lovelock is a proponent of nuclear energy is that he sees the alternative of biofuels as untenable—a view with which more and more scientists and experts seem to agree. There may not be enough farmland to feed a growing human population, let alone to provide large amounts of energy.

But what about the energy needed *for*, not provided *from*, modern agriculture? This looms as a limiting factor potentially as serious as global warming or energy policy. Modern food systems require huge inputs of energy for field preparation, fertilization, harvest, transport, and storage. All these are extremely tenuous because small disruptions or shortages (especially during critical stages of the farming process) could result in enormous food shortages. Even if energy considerations were somehow neutralized, the transformation of forest, marsh, bog, and other ecosystems into farmland is a significant and unsustainable impact in itself. In the context of the Earth's living system, these ecosystems play roles analogous to organs in a body: providing crucial functions such as filtering, nutrient transfer, and gas exchange. By comparison, farmland is relatively sterile and non-diverse with less capacity to "control its own climate and chemistry" (Lovelock 2000). Either experiencing longer term fuel shortages or reaching tipping points in the loss of functioning of the Earth's ecosystems could be catastrophic without significant prior planning because the skills necessary for local food production are not possessed by the population at large.

Fortunately, these skills are not lost to all. The imperative of accelerating local food production, using low-energy, ecological inputs, may be as important and time-sensitive as that of reducing greenhouse gases. A Gaian viewpoint compels knowledge of place because understanding local ecosystems provides a microcosm for understanding Earth as a whole, and vice versa. Individual places on Earth hold different potentials for all aspects of human existence—from climate to the availability and types of energy, water, and soil. The homogenized and mechanized agro-industrial approach does not take into account local knowledge and relies on massive inputs of energy, fertilizer, pesticides, and water usually from places far away. In *Deep Economy* (2007), Bill McKibben examined the challenges and rewards of local economies, especially for local agriculture. He charted the trends of farm and food businesses and noted that while food has become cheap and plentiful, much of this gain has come at the expense of the environment, local communities, and the poor.

A Gaian approach is needed. We must design and run farms as intricate ecosystems that are part of larger systems up to and including the entire Earth. This will enable farmlands to mimic, to the greatest extent possible, local and wild ecosystems rather than simply displace them. Many robust, exciting, and successful examples of this kind of farming can be found. For instance, the Land Institute (based in Salina, Kansas) is developing diverse perennial grain production systems that closely mimic the form and function of its native ecosystems. The organization has promoted the "big idea that humans can make conservation a consequence of production—in any region on the planet—if we use as our standard the ecosystems that existed in that region before it was utilized by humans." Part of the Land Institute's mission states that "when people, land, and community are as one, all three members prosper (Land Institute)."

Joel Salatin's family farm (named Polyface) is unique in Virginia's Shenandoah Valley. Attention to place and local natural processes is an intricate part of every operation on the farm, from its "pigaerator" system for producing compost to its cyclical system of running livestock and chickens on fields. The Salatins maintain that "mimicking natural patterns on a commercial domestic scale ensures moral and ethical boundaries to human cleverness." They do not ship food because they believe that "we should all seek food closer to home, in our foodshed, our own bioregion" (Polyface, Inc.). Although initially Joel Salatin was not aware of Gaia theory, his farming methods are so intensely "systems-based" that he was invited to speak at an October 2006 conference outside Washington, DC, that centered on Gaia theory. His talk was one of the most popular presentations, and the question-and-answer session that followed extended for hours as attendees sought to draw parallels between our understanding of natural systems and agriculture. Michael Pollan featured Salatin's work in his book *The Omnivore's Dilemma* (2006), and juxtaposed it with farming techniques used by large-scale (industrial) corn and beef production. Experimentation and inquiry, such as those outlined above, should be ramped up in both the private and public sectors, for agriculture is essential to our own survival and has tremendous impacts on the living system as a whole.

## Conclusion

The land is alive both metaphorically and in a robust scientific sense. Gaia thinking allows us to apply this worldview to all aspects of human life. No matter what our endeavor—whether food production, energy choices, or general economic activity (including our modes of recreation and leisure) —we must not push the living system to new equilibrium points that are not conducive to human life and healthy ecosystems. Gaia theory can be the model and the metaphor that guides us through the twenty-first century's most pressing problems, letting us emerge with a greater understanding of ourselves and the Earth of which we are a part. As Elisabet Sahtouris (1989: 23) offered, "once we truly grasp the scientific reality of the Gaian organism and its physiology, our entire worldview and practice are bound to change profoundly, revealing the way to solving what now appear to be our greatest and most insoluble problems."

With a Gaian worldview we may be able to transcend misleading divides between disciplines, as well as transcend any false dichotomy between humans and nature. We can celebrate the incredible beauty of the Earth with a newfound sense of joy born of the realization that we belong to it. We can blend a powerful scientific understanding of our planet as a living entity with rediscovered metaphors and stories of our ancestors to best understand our relationship to our living planet and to promote decisions in a conservation state of mind.

## References

American Council on Renewable Energy (ACORE). 2007. *The Outlook on Renewable Energy in America: Joint Summary Report*, vol. 2. Washington, DC: ACORE.

Campbell, J. 1972. *Myths to Live By*. New York: Bantam Books.

Diamond, J. 2005. *Collapse: How Societies Chose to Fail or Succeed*. New York: Viking Penguin.

Dyson, F. 1988. *From Eros to Gaia*. New York: Pantheon Books.

Flannery, T. 2006. *The Weather Makers*. New York: Atlantic Monthly Press.

Flowers, B. S., ed. 1988. *The Power of Myth*. New York: Doubleday.

Harding, S. 2006. *Animate Earth: Science, Intuition and Gaia*. White River Junction, VT: Chelsea Green.

Havel, V. 1994. The miracle of being: Our mysterious interdependence. Acceptance speech for the Liberty Medal presented at Independence Hall, Philadelphia, July 4.

Joseph, L. E. 1990. *Gaia: The Growth of an Idea*. New York: Saint Martin's Press.

Kutscher, C. F. 2007. *Tackling Climate Change in the U.S.: Potential Carbon Emission Reductions from Energy Efficiency and Renewable Energy by 2030*. Boulder, CO: American Solar Energy Society.

Land Institute. www.landinstitute.org.

Lenton, T. M. 2004. Clarifying Gaia: Regulation with or without natural selection. In S. T. Schneider, J. R. Miller, E. Crist, and P. J. Boston, eds., *Scientists Debate Gaia*. Cambridge: MIT Press, pp. 15–25.

Leopold, A. 1979. Some fundamentals of conservation in the southwest. *Environmental Ethics* 1: 131–41.

Lotka, A. J. 1956. *Elements of Mathematical Biology*. New York: Dover.

Lovelock, J. E. 1979. *Gaia: A New Look at Life on Earth*. Oxford: Oxford University Press.

Lovelock, J. E. 1991. *Healing Gaia: Practical Medicine for the Planet*. New York: Harmony Books.

Lovelock, J. E. 2000. *Gaia: The Practical Science of Planetary Medicine*. Oxford: Oxford University Press.

Lovelock, J. E. 2004. Reflections on Gaia. In S. T. Schneider, J. R. Miller, E. Crist, and P. J. Boston, eds., *Scientists Debate Gaia*. Cambridge: MIT Press, pp. 1–5.

Lovelock, J. 2006. *The Revenge of Gaia: Earth's Climate Crisis and the Fate of Humanity*. New York: Basic Books.

Margulis, L. 1998. *Symbiotic Planet: A New Look at Evolution*. New York: Basic Books.

Margulis, L., and D. Sagan. 1997. *Microcosmos: Four Billion Years of Microbial Evolution*. Berkeley: University of California Press.

Margulis, L., and D. Sagan. 2002. *Acquiring Genomes: A Theory of the Origins of the Species*. New York: Basic Books.

McCluney, R. 2003. *Renewable Energy Limits*. Cocoa: University of Central Florida.

McKibben, B. 2007. *Deep Economy: The Wealth of Communities and the Durable Future*. New York: Holt.

NASA. http://www.nasa.gov/centers/glenn/about/history/apollew.html.

Polyface, Inc. http://www.polyfacefarms.com.

Pollan, M. 2006. *The Omnivore's Dilemma: A Natural History of Four Meals*. New York: Penguin Press.

Sahtouris, E. 1989. *Gaia: The Human Journey from Chaos to Cosmos*. New York: Pocket Books.

Schwartzman, D. W., and T. Volk. 2004. Does life drive disequilibrium in the biosphere? In S. T. Schneider, J. R. Miller, E. Crist, and P. J. Boston, eds., *Scientists Debate Gaia*. Cambridge: MIT Press, pp. 129–35.

Thomas, L. 1974. *The Lives of a Cell: Notes of a Biology Watcher*. Toronto: Bantham Books.

Tickell, C. 2004. Gaia and the human species. In S. T. Schneider, J. R. Miller, E. Crist, and P. J. Boston, eds., *Scientists Debate Gaia*. Cambridge: MIT Press, pp. 223–27.

Vernadsky, V. I. 1998. *The Biosphere*. New York: Copernicus.

Volk, T. 1998. *Gaia's Body: Towards a Physiology of Earth*. New York: Springer.

Volk, T. 2003. Seeing deeper into Gaia theory. *Climate Change* 57: 5–7.

# 17

# Neocybernetics of Gaia: The Emergence of Second-Order Gaia Theory

Bruce Clarke

Systems theory often seems counterintuitive. The problem is not with the behavior of systems but with the conceptually antiquated nature of our intuitions. For instance, typically "negative" stands to "positive" as deleterious stands to desirable. In the operation of systems, however, negative functions can be desirable and positive ones deleterious. Take feedback: negative feedback generally produces beneficial self-regulation, positive feedback destructive runaway amplification. Closely related to circular functions such as feedback is a distinction between "openness" and "closure." Most of us are politically programmed to laud all things "open" and shun that which is "closed." But when it comes to the self-regulation of systems through negative feedback, only a "closed loop" will do. Again, "top-down" typically connotes a dictatorial, hierarchical, or undemocratic power structure, whereas "bottom-up" connotes participatory and egalitarian arrangements. However, in a wider analysis of systems, "top-down" names a holistic perspective attuned to emergent behaviors and protective of the integrity of what is being observed, whereas "bottom-up" names a reductionist perspective that takes things to pieces and considers them to be nothing more than the sum of their parts.

## Gaia as System

As a systems theorist of global proportions, James Lovelock is still misperceived, taken as erroneous because counterintuitive. "At last, but maybe too late," Lovelock (2006: 8) writes in *The Revenge of Gaia*, "we begin to see that the top-down holistic view, which views a thing from outside and asks it questions while it works, is just as important as taking the thing to pieces and reconstituting it from the bottom up."

Gaia theory has gathered biology, geology, geochemistry, geophysics, and meteorology into a mature systems science that contains, while surpassing, the reductivist scientific programs dominant since the seventeenth century. It is not that Lovelock is responsible for the rise of the systems paradigm, which has its roots in multiple developments that coalesced in the emergence of cybernetics at mid-twentieth century. But Gaia theory draws systems theory to a millennial head with global environmental consequences.

The Gaia concept coalesced in the 1970s at the intersection of two now-classical streams of systems theory—the thermodynamics of mechanical and natural systems first developed in the mid-nineteenth century, and the cybernetics of self-regulating control systems first developed in the mid-twentieth century. "There is little doubt that living things are elaborate contrivances," Lovelock and Margulis (1974: 3) wrote in one of their first co-authored papers. "Life as a phenomenon might therefore be considered in the context of those applied physical sciences that grew up to explain inventions and contrivances, namely thermodynamics, cybernetics, and information theory." The Gaian system was originally conceived as a natural contrivance produced in the co-evolution of the biota with their abiotic environment. Lovelock's earliest presentations of Gaia as a control system were fully informed by the engineering discourse of homeostatic feedback mechanisms.

"The primary function of many cybernetic systems is to steer an optimum course through changing conditions towards a predetermined goal," Lovelock (1979: 48) wrote in chapter 4, "Cybernetics," of his first book on Gaia.[1] His first example of self-regulation around a set point was drawn from the organic or physiological side of the classical cybernetic metaphor heuristically equating organisms and machines. It is *proprioception*—the bodily or neurological faculty of homeostatic self-perception applied to the maintenance of locomotive balance. In this analogy Gaia is the proprioceptive or internal self-balancing feedback system of the biosphere as a planetary body. Lovelock then reverted to the mechanical realm to offer the homeliest of cybernetic analogies: Gaia in operation performs like the thermostat of a kitchen oven. The Earth is the oven box, the Sun is the heat coil, and Gaia is the regulator that keeps the temperature at an "optimum."

This chapter reviews and assesses several different dialects of systems theory as they have been brought to bear on the discourse of Gaia. My main interest will center on the further development of cybernetic systems theories, particularly as these discourses first approach and then overcome the heuristic equation of mechanical contrivances and biological systems. An important but under-recognized history will be related here, one that plays a crucial role for the biological side of Gaia theory. Tracing this history will unfold the benefits of marking a distinction between *first-order* and *second-order* cybernetics. Any contemporary discussion of systems theory is incomplete without taking this distinction into account. What we learn from this finer history, to begin with, is that despite its silicon mainstreaming as all things computer-scientific, the concept of cybernetics is properly delimited neither to its point of origin at the machine–organism interface, nor to its nominal relations with later computational developments and their popularizations. Heylighen and Joslyn (2001) note how certain systems theorists "felt the need to clearly distinguish themselves from these more mechanistic approaches, by emphasizing autonomy, self-organization, cognition, and the role of the observer in modeling a system. In the early 1970s this movement became known as *second-order cybernetics*."

## Second-Order Cybernetics: Autopoiesis

The notion of recursion, or circular causality, was present in the earliest phases of cybernetic thinking around feedback, homeostasis, and related mechanisms for systemic self-correction.[2] In first-order or classical cybernetics, circular functions are instrumental for the self-regulation of the system, but the system in question may not be wholly recursive. Like the circular operation of the governor of a linear (input–output) steam engine, or the thermostat of an electric oven, the feedback mechanism may be a contrivance coupled onto a larger, more straightforward system.[3] As with any mechanical contrivance, for instance, an oven is heteronomous: the input to and outcome of the internal control is determined outside the system, by another, external system, or by the environment to be controlled. The particular temperature of an operating oven is first set by a user and only then maintained within range of that set point by the thermostat.

Simply put, first-order cybernetics is about control; second-order cybernetics is about autonomy. Second-order cybernetics presses the analysis of recursive processes beyond mechanical and computational control processes toward the formal autonomy of natural systems. Unlike a thermostat, Gaia—the biosphere or system of all ecosystems—sets its *own* temperature *by* controlling it. A decade after Lovelock's first Gaia book, Margulis and Lovelock (1989) restated the Gaia concept accordingly: "Cybernetic systems are 'steered'; biological cybernetic systems are steered *from the inside*. ... The Gaia hypothesis postulates a planet with the biota actively engaged in environmental regulation and control *on its own behalf*" (9, 11; emphasis added). In second-order parlance, Gaia has the operational autonomy of a self-referential system. Second-order cybernetics is aimed, in particular, at this characteristic of natural systems where circular recursion *constitutes the system* in the first place. This finer appreciation of recursive self-constitution refines systemic observation: natural systems—both biotic (living) and metabiotic (superorganic, psychic, and social)—are now described as at once *environmentally* open (in the nonequilibrium-thermodynamic sense) and *operationally* (or organizationally) closed, in that their dynamics are autonomous, that is, self-maintained and self-controlled.

This recursive interplay of external openness and internal closure is precisely the burden of *autopoiesis* as Humberto Maturana and Francisco Varela brought that concept forward at the outset of second-order cybernetics. Evan Thompson (2004: 389) has recently rehearsed the complex coupling of openness and closure at the basal level of the biological autopoiesis of the living cell: "Metabolism is none other than the biochemical instantiation of the autopoietic organization. That organization must remain invariant, otherwise the organism dies, but the only way autopoiesis can stay in place is through the incessant material flux of metabolism. In other words, the operational *closure* of autopoiesis demands that the organism be an *open system*" (italics in the original). And it is Margulis who has most directly brought autopoietic theory into Gaian science, to the extent of presenting Gaia as "the autopoietic planet": "The biosphere as a whole is autopoietic in the sense that it maintains itself.... In our view, autopoiesis of the planet is the aggregate, emergent property of the many gas-trading, gene-exchanging, growing, and evolving organisms in it" (Margulis and Sagan 2000: 20, 23).

Lovelock himself seldom mentions autopoiesis.[4] An admirable participant in that main line of engineering discourse, Lovelock's own Gaia discourse has continued to develop along primarily first-order cybernetic lines. As will be suggested later, Margulis's expositions of the Gaia concept through autopoietic systems theory mark a generational shift between first-order and second-order cyberneticians. Occasionally she lapses into an idiom that restricts cybernetics to mechanical or computational applications.[5] Nevertheless, by bootstrapping the concept of autopoiesis to her work on symbiosis and environmental evolution, Margulis has complemented and extended Lovelock by taking the science of Gaia on a distinctly second-order or neocybernetic path. To bring autopoiesis up to the level of Gaia, moreover, is to bind the biological microcosm to the geophysiological macrocosm in a positively fractal or holographic way, which is to say that in this vision of life on Earth, isomorphic structures and operations *recur* at many different scales. Meditating on Margulis's presentations of the bacteria spirochetes in symbiotic association with the eukaryotic protist *Mixotricha paradoxa*, William Irwin Thompson (1998: 30) set this vision down as an imbricated form of multidimensional recursion: "So we have a nested universe: the spirochete is in the protist, the protist is in the termite, the termite is in the log, the log is in the forest, the rain forest is in Gaia, and Gaia is inside the solar system, and on and on it goes, and where it stops, nobody knows."

## Autopoiesis in the Thinking of Lynn Margulis

Autopoietically considered as a self-referential cross-coupling of environmental openness and operational closure, every cell is a little Gaia, and Gaia, in its planetary autonomy—as intimated *avant la lettre* by Lewis Thomas in the lead article collected in *The Lives of a Cell*—"is *most like a single cell*" (1974: 4). Also appearing in 1974, the same year as Lovelock's and Margulis's first co-authored Gaia papers in *Icarus* and *Tellus*, was the first English-language publication of the concept of autopoiesis:

The autopoietic organization is defined as a unity by a network of productions of components which (i) participate recursively in the same network of productions of components which produced these components, and (ii) realize the network of productions as a unity in the space in which the components exist. Consider, for example, the case of a cell: it is a network of chemical reactions which produce molecules such that (i) through their interactions [they] generate and participate recursively in the same network of reactions which produced them, and (ii) realize the cell as a material unity. (Varela et al. 1974: 188)

*The Tree of Knowledge* gives a diagram of autopoietic recursion in the cell, emphasizing that this systemic organization is self-binding. As the name was coined to signify, an auto- (self-) poietic (making) system is the product of its own production—a production that can occur only because it produces for itself the conditions (here, a semi-permeable membrane) that create and maintain the operational closure that ensures the autonomy of the process (here, cellular metabolism). This is a paradigmatic case of the second-order circumstance in which recursion constitutes the system (see figure 17.1).

By the time Margulis and Sagan (1986a) coauthored their first collaborative work of scientific popularization, *Microcosmos*, the concept of autopoiesis was a prominent part of her biological vocabulary: "To be alive, an entity must first be *autopoietic*—that is, it must actively maintain itself against the mischief of the world.... This modulating, 'holistic' phenomenon of autopoiesis, of active self-maintenance, is at the basis of all known life" (1986a: 56). As often in Margulis and Sagan's later reprises of the concept, the emphasis here was more, in Maturana and Varela's terms, on the *unity* of the *network*—the self-maintenance of the autopoietic identity—than on the recursive operationality that sustains it. In this passage that aspect of autopoietic dynamics was implicitly tucked into the scare-quoted term "holistic."

In *Microcosmos* the matter of autopoietic recursion became explicit only in the final chapter, "The Future Supercosm," at the same time that, while speculating about the possibility of taking terrestrial life successfully into extraterrestrial environments, Margulis and Sagan elicited their most extensive rehearsal of the Gaia concept. This passage began with a sketch of the classical mechanistic physics to be superseded by the neocybernetic autopoietic view. The gist of the contrast was that classical-physical views of life based on the science of Descartes and Newton were *linear*, and to the detriment of both popular and professional scientific ideas, this linear bias or hangover was still the case with a lot of the mechanistic (or first-order) cybernetics and information theory then fashionably being applied to living systems through "computer-age analogies: amino acids are a form of 'input,' RNA is 'data-processing,' and organisms are the 'output,' the 'hard copy' controlled by that 'master program,' that 'reproducing software,' the genes" (1986a: 264).

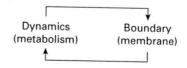

**Figure 17.1**
Autopoietic recursion in the cell. Image from Maturana and Varela (1998).

Autopoiesis first entered this section of the discussion precisely as a counterpoint to mainstream mechanistic control-oriented cause–effect or input–output linearity: "In this book we have held to a somewhat different and more abstract view....Life, a watery, carbon-based macromolecular system, is reproducing autopoiesis. The autopoietic view of life is circular" (1986a: 264). Then the Gaia hypothesis was offered as an example of the autopoietic view, in which the environmental "membrane" of the atmosphere is in co-dependent evolution with the "metabolism" of the sum of the biota: "Lovelock sees life best represented by a self-supporting environmental system which he calls Gaia..., the superorganismic system of all life on Earth" (1986a: 265). This way Margulis interlocked Gaian and autopoietic circularity: "On earth the environment has been made and monitored by life as much as life has been made and influenced by the environment....The biota itself, which includes *Homo sapiens*, is autopoietic. It recognizes, regulates, and creates conditions necessary for its own continuing survival" (1986a: 265–66).

In "Autopoiesis: A Review and a Reappraisal," Pier Luigi Luisi (2003: 49) discusses "why the theory of autopoiesis had, and still has, a difficult time being accepted into the mainstream of life-science research." Even though "eminent biologists such as Lynn Margulis accepted it as an integral part of the description of the living," and even given "the impact of autopoiesis in the social sciences... [t]he term is neither well-known nor frequently cited" (2003: 50).[6] The prominence Margulis has given to the concept of autopoiesis, then—not only in her particular expositions of Gaia theory but running the ecological gamut from the cell to the biosphere—bears examination.[7] It further reinforces her fabled maverick status as a scientist willing to stake her reputation on controversial ideas (McDermott 1991). But it also underscores her conceptual consistency, for Gaia and autopoiesis are both profoundly systems-theoretical concepts. If in Lovelock's main conception Gaia is a planetary instantiation of first-order cybernetic homeostasis, in Margulis's treatment as the autopoietic planet, the Gaia concept is unfolded into second-order cybernetic self-referential recursion. But how did it come about that Margulis incorporated the concept of autopoiesis into her biological and Gaian discourses in the first place? In fact her ties to autopoiesis have been complex. But they are also revelatory and thoroughly interwoven with a collegial network developing out of the early publication of the Gaia hypothesis, not just in the scientific press but also in the pages of a countercultural (but really uncategorizable) intellectual journal, *CoEvolution Quarterly*.

## Gaia and Second-Order Cybernetics in the *CoEvolution Quarterly*

In 1974 Stewart Brand spun *CoEvolution Quarterly* (or *CQ*) off from the first incarnations (1968–1971) of the *Whole Earth Catalog* (or *WEC*). Every issue of the *WEC* began with a section titled "Understanding Whole Systems." Every number of *CQ* contained one as well, occasionally shortened just to "Whole Systems." From the late 1960s on, these remarkable publications were the virtual house organ on the world stage for the popular discussion of the breadth of cybernetic complexities. Put another way, beginning in the planetary banner year of 1968, Brand's Whole Earth publishing collectives were Gaian before Gaia was cool. In a way it is shocking to look back at these publications from the present moment of our gathering planetary emergency and see so many superb thinkers laying out detailed global ecological perspectives on local practices for what they perceived then as *their* emergency—explicit premonitions of global warming, the imminence of environmental devastation by nuclear war, rampant monoculture, and/or unsustainable population explosion—and to admit that they already had it just about right. The ecological sciences explored, the technological and political solutions debated, and the cultural and spiritual practices recommended there have hardly aged a bit, if at all. They are as relevant as ever to our current abysmal quandaries.

A case in point is our particular topic—Gaia, the hypothesis from which it originated, the theoretical variants into which it has evolved, and the wider cultural ramifications taken from the whole of these developments. What follows is an abbreviated sampling of the contents of *CQ* as they pertain to the emergence there of interrelations among Gaia, autopoiesis, and second-order systems theory. To begin with, appearing several times in the early pages of *CQ* were various items associated with Heinz von Foerster. Founder and director of the Biological Computer Laboratory at the University of Illinois from 1957 to 1975, von Foerster coined the phrase "second-order cybernetics" in the early 1970s. Its advent marked the moment when cybernetic theory explicitly factored self-reference into its own discourse, precisely as a "cybernetics of cybernetics," that is, as an effort to observe (at "second-order") its own processes of system observation. A formidable authority in cybernetic matters, von Foerster's relevance to this narrative will soon become clear.

As the purveyor of the *Whole Earth Catalog*, clearly Stewart Brand was primed to respond to a bona fide cybernetic systems theory taking the Earth altogether as a "whole system." For the sixth number of CQ, appearing in Summer 1975, Brand arranged for the first presentation of the Gaia hypothesis in a nonspecialist journal. He observed in his head note that this serious scientific fare might appear anomalous in such a venue, to the detriment of the authors: "Margulis and Lovelock will doubtless take some flak for appearing in suspect company— condom evaluations, poetry, and such." But its inclusion was an inspired intervention on the part of all concerned. "The Atmosphere as Circulatory System of the Biosphere: The Gaia Hypothesis" (Margulis and Lovelock 1975) led its reader into the topic with a seventeenth-century engraving and a discussion of Harvey's demonstration of the circulatory system of the body, as an analogy for the atmosphere's Gaian role in relation to the planetary "body." Fatefully, however, Brand (or someone) decided to append to this more popular article a separate section titled "Gaia and cybernetics," an excerpt from a more technical piece that had appeared the year before, "Atmospheric Homeostasis by and for the Biosphere: The Gaia Hypothesis" (Lovelock and Margulis 1974), invoking first-order cybernetics and giving mathematical formulas for the application of Shannon and Weaver's information theory to the thermodynamics of living systems.

In the next, seventh number (Fall 1975), an entire page of CQ was devoted to a long letter to the editor from von Foerster, asserting defects in the mathematical formulas offered in "Gaia and cybernetics." Von Foerster's response was in no way dismissive of the main ideas in Margulis and Lovelock's presentation. His tone was collegial and colloquial: "I have no way to find out who is to be charged with these booboos: CQ who misprints Lovelock and Margulis; Lovelock and Margulis who misquote Denbigh (1951) and Evans (1969); or Denbigh and Evans who misunderstand. But this is not my job." Indeed, "I found Lovelock's and Margulis's ideas too important to see them becoming vulnerable because of deficiencies of a different kind. As a comment on their—or anybody else's—classification of Life, I suggest that you reproduce 'Autopoiesis: The Organization of Living Systems, its Characterization and a Model'" (von Foerster 1975). Clearing incidental away from important matters, von Foerster's constructive criticism is to my knowledge the first and original suggestion of a relation between the cybernetics of Gaia and the theory of autopoiesis as a description of the operational organization of living systems.[8]

The following summer *CQ* published a long interview, "On Observing Whole Systems," with the co-inventor of the concept of autopoiesis, Francisco Varela. In this I detect the continuing hand of von Foerster, who had a longstanding relationship with Stewart Brand, and an even longer one with Maturana and Varela. In his head note to the Varela interview, Brand rehearsed von Foerster's own articulation of the distinction between first-order and second-order cybernetics: "This sounds abstruse, but I share the opinion of Ludwig Wittgenstein, Gregory Bateson, G. Spencer Brown, Heinz von Foerster and others that failure to understand self-reference is the poison in the brain of most Western misbehavior, public and personal. In his recent landmark paper, 'A Calculus of [sic] Self-Reference' [Varela 1975] and in this interview, Francisco is helping build what von Foerster calls 'a cybernetics of *observing*-systems,' which is the rest of the story after 'the cybernetics of *observed*-systems'—feedback, goal-seeking, and such" (Varela with Johnson 1976: 26).

First-order cybernetics is "the cybernetics of observed systems," that is, of objects such as natural or technological systems, while second-order cybernetics is "the cybernetics of observing systems," that is, of subjects, the cognitive systems capable of producing observations in the first place. In *Autopoiesis and Cognition* (1980) Maturana and Varela make their definitive case for considering autopoietic systems, such as living cells, as *cognitive*—not merely as observed but more fundamentally as *observing* systems producing life-maintaining self-making cognitions of their environments. One inference following from the recursive logic of observing systems in second-order systems theory is that the traditional distinction between objects and subjects is untenable, or more to the point, ungrounded. For instance, we cannot really look at Gaia as a planetary whole without looking, self-referentially, at ourselves, a part of Gaia, looking at Gaia. "Objectivity" is surpassed by participation.

Varela returned to the pages of *CQ* a year later with the publication of selected materials presented at the 1976 Mind-Body Dualism Conference organized by Gregory Bateson and attended by von Foerster and Varela. In following years *CQ* printed further articles separately on, and some written separately by, Lovelock and Margulis. In 1981 it published one of the first major critiques of Gaia theory, W. Ford Doolittle's review-essay on *Gaia: A New Look at Life on Earth* (Lovelock 1979), "Is Nature Really Motherly?" along with responses from both Lovelock and Margulis. At the end of this run of materials on Gaia and second-order cybernetics was a 1981 Lovelock article, "More on Gaia and the End of Gaia." Brand's head note reads: "The Gaia Hypothesis, you may recall, is the notion proposed by British chemist Lovelock and American microbiologist Lynn Margulis, that the chemical composition of the Earth's atmosphere is a highly subtle buffering device maintained by all of the planet's life—making the Earth as a whole in effect one life. The following recent thoughts by Lovelock on the subject are the closing third, or so, of a talk he gave at The Lindisfarne Fellows Conference, June 4, 1981" (Lovelock 1981: 36).

## Gaia at Lindisfarne

From the mid-1970s to the mid-1980s, mixed in with *CQ*'s occasional presentations of the science of Lovelock and Margulis, von Foerster and Varela, were occasional references to cultural historian William Irwin Thompson. Begun within a year of the start of *CQ*, his brainchild the Lindisfarne Association has been gathering notable workers in the arts and humanities, in politics and economics, in green technologies and ecological sciences, with "the esoteric teachings and practices of the great universal religions" (summer 1974: 130). Such holistic purposes ran parallel, most of the time, with the ecological vision of *CQ*, particularly in its support for second-order cybernetic epistemological issues and systems thinkers such as von Foerster and Bateson. Thompson relates that it was in the pages of *CQ* that he first came to hear about the authors of the Gaia hypothesis and the work of Varela (personal communication). As Lindisfarne took on form throughout the 1970s, Bateson became its first scholar-in-residence in 1977, followed by Varela in 1978 and 1979. A roster of Fellows developed from which an annual, moveable Fellows Conference has been drawn.[9] Within this milieu, for several decades Thompson himself has been elaborating provocatively on the political, economic, and cognitive implications of Gaia for a "planetary culture."

*CoEvolution Quarterly*, then, left an appreciable imprint on the Lindisfarne group. Moreover, I would suggest, the Lindisfarne Association shaped the primary collegial milieu within which Margulis absorbed—in relation to Gaia theory—second-order cybernetics in general and the concept of autopoiesis in particular. From the Lindisfarne Fellows meetings of 1981 and 1988 Thompson published two Gaia-centered essay collections. *Gaia—A Way of Knowing: Political Implications of the New Biology* (Thompson 1987) gathered papers from the 1981 event. The first six chapters comprised an international tour of predominantly biological systems theory before and after von Foerster (who also spoke at the meeting) and his Biological Computer Laboratory, in the following order: Bateson (posthumously), Varela, Maturana, Lovelock, Margulis, and Henri Atlan.[10] However, the most sustained discussion of autopoiesis in this volume was not biological but socioeconomic, "Gaia and the Politics of Life," section II, "Toward an Autopoietic Economy," treating the proliferation of "shadow economies" as emergent autonomous formations redolent of the Gaian interconnectedness and evolutionary mobility of microbial symbioses (Thompson 1987). And on the evidence of *Microcosmos*, which we reviewed above, and also of the more technical and focused volume *Origins of Sex*,[11] both first published in 1986, it is clear that Margulis had appropriated the concept of autopoiesis for her biological and Gaian thinking and writing by the mid-1980s.[12]

Meanwhile the repercussions of the Gaia concept continued at Lindisfarne, as documented by *Gaia 2: Emergence, The New Science of Becoming* (Thompson 1991) drawn from the May 1988 Fellows meeting. Thompson's introduction for this second Gaia volume made explicit and emphatic his envisioning of Gaia as a planetary myth or global imaginative structure for a new world culture in the making, based on the specific convergence of "Lovelock, Margulis, and Varela" (Thompson 1991: 22). However, between the late 1980s recorded by this second volume and the early 1980s of its predecessor, a new set of systems discourses, somewhat forgetful of their cybernetic incubations, had burst upon the scene, dynamical systems theory and complexity theory, developments encapsulated under the names of "chaos theory" and "emergence."[13] Capturing the ferment of this new conceptual yeast in the Lindisfarne dough is the transcript of a vigorous concluding symposium (Thompson et al. 1991). This long and spirited conversation begins with Varela's lengthy assessment and critique of Lovelock's Gaia theory at the point to which it had then arrived, especially with his recent addition of the Daisyworld computer models.[14] I will excerpt Varela's tour de force of scientific conversation at some length, on two points in particular, for it provided a definitive second-order cybernetic perspective on Lovelock's first-order orientation.

Varela's first point was absolutely crucial. It concerned Lovelock's continued use of phrasings that hypostatize the "life" of Gaia. Varela implicitly suggested for that complex coupling of biotic and abiotic component systems, instead, an adaptation of the discourse of autopoiesis centered on the operational autonomy of *meta*biotic systems:

Jim has made it very clear...that Gaia cannot be described as other than having the quality of life....But it seems to me that this difficult issue can perhaps be helped and clarified by making a distinction....It is the difference between being alive, which is an elusive and somewhat metaphorical concept, and a broader concept, which is perhaps easier to tackle, that of autonomy. The quality we see in Gaia as being living-like, to me is the fact that it is a fully autonomous system...whose fundamental organization corresponds to operational closure....

Operational closure is a form, if you like, of fully self-referential network constitution that specifies its own identity....Autonomy, in the sense of full operational closure, is the best way of describing that living-like quality of Gaia, and...the use of the concept of autonomy might liberate the theory from some of the more animistic notions that have parasitized it. (Thompson et al. 1991: 211)

Margulis and Sagan (2000: 20) have been echoing Varela's point in the way that their remark in *What Is Life?* stresses Gaia's participation in, rather than identity with, the form of life per se: "The biosphere as a whole is autopoietic in the sense that it maintains itself....As an autopoietic system, Gaia therefore shares an essential quality with individual living systems." Although Varela would not have put it precisely this way, the recognition that there are metabiotic modes of autonomy based on autopoietic closure—that autopoiesis broadly considered describes a general mode of systemic self-reference, *one* form of which is biological—resolves the central problem with the overly "strong" form of the Gaia concept.[15]

Varela's second point was more problematic. He critiqued the heuristic limitations of the Daisyworld model in claiming that its feedback mechanism is "linear." By this I take Varela to have meant that the operational *outcome* of a first-order cybernetic feedback loop may be "linear" in the sense of hovering around an invariant homeostatic set point. Varela's second-order thrust concerned the complex adaptability of Gaia's ongoing emergence as a globally distributed network of systems, a planetary network that, like an immune system, continues to learn on the job:

Daisyworld, in the best tradition of feedback engineering, which Jim has referred to, is not the same thing as a fully plastic network, that is, a network which has some way of changing itself....Here there is a distinction between a single, linear feedback mechanism, or circumstance where you have one, two, or three feedback loops, and a network. A whole bunch of feedback mechanisms added together does not amount to the same thing as a network, for a network has a distributive quality and has its own dynamic....So I propose, I hope not too boldly to its own inventor, that the best model for Gaia is not one of the old tradition of feedbacks added together, but one of a fully distributed network....In the same way that you will not get a cell by just adding together the regulatory circuits of enzymes and substrates, you will not get Gaia out of the regulatory circuits of Daisyworld. I believe that one will not have a fully convincing argument for Gaia until the full plastic network qualities of Gaia become apparent. For then, you see, you will actually be able to put your finger on the learning capacity of Gaia to show just how it becomes adaptive. (Thompson et al. 1991: 212)

After a fair amount of further exchange between Varela and Lovelock, Thompson's discreet moderation articulated once more the distinction this essay has been working to clarify, between second-order cybernetic constitutive recursion and the first-order homeostatic paradigm. Speaking directly to Varela, Thompson remarks:

I see your comment on Jim's talk as a generational development. The first and founding generation of cybernetics, names associated with the famous Macy Conferences, such as McCulloch, von Neumann, von Foerster, and Bateson, gave us basic concepts for systems guidance and correction, the feedbacks you're talking about. Now your generation comes along with its connectionist language of "Net Talk," "Hopfield neural nets," or your own "autopoiesis," and says, "Our generation wants to take it another step, from feedback to the metadynamics of the system as a learning one." (Thompson et al. 1991: 214–15)

Margulis's Gaia discourse has not gone as far as Varela did in this direction of distributed cognition. It still occupies a conceptual position mediating Lovelock's first-order and Varela's second-order orientations. It appears that a paper Margulis co-authored with Lovelock, "Gaia and Geognosy" (published a year after the 1988 Lindisfarne Fellows meeting), represents their defense of Lovelock's cybernetic idiom against Varela's critique, although with a terminological refinement indicating that Gaian self-regulation, in itself *and* as modeled by Daisyworld, is *homeorrhetic* as well as homeostatic. Margulis and Lovelock (1989: 10, 27) explained: "The Gaian system, unlike any engineered one, is not controlled by a 'steersman' or governor from the outside. Rather, like any living system, Gaian control is homeorrhetic. Set points change through time: the apparent stability is dynamically maintained by organization inside the system itself." Nonetheless, self-maintenance under organizational closure "inside the system itself" is another way to say autopoiesis. And in its autopoietic commitments, as we have already examined, Margulis's Gaia discourse also joins with Varela's biological metadynamics.[16]

In *What is Life?* the upshot of the merger of Gaian autopoietic closure and symbiogenetic openness is the recognition that "organisms are less self-enclosed, autonomous individuals than communities of bodies exchanging matter, energy, and information with others" (Margulis and Sagan 2000: 23). From the bacterial bioplasm onward, *pure* individuality does not exist: autonomy does not equal isolation. And the sym- in symbiosis stands not just for the enduring aggregation of different living systems but for the coupling of life altogether with its abiotic thermodynamic environments and, eventually, its metabiotic elaborations of psychic and social observing systems, which are themselves built up from life's own self-referential sentience. In drawing these various conceptual lines together, the following passage quietly lays in the breadth of the second-order systems-theoretical framework of Margulis's Gaia: "Mind and body, perceiving and living, are equally self-referring, self-reflexive processes already present in the earliest bacteria. Mind, as well as body, stems from autopoiesis" (2000: 31). From the Gaian perspective of sentient biospheric interconnectedness, a level to which Maturana and Varela never take the concept of autopoiesis, the paradox of operational boundaries that are at once both thermodynamically open and organizationally closed is resolved in the form of a global *environment* that is also and in itself, not precisely a *living* system but a higher order autopoietic consortium of living autopoietic systems.

## Neocybernetic Mythopoesis

Second-order systems theory in the key of Gaia opens a cognitive passageway between hard science and communal vision. At the end of *What is Life?* Margulis and Sagan (2000: 218) articulate their own vision of a science-inspired postreligious spirituality: "The facts of life, the stories of evolution, have the power to unite all peoples.... The most meaningful story of existence for future humanity is more likely to come from the evolutionary worldview of science than from Hinduism, Buddhism, Judeo-Christianity, or Islam. The dual understanding of scientific enquiry and creation myth could become a single view: a science tale rich both in verifiable fact and personal meaning."

But even as recently as 1995, when these sentiments were first published, one was not unduly uncertain whether there would even *be* a "future humanity" to worry about such things. Stephan Harding's 2006 volume *Animate Earth: Science, Intuition and Gaia* is composed in full confrontation with the dire climate trends that have been confirmed beyond reasonable doubt only in the last decade. In this context Gaia theory becomes much more than a systems-theoretical curiosity, it becomes a lifeline to the very possibility of a future. Broadly parallel with Margulis and Sagan's vision of neocybernetic mythopoesis, Harding's (2006) concern throughout *Animate Earth* is to balance the explanation of scientific information, "cycles and feedback loops," with the ethical understanding born of "intimacy and connection with what has been explained" (2006: 13).

*Animate Earth* offers a powerful development of Lovelock's brand of Gaian systems science. Gaia is precisely, like any individual organism but more so, a supersystem, a system of systems of systems. This is both hard-headed natural science and the key to the way that all things human—from individual minds and feelings to every artifact of social commerce and communication—are always already more than human, are looped into the global and cosmic whole. Harding illuminates the science of emergence (cycles, feedback loops, structural couplings, emergent behaviors arising out of nested interdependencies) specifically as that science is exemplified by Gaian interrelationships—the carbon cycle, the sulfur cycle, the phosphorus cycle, the atmospheric and oceanic loops that course through the geological continents and play upon the tectonic currents. He also provides meditative instructions guiding the Gaian imagination through every loop of each complex cycle, and as a result, systems science comes out of discursive abstraction and takes on flesh and blood.

With global ecological disaster closer than anyone had thought up till the end of the last millennium, it may seem beside the point to quibble about the epistemological niceties of anyone's visionary stance toward a possible future. Nevertheless, I would suggest that the force of Harding's evocations of Gaian sanctity are constrained by his standing as yet on the first-order cybernetic or Lovelockian side of the Lindisfarne debate over autopoietic metadynamics as opposed to homeostatic correction. Harding (2006: 30) invests in a phenomenological rhetoric of immediacy, whereby "sensation and intuition are perceptive in that they make us aware of what is happening without interpretation or evaluation." From the standpoint of autopoietic systems theory, key developments in the same systems sciences that provide access to Gaian insights also indicate, on the contrary, the operational closure of any observing system. Sensation, perception, intuition—these are in all cases internal constructions of psychic systems for which the only possible kinds of relations to their environments and to the other systems they contain are *mediated* ones, and the structure of those mediations is such as ineluctably to "interpret or evaluate" whatever noumena manage to arise in our awareness as phenomenological constructions.

Harding uses meditative and literary techniques to embody and enrich the cognitive absorption of scientific information, a confluence of astute social communication and guided epistemological construction, for which the crucial consideration is not the absence (which is not possible) but the *quality* of the mediations at work. So I think that the mythopoetic force of Harding's Gaian panpsychism would ultimately be better served by couching itself within the epistemological constructivism developed in the second-order systems theories associated with Gregory Bateson, Heinz von Foerster, Francisco Varela, and Niklas Luhmann. As a systemic totality of the biota, rocks, air, and seas, one can now state with conceptual precision, the biosphere is neither literally "alive" nor merely figuratively "life-like": it is *autopoietic*—operationally closed, environmentally open, structurally coupled, and complexly interpenetrated. Through this conceptual refinement Gaian science provides both the hard explanation *and* the visionary understanding of systematic global interconnectedness through which the grim predictions of current climate models gain both rational conviction and persuasive force.

For the crisis we are already in right now, only a formidable cultural change of heart will do. And yes, while the right kind of science is indispensable, it is also not enough just to know it, we have to feel it. For that, Harding's particular constructions of ecological epiphany do have the merit of a certain traditional familiarity. As Harding winningly puts it, we need to be *Gaia'ed*. But I would add, we humans need to be posthumanized, and for that, autopoietic systems theory is an indispensable discourse (see Clarke 2008). We need to be interpenetrated spiritually *and* conceptually with the more-than-human geological and biological "cycles and feedback loops" that keep the Earth fit for life as we know it.

## Notes

My thanks to Lynn Margulis for hosting an early version of some of these arguments in her graduate seminar at the University of Massachusetts/Amherst. I have incorporated many helpful suggestions from Eileen Crist, Stephan Harding, Hans-Georg Moeller, Steven Norwick, Dorion Sagan, and William Irwin Thompson. Any errors left standing are strictly my own.

1. Gaian homeostasis is further codified in its Daisyworld computer simulations. Lovelock remarks in a documentary interview, "To my delight, Daisyworld turned out to be a most magnificent thermostat. So good that I thought of patenting it for engineering purposes" (Suzuki 1991).

2. The inaugural 1946 meeting of the famed Macy conferences on cybernetics was titled "Feedback Mechanisms and Circular Causal Systems in Biological and Social Systems." See www.asc-cybernetics.org/foundations/history/MacySummary.htm.

3. "Early cybernetics is essentially concerned with feedback circuits, and the early cyberneticists fell short of recognizing the importance of circularity in the constitution of an identity. Their loops are still inside an input/output box" (Varela 1995: 212).

4. The one example I have found is arguably drawn not from Maturana and Varela but from Jantsch's loose reformulation of autopoiesis toward Ilya Prigogine's prebiotic dissipative structures: "The tightly coupled evolution of the physical environment and the autopoietic entities of pre-life led to a new order of stability" (Lovelock 1988: 219–20; see also Jantsch 1980: 29–35 *et passim*).

5. For instance, in a paper Margulis co-authored with Hinkle (1997: 216), remarking on "features that make autopoietic (living) systems different from cybernetic ones."

6. Luisi (2006) provides an extended update of this discussion.

7. The absence of the term *autopoiesis* from Sagan's own recent writings—it occurs neither in Schneider and Sagan (2005), nor in Sagan (2007)—suggests that its presence in texts co-authored with Margulis is due to her particular contributions.

8. Von Foerster was also instrumental in placing Varela et al.'s 1974 paper in *BioSystems*. See Clarke (2009). For more on the publication of the Gaia hypothesis in *CQ*, see Kirk (2007).

9. A detailed history of Lindisfarne and a current roster of Fellows is available at williamirwinthompson.org/lindisfarne.html. The Fellows Conference reconvened in the summer of 2007 after a decade in abeyance.

10. Atlan is peripheral to the main narrative of this essay, but squarely within the conceptual lineage descending from von Foerster, in particular, the latter's earlier work on self-organization. See Atlan (1981, 1987).

11. See, in particular, chapter 1, "What Is Life? DNA, Autopoiesis, and the Reproductive Imperative," in Margulis and Sagan (1986b: 9–15).

12. "Autopoiesis" does not appear in the index of Margulis (1981). Multiple entries for "autopoiesis" and "autopoietic systems" appear in the second edition, Margulis (1993).

13. For differing recent assessments of these newer trends in Gaia theory, see Lenton and van Oijen (2002), Klinger (2004), and Sagan and Whiteside (2004).

14. In his volume recently finished at the time of the 1988 Lindisfarne meeting, he contrasted Daisyworld specifically to the chaotic population-ecology models of Robert May et al. It "differs profoundly from previous attempts to model the species of the Earth. It is a model like those of control theory, or cybernetics, as it is otherwise called. Such models are concerned with self-regulating systems; engineers and physiologists use them to design automatic pilots for aircraft or to understand the regulating of breathing in animals, and they know that the parts of the system must be closely coupled if it is to work. In their parlance, Daisyworld is a closed-loop model" (Lovelock 1988: 60).

15. The most fully developed neocybernetic work on the metabiotic application of autopoiesis has been done by social systems theorist Luhmann (1995).

16. For a recent discussion of autopoiesis in relation to Gaia theory, see Bourgine and Stewart (2004: 336–37).

# References

Atlan, H. 1981. Hierarchical self-organization in living systems: Noise and meaning. In M. Zeleny, ed., *Autopoiesis: A Theory of Living Organization*. New York: Elsevier North Holland, pp. 185–208.

Atlan, H. 1987. Uncommon finalities. In W. I. Thompson, ed., *Gaia—A Way of Knowing: Political Implications of the New Biology*. Great Barrington, MA: Lindisfarne Press, pp. 110–27.

Bourgine, P., and J. Stewart. 2004. Autopoiesis and cognition. *Artificial Life* 10: 327–45.

Brand, S., ed. *CoEvolution Quarterly*. 1974–1984. Sausalito, CA: Point.

Clarke, B. 2008. *Posthuman Metamorphosis: Narrative and Systems*. New York: Fordham University Press.

Clarke, B. 2009. Heinz von Foerster's demons: The emergence of second-order systems theory. In B. Clarke and M. Hansen, eds., *Emergence and Embodiment: New Essays on Second-Order Systems Theory*. Durham: Duke University Press, pp. 34–61.

Harding, S. 2006. *Animate Earth: Science, Intuition and Gaia*. White River Junction, VT: Chelsea Green.

Heylighen, F., and C. Joslyn. 2001. Cybernetics and second-order cybernetics. *Encyclopedia of Physical Science and Technology*, 3rd ed. San Diego: Academic Press.

Jantsch, E. 1980. *The Self-organizing Universe: Scientific and Human Implications of the Emerging Paradigm of Evolution*. New York: Pergamon.

Kirk, A. G. 2007. *Counterculture Green: The* Whole Earth Catalog *and American Environmentalism*. Lawrence: University Press of Kansas.

Klinger, L. F. 2004. Gaia and complexity. In S. H. Schneider, J. R. Miller, E. Crist, and P. J. Boston, eds. 2004. *Scientists Debate Gaia: The Next Century*. Cambridge: MIT Press, pp. 187–200.

Lenton, T. M., and M. van Oijen. 2002. Gaia as a complex adaptive system. *Philosophical Transactions of the Royal Society*, B357: 683–95.

Lovelock, J. E. 1979. *Gaia: A New Look at Life on Earth*. New York: Oxford University Press.

Lovelock, J. E. 1981. More on Gaia and the end of Gaia. *CoEvolution Quarterly* 31 (Fall): 36–37.

Lovelock, J. E. 1988. *The Ages of Gaia: A Biography of Our Living Earth*. New York: Norton.

Lovelock, J. E. 2006. *The Revenge of Gaia*. London: Allen Lane.

Lovelock, J. E., and L. Margulis. 1974. Atmospheric homeostasis by and for the biosphere: The Gaia hypothesis. *Tellus* 26: 2–10.

Luhmann, N. 1995. *Social Systems*, trans. John Bednarz Jr. with Dirk Baecker. Stanford: Stanford University Press.

Luisi, P. L. 2003. Autopoiesis: A review and a reappraisal. *Naturwissenschaften*, 90: 49–59.

Luisi, P. L. 2006. Autopoiesis: The logic of cellular life. In *The Emergence of Life: From Chemical Origins to Synthetic Biology*. Cambridge: Cambridge University Press, pp. 155–81.

Margulis, L. 1981. *Symbiosis in Cell Evolution: Life and Its Environment on the Early Earth*. San Francisco: Freeman.

Margulis, L. 1993. *Symbiosis in Cell Evolution: Microbial Communities in the Archean and Proterozoic Eons*, 2nd ed. New York: Freeman.

Margulis, L., and D. Sagan. 1986a. *Microcosmos: Four Billion Years of Evolution from Our Microbial Ancestors*. New York: Summit Books.

Margulis, L., and D. Sagan. 1986b. *Origins of Sex: Three Billion Years of Genetic Recombination*. New Haven: Yale University Press.

Margulis, L., and D. Sagan. 1997. *Slanted Truths: Essays on Gaia, Symbiosis, and Evolution*. New York: Springer.

Margulis, L., and D. Sagan. 2000. *What Is Life?* Berkeley: University of California Press.

Margulis, L., and G. Hinkle. 1997. The biota and Gaia: One hundred and fifty years of support for environmental sciences. In L. Margulis and D. Sagan, *Slanted Truths: Essays on Gaia, Symbiosis, and Evolution*. New York: Springer, pp. 207–20.

Margulis, L., and J. E. Lovelock. 1975. The atmosphere as circulatory system of the biosphere—The Gaia hypothesis. *CoEvolution Quarterly* 6: 31–40.

Margulis, L., and J. E. Lovelock.1989. Gaia and geognosy. In M. B. Rambler, L. Margulis, and R. Fester, eds., *Global Ecology: Towards a Science of the Biosphere*. San Diego: Academic Press, pp. 1–30.

Maturana, H. M., and F. J. Varela. 1980. *Autopoiesis and Cognition: The Realization of the Living*. Dordrecht: Reidel.

Maturana, H. M., and F. J. Varela. 1998. *The Tree of Knowledge: The Biological Roots of Human Understanding*, rev. ed. Boston: Shambhala.

McDermott, J. 1991. Lynn Margulis: Vindicated heretic. In C. Barlow, ed., *From Gaia to Selfish Genes: Selected Writings in the Life Sciences*. Cambridge: MIT Press, pp. 47–56.

Sagan, D. 2007. *Notes from the Holocene: A Brief History of the Future*. White River Junction, VT: Chelsea Green.

Sagan, D., and J. H. Whiteside. 2004. Gradient reduction theory: Thermodynamics and the purpose of life. In S. H. Schneider, J. R. Miller, E. Crist, and P. J. Boston, eds., 2004. *Scientists Debate Gaia: The Next Century*. Cambridge: MIT Press, pp. 173–86.

Schneider, E. D., and D. Sagan. 2005. *Into the Cool: Energy Flow, Thermodynamics, and Life*. Chicago: University of Chicago Press.

Schneider, S. H., J. R. Miller, E. Crist, and P. J. Boston, eds. 2004. *Scientists Debate Gaia: The Next Century*. Cambridge: MIT Press.

Suzuki, D. 2002. Journey into new worlds. In *The Sacred Balance*. VHS.

Thomas, L. 1974. *The Lives of a Cell: Notes of a Biology Watcher*. New York: Bantam.

Thompson, E. 2004 Life and mind: From autopoiesis to neurophenomenology. A tribute to Francisco Varela. *Phenomenology and the Cognitive Sciences* 3: 381–98.

Thompson, W. I. 1987. Gaia and the politics of life: A program for the nineties? In W. I. Thompson, ed., *Gaia—A Way of Knowing: Political Implications of the New Biology*. Great Barrington, MA: Lindisfarne Press, pp. 167–214.

Thompson, W. I. 1991. The imagination of a new science and the emergence of a planetary culture. In W. I. Thompson, ed., *Gaia 2: Emergence, The New Science of Becoming*. Hudson, NY: Lindisfarne Press, pp. 11–29.

Thompson, W. I. 1998. *Coming into Being: Artifacts and Texts in the Evolution of Consciousness*. New York: St. Martin's Griffin.

Thompson, W. I., ed. 1987. *Gaia—A Way of Knowing: Political Implications of the New Biology*. Great Barrington, MA: Lindisfarne Press.

Thompson, W. I., ed. 1991. *Gaia 2: Emergence, The New Science of Becoming*. Hudson, NY: Lindisfarne Press.

Thompson W. I., et al. 1991. From biology to cognitive science: General symposium on the cultural implications of the idea of emergence in the fields of biology, cognitive science, and philosophy. In W. I. Thompson, ed., *Gaia 2: Emergence, The New Science of Becoming*. Hudson, NY: Lindisfarne Press, pp. 210–48.

Varela, F. J. 1975. A calculus for self-reference. *International Journal of General Systems* 2: 5–24.

Varela, F. J. 1995. The emergent self. In J. Brockman, ed., *The Third Culture*. New York: Touchstone, pp. 209–22.

Varela, F., with D. Johnson. 1976. On observing natural systems. *CoEvolution Quarterly* 10 (summer): 26–31.

Varela, F. J., H. M. Maturana, and R. Uribe. 1974. Autopoiesis: The organization of living systems, its characterization and a model. *BioSystems* 5: 187–96.

Von Foerster, H. 1975. Gaia's cybernetics badly expressed. *CoEvolution Quarterly* 7 (fall): 51.

# 18

## Intimations of Gaia

Eileen Crist

The most compelling contribution of Gaian science, which has complemented the evolutionary and ecological perspectives on life, is that organisms do not merely adapt to the environmental conditions they find themselves in but actively shape them. In his four decades of Gaian thought, James Lovelock (2004: 1) has always insisted that "organisms are not mere passengers on the planet"—they are more like pilots. On the basis of the Earth case, Gaia theory postulates that once life becomes abundant enough to have considerable environmental effects, it takes over its planet home: life in the universe, in general, is likely to be a planetary phenomenon. Besides existing everywhere and mostly profusely on the Earth's surface, life is also found, albeit more sparsely, fifty kilometers above the surface and at depths a few kilometers into the crust.

### What Gaia Taught Us

On Earth the composition of the air is 99 percent biogenic, life has a strong influence on global ocean chemistry and possibly on the retention of the planet's water, and soil is a Gaian phenomenon in which living and nonliving components are thoroughly hybridized. Atmosphere, hydrosphere, and upper lithosphere are markedly different from what they would have been were life absent from Earth. The Gaian perspective thus submits that a planet with life becomes more akin to a biological composition than a geophysical body: biological and geological forces merge, and a new kind of entity—a geophysiology—is born. Lovelock called that entity "Gaia." W. E. Krumbein and A. V. Lapo (1996) offered the neologism "bioid" to describe Gaian-type planets at large in the universe as opposed to "geoids" that support no life.

As life spreads and becomes abundant so do the effects of its activities and metabolisms, until those effects become first large scale then global in scope. Inevitably, life-driven environmental modulations feed back on life itself—on the groups of organisms that caused the particular changes as well as on other groups. Environmental changes that turn out detrimental to the life that generated them tend to be self-limiting by boomeranging on the creatures that instigated them (Lenton 2004). Effects that happen to be beneficial, on the other hand, may tend (other things being equal) to be self-enhancing by promoting the instigating organisms (ibid.). Of course, there is nothing to stop organisms that foul their environments (in a way that rebounds upon them) from arising, and even thriving for a while, but they are less likely to persist.

Gaian inquiry has offered a renewed perspective on life's grandeur by foregrounding what Lovelock has evocatively called life's "cosmic lifespan." Life has existed on Earth for about 3.8 billion years—a quarter of the age of the universe. Conditions on the planet have varied greatly through life's eons, yet all have been habitable for life. (Conditions that have been extreme—too hot or too cold, for example—have been tolerated by a narrower spectrum of life, which then may have contributed to the emergence of environmental parameters viable for a broader spectrum; see McMenamin 2004.) Although it is a point of debate within the Gaian community, Gaia theory proposes that the endurance of life through vast passages of time has not been accidental. To be sure, life's longevity could have been simply a matter of luck. The shortcoming of this idea, however, is that it does not encourage interesting thinking nor research into the possibility of life's participation in securing its own survival. Gaian inquiry, on the other hand, keeps alive the fascinating question of how an ever-changing biota can have the power to contribute to sustaining an ever-changing yet always viable world.

The Gaian logic proceeds as follows: when organisms drive environmental variables toward uninhabitable conditions, the growth of those organisms is likely to be eventually suppressed while organisms that enhance habitability, especially for themselves, are selected for (Lenton 1998). Further, if the community models of Tim Lenton and his colleagues apply to real Earth conditions, then "ecosystems or communities that 'foul their nest' [may tend to] lose out to those that improve their local conditions" (Lenton and Williams, chapter 5 in this volume). Scaling up these insights to the level of the biosphere—and adding the reinforcement of viable effects via strong linkages between successful, pervasive groups of organisms that are metabolically complementary with one another (more on this shortly)—we might conjecture that environment-enhancing life, as a networked and emergent whole, has a hand in maintaining a gradient of habitability for a broad phenotypic gamut of organisms, from tropical butterflies to penguins (to borrow from Volk, chapter 3 in this volume).

Once life got seriously underway—which is to say after microbes became pervasive on Earth—organisms have always evolved under conditions formed by life. As Lovelock has poetically put it, organisms "live in a world that is the breath and the bones and the blood of their ancestors and that they themselves are now sustaining" (1996: 19). After four decades of Gaian inquiry, the view that life has a (trans)formative impact on the environment now enjoys broad consensus—even as many scientists may shy away from the idea that a tightly coupled living and nonliving world "regulates" conditions on the planet.

Earth's average surface temperature is a key example of life's enormous influence on environment—"influence," in this context, seeming too weak a concept, even as "regulation" may sound too strong. Quantifying life's impact on the planet's surface temperature—via its role in removing $CO_2$ from the atmosphere (biotic enhancement of weathering) and seeding clouds through the biological production of dimethyl sulfide (DMS)—Tyler Volk and David Schwartzman have estimated that biotic processes may presently cool the planet by 35 to 45 degrees centigrade (Schwartzman and Volk 1989; Schwartzman 1999; Volk 1998). In other words, an abiotic Earth in present time would be at least 35 degrees hotter. From this example alone we can discern why Gaian scientists maintain that "the global environment is being transformed by life into a state very different from a planet without life" (Volk 2004: 27).

Early Gaian literature offered the metaphor of "superorganism" for the Earth. On this metaphor the biosphere became comparable, for example, to a beehive whose ranges of temperature, humidity, and other conditions suitable for the bees happen to be created by the bees. (Subtract the bees, and conditions in and out of the hive swiftly equilibrate.) Similarly, Gaians argued, the biota as a whole shapes environmental parameters to be suitable for life. Such was the early formulation of Gaia (which generated a storm of protest): this is a planet constructed by and for the biota. For example, in an evocative, albeit contentious, metaphoric description of how life molds the Earth's atmosphere, Lovelock submitted that "the interaction between life and environment, of which the air is a part, is so intense that the air could be thought of as being like the fur of a cat or the paper of a hornet's nest: not living but made by living things to sustain a chosen environment" (1987: 88).

The superorganism metaphor enjoyed only brief popularity in Gaian literature: eventually it was conceded that unlike the members of a bee or ant colony, organisms on Earth are not genetically similar enough to participate in co-creating a home. (Also the mechanism for the evolution of beehives is kin—possibly along with group—selection, with no analogues for Gaia.) In the beehive, what benefits one bee is likely to benefit all, since they can be regarded as extended phenotypes of one genetic blueprint; in the biosphere, on the other hand, organisms are genetically very different and often adapted to widely divergent conditions.

Yet even as the superorganism metaphor has fallen into disuse, it is conceivable that the Earth is more like a beehive than we are able to grasp or formulate rigorously. For while the biosphere's organisms are not genetically identical, they *are* genetically related, all having descended from a single common ancestor. There is only one form of life on Earth, a form of life possessing a shared genetic mechanism, cellular infrastructure, and (to a large extent) biochemical language. All life forms are evolved expressions of an ancestral form, and that they may participate in co-creating and sustaining a particular range of environmental conditions—within which they survive and often flourish—seems intuitively probable, even if a scientific specification of how exactly this emerges is, now or perennially, elusive.

A new generation of Gaian scientists—Tyler Volk, Tim Lenton, David Wilkinson, and David Schwartzman, among others—have sharpened Gaian thinking while at the same time responding to neo-Darwinian critiques of Gaia as a teleological concept. They have argued that organisms do not evolve by-products (or traits) *in order to* control their physical and chemical environments; rather, the by-products of organisms end up having side effects that are both inevitable and potentially consequential. As previously described, the side effects will feed back (one way or another) on their creators and also delimit the evolution of other organisms that must, on the extremes of a continuum, either adapt or perish. To revisit the example of Earth's viable surface temperature: it is not so much that life has contributed to creating a global climate regime that is habitable, as that it has contributed to creating a global climate regime that became inhabited by life forms that were able to evolve within it, having found it at least tolerable if not ambient. (Speaking only partly tongue in cheek, Lynn Margulis likes to quip that the

Earth is "room temperature.") The organisms that evolved within such partly life-driven conditions often reinforce the effects of their ancestors, being relatives of these predecessor life forms with compatible biochemistries and metabolic outputs. The maintenance of certain environmental effects—whether or not we choose to call this outcome "Gaia's self-regulation"—thus becomes a self-reinforcing phenomenon.

Once Gaia gets going, in other words, it keeps itself going. (At the dawn of life, however, and perhaps after major setbacks like mass extinctions, the starting physical and chemical conditions are critical for life's (re)ascent into prominence.) The biosphere, or Gaia, is thus an emergent phenomenon of life's *abundance*, because it is only through being abundant that life can (chemically and physically) shape its surroundings consequentially enough to generate feedback for itself and constraints (enabling or limiting) for other organisms.

Teleology, as has been duly and often noted by Gaians, is redundant. The biota does not need to be purpose-driven or cooperative to shape life-sustaining surroundings, but instead environmental feedback takes care of that end result. Charles Darwin recognized such a feedback mechanism in the case of earthworms that, by moving through, chemically processing, and physically triturating earth, create soil on which their food—plants—grows (Darwin 1881).[1] (Earthworms thus cultivate their food; Darwin called them "gardeners.") This feedback process that Darwin identified for one group of soil invertebrates can be generalized to all soil organisms (Carson 1962; Lavelle 1996; Volk 1998). Soil, created by and virtually composed of life, is good for every creature that lives in it. Furthermore none of the creatures that collectively make the commonwealth of soil through their behaviors, their excretions, their body parts, and ultimately their corpses are being either selfish or collaborative in so doing.

Groups of organisms not only impact their surroundings by putting out metabolic byproducts the effects of which feed back upon them; their abundant by-products also create opportunities for the evolution of other kinds of organisms that can metabolize or utilize those by-products (Volk 1998). (By-products include such things as feces, urine, leaf litter, corpses, oxygen, and nitrogen compounds.) The latter organisms will then create other by-products that change the surroundings in ways that will reverberate back upon them, and also create opportunities for yet other groups to evolve and grow. Feedback cycles arise and exchanges are created, some of which may involve the reciprocal consumption of each other's excretions. Large-scale partnerships between successful, pervasive groups with complementary metabolisms, feeding on one another's "wastes," may emerge and stabilize: for example, between land organisms and marine creatures (Harding 2006), between autotrophs and decomposers (Rinker, chapter 6 in this volume), as well as between photosynthesizers and respirators or, more colloquially if less exactly, plants and animals (Volk, chapter 3 in this volume). Organisms become interlinked in matrices of chemical exchanges. For Gaian scientist Scott Turner (2004), groups of organisms that reciprocally enhance each other's survival and growth—creating what he calls "closed loops of nutrient flows"—can form enduring, mutually favorable, and environmentally dominant associations.

Life is a fundamentally imbricating phenomenon, an elaborate edifice of nestings that stabilize for extended periods into complex, in-flux equilibrated states (e.g., of atmospheric composition, climate regimes, life-forms, or ecosystems). The intricate webbing of life has happened, and continues to occur, from the most intimate dimension of endosymbioses that created and sustain complex life, to the exchange of nutrients via the trading zones of air, water, and soil (Margulis 1981; Lovelock 1988). Gaia is this interconnected flux, what Darwin called the "entangled bank." Strictly speaking, there is no selfishness or cooperation in life's activities within the biosphere—only a whirl of obligate interconnectedness. For some, the absence of selfishness and cooperation may be testimony to a morally indifferent natural world, one neutral to direction, outcome, and relationship. And yet, the obligate interconnectedness of life, from which mutual benefit is constantly flowing to all, can also be interpreted as evidence that goodness is profoundly rooted in a primordial and objective condition of being (see Kropotkin 1902; Bookchin 1996). Goodness, in other words, can be understood as the distilled concept and conscious practice of what life does simply as a matter of fact. In this light, human ethics (for which service and benevolence are universal ideals) are continuous with, not extrinsic or epiphenomenal to, Nature's ways.

For Gaian thought "the environment" is not a range of external conditions that sets the stage for life's "struggle for survival," but more like a physiological matrix or co-created interface that eases the flow of matter and energy between an abundance of organisms. Elsewhere I called atmosphere, hydrosphere, and upper lithosphere "the commons" of the biota (Crist 2004a). By playing such a significant role in creating their surroundings, organisms are essentially protagonists in creating and changing themselves, a process that Gaians call "autopoiesis" or self-creation (Margulis and Sagan 1995; Clarke, chapter 17 in this volume). But the fact that organisms modify their interfaces in ways that turn out to facilitate vital communication in a variety of biochemical dialects does *not* mean that life is in control of its environment or its own destiny. Life's exquisite powers offer absolutely no guarantee for the persistence or even the resilience of the biota that constitutes it.

Despite life's awesome ability (thus far in its long history) to renew itself and rebound with the passage of time, the biota has been exceedingly vulnerable to the human onslaught. And what remains of Earth's biological diversity today is all too clearly as fragile and ephemeral as fireballs over marshes.

## The Danger of Overtheorizing the Earth as "System"

Life's unimaginably long tenure on Earth, along with its capacity to survive massive blows (e.g., asteroid strikes and extreme climatic shifts), so enthralled Gaian thinkers that many, especially in the early days of the Gaia hypothesis, tended to overemphasize Gaia's intrinsic toughness. The discovery that life has the "power to tame the forces of the universe," as Eugene Linden puts it (chapter 19 in this volume), led Gaians to privately underestimate and publicly understate the human-driven devastation of the biosphere. How much damage could arrogant but puny humankind inflict on tough Gaia? Such questioning put Gaian thought at loggerheads with environmental sensibilities.

Indeed, if we focused solely on cultural appropriations of Gaia—on the New Age enthusiasm with which the goddess-Earth concept has been embraced—the paradox of a tense relationship between environmentalism and a Gaian perspective would be missed. Yet tensions between the two have existed from the early days of the Gaia hypothesis. In comparison to Gaia's self-regulating power, to her robustness and longevity, the Gaian paradigm seemed to dismiss human beings as a relatively trivial force—hell-bent, perhaps, on our own destruction, but incapable of jeopardizing the Gaian system. "The environmentalist," Lovelock averred in the 1980s, "who likes to believe that life is fragile and delicate and in danger from brutal mankind does not like what he sees when he looks at the world through Gaia. The damsel in distress he expected to rescue appears as a robust man-eating mother" (1987: 96). The ecologically minded naturally worried about the implications of this position. Environmental ethicist Anthony Weston, for example, challenged the emphasis on Gaia's "powers, not our responsibilities," on ground that it could "undercut rather than reinforce many of the legal safeguards that environmentalists have established." For, he continued, "it can be argued that nature is not fragile, on the whole, and therefore, that many of the protections we have enacted based on our fears of its fragility are probably unnecessary" (1987: 219–20).

Underestimating the human impact went hand-in-hand with the emphasis on *homeostasis*, a term transposed from systems theory and cybernetics to conceptualize Gaia. For example, Lovelock and Margulis (1989) described "the world system, which is Gaia," as having "the thermostat-like capacity to maintain the earth temperature constant, in spite of an increase of heat from the sun, and also to maintain the chemical composition relatively stable." Homeostasis, sustained through built-in or autopoietic negative feedback mechanisms, refers to a system's tendency to keep its basic states relatively stable in response to perturbation. The early Gaia literature, in particular, found evidence for homeostasis in the biosphere's "maintenance of relatively constant conditions by active control" having prevailed for thousands of millions of years (Lovelock and Margulis 1974). The ostensible capacity of Gaian system to self-regulate, maintaining stable parameters of temperature, chemical composition, and other variables, was unfortunately often taken to imply that human beings have considerable latitude to perturb the Earth with impunity: not only was the Earth system thought potent enough to withstand human perturbation, it was sometimes also deemed capable of automatically countering it.

Whether Gaian scientists intended such erroneous inferences to be drawn from an Earth systems perspective, commentators did in fact draw them. In his review of Lovelock's *Homage to Gaia*, for example, Adolfo Olea-Franco (2002: 602) wondered: "Since Gaia is resilient and homeostatic, why should we care about pollution and global warming?" The Earth system's supposed ability to handle disturbance could thus be glibly interpreted as proverbial "license to pollute." This implication was also picked up by ecofeminist Val Plumwood (1992: 63). "It does not matter," she noted about the potential Gaian environmental message, "if we don't wash our dishes and throw our dirty linen on the floor because Gaia, a sort of super housekeeping goddess operating with whiter than white homeostatic detergent, will clean it all up for us. In this form the concept...denies the need for any reciprocal human responsibilities towards Gaia. Such a Gaia may have the trappings of a goddess but is really conceived as a sort of super-servant." This was fair if caustic commentary on problematic environmental "uses of the Gaia concept" (ibid.).

These problematic uses of Gaia, far from reflecting indifference toward the fate of an overexploited biosphere within the Gaian community, stemmed from an overzealous application of systems theory. Systems theory, too rigidly or literally applied, may have propped the underestimation of humans' disruptive power, on the one hand, and encouraged an irrelevant emphasis on Gaia's capacity to survive and triumph in the long haul, on the other.

When the idea of system is applied loosely to the biosphere—simply to highlight the physical and energetic interconnectedness of elements within a whole—then no overwrought conceptual repercussions follow. A loose use of "system" resonates with its connotations as a suffix in the word *ecosystem*. Indeed Lynn Margulis (1996), who has resisted the notion of the Earth as a "singularity," has often preferred to describe Gaia as "a set of interacting ecosystems" rather than in cybernetic terms. But when systems theory is literally and vigorously applied to the Earth, the emergent perspective suffers from enormous flaws. For one, it props technological metaphors for describing the Earth—the most widely used having been the comparison of Gaia to "thermostat." The technological idiom tends to reinstate a mechanistic conception of the biosphere—the worldview most implicated in ecological destruction, as historians, philosophers, and scientists have compellingly argued.[2] Ironically the resurrection of a vision of *anima mundi*, in the scientific form of Gaia, intended to supersede deadened concepts of Earth and cosmos (Abram 1996; Harding 2006).

Moving forward, the most insidious repercussion of a newly minted mechanistic biosphere (of Earth as cybernetic system, thermostat, etc.) is taking shape in the increasingly aired proposals to solve anthropogenic climate change via so-called geoengineering methods. The most widely discussed possibility is the idea of shooting aerosols into the stratosphere to mask global warming via the effect of global dimming (see Crutzen 2006). Geoengineering schemes constitute dangerous strategies for addressing our ecological predicament, at both ideological and real-worldly levels.[3] But geoengineering schemes are also profoundly dubious for being premised on the assumption that the Earth is *literally* a single gigantic cybernetic contraption that we might manipulate as a whole. As David Abram (1996: 238) has pointed out about the repercussions of mechanistic thinking, "the mechanical metaphor...not only makes it rather simple for us to operationalize the world..., it also provides us with a metaphysical justification for any and all such manipulations." An emergent technological metaphor of Earth as "cybernetic system" is the hidden empirical assumption, and the underlying metaphysical justification, of geoengineering proposals: both the empirical *and* metaphysical underpinnings are highly suspect.

A stringently applied systems-theoretic framework for Gaia is unfalsifiable—another gaping cognitive flaw. As was noted in the Preface to the volume *Scientists Debate Gaia*, "the direction of feedback [in the Gaian system] is not clear and is likely to be destabilizing as stabilizing at different times and scales" (Schneider et al. 2004: xv). Thus any planetary state (whether a consequence of stabilizing or destabilizing feedback) can be rendered conformable to systems theory: from the extreme climatic episodes of snowball Earth or the Eocene's runaway heating, to oscillations between glacial periods and interglaciers, and the onslaught of feedbacks that are currently strengthening rather than offsetting global warming—all can be rationalized within a systems perspective. This is partly because systems theory offers a range of concepts able to account for *both* stability and change. While early Gaian thinking emphasized the former (negative feedback and homeostasis), the current documentation of anthropogenic climate change has set in motion the marshalling of the latter (positive feedback and chaos).

Gaian scientists began to steer away from stressing homeostasis (negative feedback), both because of growing knowledge that global environmental parameters have ranged widely in geological time (sometimes settling into extreme regimes dangerous for life's tenure) and because of a growing understanding of what is unfolding with global warming. Indeed, if the hope that "homeostatic Gaia" might counter the human-driven amplification of the greenhouse effect ever induced comfort— namely that the system would kick in with negative feedback to offset adverse heating—such comfort has evanesced in the face of steadily increasing temperatures, melting ice and glaciers, and rising sea levels (Lovelock 2006; Flannery 2006; Aitken, chapter 8 in this volume). Yet the apparent failure of homeostatic mechanisms to emerge in response to the human-driven $CO_2$ forcing has not (necessarily) inspired the abandonment of Gaia-qua-system. Instead, an alternate panoply of system concepts is being marshaled—in particular, those of threshold, tipping point, amplifier, and positive feedback (Lovelock 2006). Applied to our current greenhouse predicament these concepts generate an apocalyptic vision, as the Earth system becomes conceived on the verge of shifting rapidly and irreversibly (for a human time scale) to a new hot state (see Crist 2007).

Here is how Lovelock (1996: 24) summarily captured the dynamics of Earth-system shifts in an earlier publication: "Gaia theory sees the Earth as a responsive supra-organism that will at first tend to resist adverse environmental change and maintain homeostasis. But if stressed beyond the limits of whatever happens to be the current regulatory apparatus, it will jump to a new stable environment where many of the current range of species will be eliminated." As this classic formulation crisply demonstrates, systems theory reasoning is *digital*, tending to deliver a binary storyline: on the one hand, the Earth-system is assessed robust enough to withstand some disturbance through actively maintaining homeostasis; on the other hand, too much disturbance is regarded as forcing a threshold-crossing that throws the Earth-system off kilter before it stabilizes into a new state.

But this two-tiered takeaway picture, which emerges through an overly stringent application of systems theory to the biosphere, may be assessed as far too unrefined: it entirely bypasses the fact that, as a consequence of human colonization, the Earth has suffered profound losses of ecosystems and species without adverse whole system consequences. The losses that have occurred, and continue to unfurl, can only be discerned through *analogue* thinking: biodiversity destruction has been a continuous, incremental, and cumulative event—and the binary systems-theory construct of *stability* to *chaos* (whatever truth it may hold for extreme climate forcings) has not much to do. In fact the vast diminishment of life's richness has unfolded within apparently stable system conditions; it has not resulted from, nor (to our knowledge) led to, the overstepping of any global thresholds; and it has proceeded as a linear unraveling—species by species, population by population, habitat by habitat, and today (after the steady chiseling of centuries if not millennia) acre by acre.

In brief, systems theory applied to the biosphere has been conceptually unequipped to capture the import of the biodiversity crisis (which includes the mass extinction underway), except insofar as this crisis emerges as consequence (or potential cause) of a jump from one Earth-system state to another. This cognitive failure of systems theory is a straightforward consequence of systems thinking. For when the Earth is conceived as a system, inevitably quandaries about environmental troubles become posed in terms of whether *the system* is endangered; the question of whether the biosphere is being destroyed becomes coextensive with the question of whether the Earth-system, as we know it, is breaking down. For those of us, however, who understand the scope of biodepletion as the precipitous loss of life's richness, as a profound crisis for the composition of the biosphere (and not necessarily a crisis for the biosphere's stability, whatever that means), the systems view has not been a sufficiently nuanced theoretical instrument to register, and thereby bring into discursive view, the anthropogenic devastation of life.

This is not simply a normative grievance; it is a cognitive grievance as well. Systems theory, especially in its unpalatably mechanistic metaphors, is simply no match for the Earth's mystery and immensity. By laying the biosphere on the Procrustean bed of overtheorizing, systems theory ends up reducing it to fit its framework. Every theory obscures at least as much as it reveals and is therefore *reductionist* in some sense. The celebrated holism of systems theory can make it all the more deceptive an instrument of knowledge. Holistic theories are invisibly reductionist because by making the implicit validity claim of "capturing the whole" they blindside us to what their framing crops out. (I discuss what is cropped out with respect to Gaia in the next section.) And so it is with an overtheorized systems view of the biosphere. Fortunately neither Gaian inquiry nor our intuitive sense of the Earth's oneness hang on a systems-theoretic conception of the living planet.

The whirl of intensely interactive, abundant, diverse, and complex life that shapes the wondrous commons of the biosphere does not need cybernetics for clarification.

## Restoring the Holocene

The famous Daisyworld computer model, elaborated by Andrew Watson and Lovelock in the early 1980s, gave the idea of biospheric self-regulation a tremendous boost, by simulating how the differential growth of black and white daisies could tune a planet's temperature (as a kind of albedo dial) within relatively ambient zones for the daisies, even as the sun's heat output gradually increased. The model was "a splendid rhetorical asset," as Jon Turney (2003) put it, in demonstrating that systemic regulation can occur as an automatic consequence of organismal growth, life's environmental input, feedback from input, and natural selection.

The model demonstrated that the abundant, differential growth of black and white daisies can do the global work of modulating temperature over time in the face of a heating sun. Yet within the habitable climatic conditions thereby created, what is arguably *the main event* on Daisyworld starts to unfold: gray daisies evolve and spread in the life-molded, life-supporting matrix they find. It was the neo-Darwinists who brought up the "gray daisies"—a pigment-free variety that could exploit the comfortable environment without investing any work in sustaining it—as a challenge to the ultimate tenability of the self-regulation of Daisyworld. Within a neo-Darwinian framework that elevates competition to a first-line biological principle, the gray daisies were naturally construed as the so-called cheats (see Lenton and Williams, chapter 5 in this volume). (It is only by over-inflating the status of competition and struggle in living processes that such loaded language has descriptive purchase.) But in a broader evolutionary and biospheric perspective, the "gray daisies" represent something much less inflected and much more important than freeloaders—they represent *biodiversity*, the influx of evolutionary proliferation within partly life-created niches that afford living means and habitat. If what the biosphere epitomizes is a capaciousness in creating life, then the gray daisies are its very essence—even though, from a systems perspective, their regulatory functions may be auxiliary, redundant, or even nonexistent.

Systems-theory reasoning places an ontological premium on the whole. The component parts are considered important, of course, but primarily because of the functional roles they play. As a consequence systems thinking produces a discursive blind spot for those components of life, or levels of taxonomy, that do not have critical functions within the system. Importantly, thinking along such functionalist lines leads to the inference that any number of life forms may be "redundant," vis-à-vis the (adequate or even healthy) functioning of the (eco)system, for any of the following reasons: they are too rare to have a serious impact; they are fungible, which is to say replaceable with functional equivalents; and/or they are simply taking a free ride in a system that more biochemically robust groups are running.

The idea of redundancy of life forms within the Gaian system (or within ecosystems) insidiously mutates into a notion of *dispensable* life forms. But the notion that certain life forms are, or may be, dispensable is completely theory-laden: dispensability gets its façade of empirical coherence only within a theoretical framing that prioritizes the functionality of some imagined whole. Defenders of biodiversity—who naturally recoil at the implications of dispensability in a time when life's richness is critically imperiled—are often duped by the pseudora-tionality of the notion of dispensability, and driven to the weak argu-ment that we should protect all life forms because we cannot know which ones are, or are not, dispensable, or because the dispensable ones are backups—spare parts, as it were—in the system.

But it is the *very idea* of dispensability that needs to be deconstructed and jettisoned as intrinsically incoherent. Were the world's old-growth forests dispensable? What if remaining old-growth forests are replaced with oxygen-producing and carbon-absorbing fast-growing tree planta-tions—are the rest of them (also) dispensable? Was Costa Rica's golden toad, driven to extinction by climate change, dispensable? How about the Tasmanian wolf, hunted to oblivion? Seventy percent of flowering plants are endangered or threatened: what fraction of them is dispens-able? Until the recent devastation of marine ecosystems, the oceans were home to "a great abundance of whales, walruses, sea cows, seals, dol-phins, sea turtles, sharks, rays, and large fish" (Jackson 2007). That abundance is no longer: Was it dispensable?

It is the flaw of systems paradigms—be they social or biological—to lack the conceptual tools for honoring the member parts for their intrin-sic existence, their unknown (or even trivial) roles, their sheer contribu-tion to complexity, their sheer contribution to diversity and/or biomass, and their unknowable destinies. In what is a purely reductive move, systems thinking is only equipped to appreciate component parts as functional cogs in the whole. It is systems thinking that is dispensable, not life forms.

## Concluding Remarks

If life in the biosphere has an essence, it might be expressed under the rubric of three interconnected qualities: diversity, complexity, and abun-dance. These qualities have been captured peerlessly by the perspectives of Darwinian evolution, ecological science, and Gaian inquiry in their complementary paradigms that, together, hold the potential to create a Zeitgeist of deep understanding and harmonious living on Earth. The tendency of life to become increasingly diverse, increasingly complex, and increasingly abundant has, over the course of eons, created and recreated a living Earth that in the temporal and spatial unfolding of the universe can be celebrated as a *cosmos*—a world of intrinsic order and beauty. Amazingly the time frame within which *Homo sapiens* evolved and proceeded to develop a cornucopia of cultures coincided with what biologists believe may have been the most biodiverse era of Gaia's natural history (Wilson 1999).

But we are living in a time that we are daily inundated by overwhelming news about the biosphere's predicament. Wherever we turn there is a crisis: an amphibian crisis, a climate crisis, an ocean crisis, a water crisis, a coral reef crisis, a bird crisis, a rainforest crisis, a carnivore crisis—the list is endless. As sorrowful as the specifics are, the deepest tragedy lies in the scope they add up to: human beings have taken aim at the very qualities that define the living planet, dismantling, with an intent that seems paradoxically both blind and demonic, the diversity, complexity, and abundance of life on Earth. Elsewhere, I collectively called these properties *the flame of life*, because they form the matrix of Earth's life-generating creativity and of the biosphere's robustness so celebrated by Gaians (Crist 2004b).

Extinction of species is occurring at a rate thousands of times the rate of natural (also known as background) extinction. Life scientist Peter Raven (2001) has calculated (using an estimate of ten million extant species and an average species lifetime of four million years) that in the absence of the human impact, between two to three species would be going extinct each year. By contrast to this natural rate, thousands (if not tens of thousands) of species are vanishing yearly. And the biosphere is not only hemorrhaging species, it is losing its abundance of wilderness and wild creatures. The great masses of flocks, schools, and herds of animals are vanishing, and so are their migrations. The populations of top predators—tigers, lions, jaguars, wolves, grizzlies, sharks, and others—are a mere fraction of what they were even a hundred years ago. The numbers of ocean fish are rapidly shrinking, and on land the same holds for the once globally abundant forested tracts of the Holocene. Half of the world's wetlands, intensely rich in biodiversity, were lost in the twentieth century alone. On landscape and seascape levels such losses are tantamount to the accelerating dismantling of ecological complexity. The impoverishment can be large or small, depending on the particulars, but the global trend has been in the direction of simplification of one ecosystem after another. Simplification is ratcheted up by the frenetic mixing of the world's biota, brought on by globalization, that is swiftly homogenizing the biosphere. In contrast to mixing cultures, where loss of cultural diversity may be offset by the gain of greater mutual acceptance and equality between peoples, there is *no silver lining* to the biological melting-pot: the generalists win, and diversity—the unique loveliness of each place—first recedes, then vanishes.

Will the biodepletion underway generate a resource problem for humanity? I submit that fear of resource depletions is all but a red herring: it distracts attention from the fact that the transformation of the biosphere into a stock of resources is what has devastated it in the first place. We have in fact gained a world of resources by forfeiting the living Earth. Misplaced anxiety about resources occludes from view that the resourcist worldview has been, and is, destroying the beauteous wealth of the biosphere (see Foreman 2007). Fear of losing resources will not generate the vision we need to preserve and restore the Holocene's richness, but instead will encourage technological fixes (e.g., geoengineering the atmosphere or the oceans) and managerial approaches to land, water, and air. The day may not be far when, for example, instead of working toward restoring the abundance, diversity, and beauty of marine ecosystems, we start farming the oceans to produce "protein."

In Bill McKibben's opening words, this is our moment. It is the moment to face the root of the terrible trouble we have unleashed for the biosphere and for ourselves: our expansionism, arrogance, and domination within the biosphere. In acts of beauty, and without fear, this is our moment to put Gaia first.

## Notes

I would like to thank Dave Abram, Stephan Harding, Rob Patzig, H. Bruce Rinker, Tyler Volk, and David Schwartzman for their critical readings of an earlier draft, and most helpful suggestions and encouragement.

1. For an analysis of Darwin's last book—often called his "worm book"—in a geophysiological or Gaian light, see Crist (2004a).

2. Carolyn Merchant, for example, in her celebrated work *The Death of Nature* (1980: 193), wrote that "the removal of animistic, organic assumptions about the cosmos constituted the death of nature." More recently Stephan Harding (2006) echoed this assessment, condemning the "mechanistic view" as "literally killing the Earth as it was configured at the time of our birth as a species."

3. For a more elaborate critique of geoengineering that for reasons of space I cannot reiterate here, see Crist (2007).

## References

Abram, D. 1996. The mechanical and the organic: Epistemological consequences of the Gaia hypothesis. In P. Bunyard, ed., *Gaia in Action*. Edinburgh: Floris Books, pp. 234–47.

Bookchin, M. 1996. *The Philosophy of Social Ecology*. Montréal: Black Rose Books.

Carson, R. 1962. *Silent Spring*. Boston: Houghton Mifflin.

Crist, E. 2004a. Concerned with trifles? A geophysiological reading of Charles Darwin's last book. In S. Schneider, J. Miller, E. Crist, and P. Boston, eds., *Scientists Debate Gaia: The Next Century*. Cambridge: MIT Press, pp. 161–72.

Crist, E. 2004b. Against the social construction of nature and wilderness. *Environmental Ethics* 26: 5–24.

Crist, E. 2007. Beyond the climate crisis: a critique of climate change discourse. *Telos* 141: 29–55.

Crutzen, P. 2006. Albedo enhancement by stratospheric sulfur injections: A contribution to resolve a policy dilemma? *Climate Change* 77 (3/4): 211–19.

Darwin, C. [1881] 1985. *The Formation of Vegetable Mould, Through the Actions of Worms with Observations on Their Habits*. Chicago: University of Chicago Press.

Flannery, T. 2006. *The Weather Makers: How Man is Changing the Climate and What It Means for Life on Earth*. New York: Grove Press.

Foreman, D. 2007. The arrogance of resourcism. *Around the Campfire* 5 (March 1).

Harding, S. 2006. *Animate Earth: Science, Intuition and Gaia*. White River Junction, VT: Chelsea Green.

Jackson, J. 2007. When ecological pyramids were upside down. In J. A. Estes, ed., *Whales, Whaling, and Ocean Ecosystems*. Berkeley: University of California Press, pp. 27–37.

Kropotkin, P. [1989] 1902. *Mutual Aid: A Factor of Evolution*. Montreal: Black Rose Books.

Krumbein, W. E., and A. V. Lapo. 1996. Vernadsky's biosphere as a basis of geophysiology. In P. Bunyard, ed., *Gaia in Action: Science of the Living Earth*. Edinburgh: Floris Books, pp. 115–34.

Lavelle, P. 1996. Mutualism and soil processes. In P. Bunyard, ed., *Gaia in Action: Science of the Living Earth*. Edinburgh: Floris Books, pp. 204–19.

Lenton, T. 1998. Gaia and natural selection. *Nature* 394: 439–47.

Lenton, T. 2004. Clarifying Gaia: Regulation with or without natural selection. In S. Schneider, J. Miller, E. Crist, and P. Boston, eds., *Scientists Debate Gaia: The Next Century*. Cambridge: MIT Press, pp. 15–25.

Lovelock, J. 1987. Gaia: A model for planetary and cellular dynamics. In W. I. Thompson, ed., *Gaia, A Way of Knowing: Political Implications of the New Biology*. Great Barrington, MA: Lindisfarne Press, pp. 83–97.

Lovelock, J. 1988. *The Ages of Gaia*. New York: Norton.

Lovelock, J. 1996. The Gaia hypothesis. In P. Bunyard, ed., *Gaia in Action: Science of the Living Earth*. Edinburgh: Floris Books, pp. 15–33.

Lovelock, J. 2004. Reflections on Gaia. In S. Schneider, J. Miller, E. Crist, and P. Boston, eds., *Scientists Debate Gaia: The Next Century*. Cambridge: MIT Press, pp. 1–5.

Lovelock, J. 2006. *The Revenge of Gaia*. London: Allen Lane.

Lovelock, J., and L. Margulis. 1974. Biological modulation of the Earth's atmosphere. *Icarus* 21: 471–89.

Margulis, L., and J. Lovelock. 1989. Gaia and geognosy. In M. Rambler, L. Margulis, and R. Fester, eds., *Global Ecology: Towards a Science of the Biosphere*. San Diego: Academic Press, pp 1–30.

Margulis, L. 1981. *Symbiosis in Cell Evolution: Life and Its Environment on the Early Earth*. San Francisco: Freeman.

Margulis, L. 1996. James Lovelock's Gaia. In P. Bunyard, ed., *Gaia in Action: Science of the Living Earth*. Edinburgh: Floris Books, pp. 54–64.

Margulis, L., and D. Sagan. 1995. *What Is Life?* New York: Simon and Schuster.

McMenamin, M. 2004. Gaia and glaciation: Lipalian (Vendian) environmental crisis. In S. Schneider, J. Miller, E. Crist, and P. Boston, eds., *Scientists Debate Gaia: The Next Century*. Cambridge: MIT Press, pp. 115–27.

Merchant, C. 1980. *The Death of Nature*. San Francisco: HarperSanFrancisco.

Olea-Franco A. 2002. Review of *Homage to Gaia: The Life of an Independent Scientist*. *Journal of the History of Biology* 35 (3): 600–602.

Plumwood, V. 1992. Conversations with Gaia. *APA Newsletters* 91 (1).

Raven, P. 2001. What have we lost? What are we losing? In M. J. Novacek, ed., *The Biodiversity Crisis: Losing What Counts*. New York: New Press, pp. 58–62.

Schneider, S., J. Miller, E. Crist, and P. Boston, eds. 2004. Preface. In *Scientists Debate Gaia: The Next Century*. Cambridge: MIT Press, pp. xiii–xvii.

Schwartzman, D., and T. Volk. 1989. Biotic enhancement of weathering and the habitability of Earth. *Nature* 340: 457–60.

Schwartzman, D. 1999. *Life, Temperature, and the Earth: The Self-organizing Biosphere*. New York: Columbia University Press.

Turner, S. 2004. Gaia, extended organisms, and emergent homeostasis. In S. Schneider, J. Miller, E. Crist, and P. Boston, eds., *Scientists Debate Gaia: The Next Century*. Cambridge: MIT Press, pp. 57–70.

Turney, J. 2003. *Lovelock and Gaia: Signs of Life*. New York: Columbia University Press.

Volk, T. 1998. *Gaia's Body: Toward a Physiology of Earth*. New York: Copernicus.

Volk, T. 2004. Gaia's life in a wasteworld of by-products. In S. Schneider, J. Miller, E. Crist, and P. Boston, eds., *Scientists Debate Gaia: The Next Century*. Cambridge: MIT Press, pp. 27–36.

Weston, A. 1987. Forms of Gaian ethics. *Environmental Ethics* 9 (4): 217–30.

Wilson, E. O. 1999. *The Diversity of Life*. New York: Norton.

# V

## Afterword

# 19

# Gaia Going Forward

Eugene Linden

Certain ideas become invested with a metaphorical power that extends their influence far beyond their original domain. The idea of natural selection, arguably the most important unifying concept of the past 1,000 years, has been appropriated—occasionally for better or more often for worse—by diverse fields including philosophy, sociology, political science, and economics; Thomas Kuhn's notion of paradigm shifts articulated in *The Structure of Scientific Revolutions* provides a general model for understanding the evolution of the context of ideas; quantum mechanics, conceived of to explain the bizarre mechanics of the invisible world, has been dragged into realm of the visible and proposed as a model of reality that can even explain such hardy New Age perennials as time travel and ESP (Herbert 1988, 1985).

And then there's Gaia. Apart from its utility in providing a framework for understanding the integration organic and inorganic systems, this simple, elegant theory offers the possibility of reuniting science and religion. For me, Gaia provides an elegant answer for a host of questions and suppositions that have been informing my worldview throughout my career. Perhaps the most beguiling aspect of Gaia is its implication that life is not fragile and passive but active and self-protective, invested with a godlike power to tame the brutal raw forces of the cosmos (Lovelock 1979).

The interconnectedness of life and stuff was a given in many animistic societies, but the history of Western thought, as Arnold Toynbee dryly observed, has been to get the ancestors/gods out of daily life, first pushing them out of trees to mountain tops (e.g., Olympus), and ultimately exiling the Almighty in outer space (Toynbee 1972). Freed of religious fetters, Western science and society have prospered by dealing with Earth's components in isolation.

The price of this blinkered approach, however, has been that Earth's ignored systems, out of sight and largely out of mind, began to break down. It's difficult to protect interconnected systems if one neither knows of nor cares about the interconnections. As the West conquered the world, however, it was increasingly brought face to face with those ignored bonds. This happened through the normal course of science but also through simple encounters, such as the discovery of apes that were far more humanlike than the monkeys encountered previously in Europe.

Over the past century and a half, the pace of scientific discovery accelerated, ultimately creating a situation where hard-core materialists have come to look at Earth and life from a perspective awfully similar to that of animists. The evolution of the major religions, however, has lagged, with the dominant monotheistic beliefs remaining committed to the belief that humanity is intrinsically different from the rest of creation. This is a situation ripe for resolution, and the advent and growing influence of Gaia suggests that this untenable contradiction might be in the process of being resolved. Simply put, Gaia provides a gangway between science and pantheism.

This is by no means an endorsement of strong Gaia (Kirchner 2002). You don't have to believe that there is some purpose to the way in which life tends to perpetuate a stable and life-friendly environment in order to celebrate the mystery, the blind artfulness, and self-correcting balance of the myriad interactions that domesticate and stabilize geophysical processes. Nor does the fact that life has taken over this planet require you to believe that Gaia has any vested interest in any particular ensemble of life forms.

Even weaker forms of Gaia—for example, that Earth is a self-organizing system that constantly adjusts to maintain an equilibrium friendly toward life—offer a scientific construct supportive of a theology with life at its center. That represents a refreshing shift from envisioning a creator with a vested interest in humanity. If life possesses the autochthonous power to emerge from the elements and harness the cosmos to perpetuate and protect itself...well, for me at least, that's pretty godlike, and worthy of respect, deference, and, yes, worship (though not necessarily in the form of dancing naked around Stonehenge during the summer solstice).

This shift in epistemology and cosmology toward Gaia (or at least the interconnectedness that is at the center of Gaia) is not trivial. It is also probably inevitable. The contradictory worldviews of religion and science can persist in parallel for some time but probably not forever, if for no other reason than the traditional view of human dominion that flows from our view of ourselves as God's chosen species has itself proved to be unsustainable.

As the consequences of this mismatch come home in the form of dead zones, stripped oceans, extinctions, silent springs, not to mention droughts, floods, and killing heat waves, people are reacting. I remember a conversation some years back with James Parks Morton when he was dean of the Cathedral of Saint John the Divine in New York. We were discussing this very issue—whether Christianity could adapt to a world in which life and ecology rather than humanity were at the center of belief (an issue somewhat broader than the ongoing debate among theologians over whether the creation is sacred). We were in the Cathedral, and his answer was playful, eloquent, and simple, using the metaphor of the cathedral itself, which has the shape of a cross with the high alter situated at the transept. "The Cathedral is the center," he said, "and if the center moves, the Cathedral must move."

Is that possible? Can the Cathedral move? Or will the disaffected rally around some emergent prophet who articulates a set of beliefs more in accord with the world we have discovered? There is plenty in the Bible to encourage respect for God's creatures, but all of it seems to flow from a type of noblesse oblige. If we're going to remain tenants of this planet, we need a cosmology that makes it abhorrent to heedlessly impair Earth's life support systems rather than our present-day custodial attitudes, which are more focused on cleaning up after the fact. Moreover even a reformulated monotheism could hardly be expected to do a better job of ensuring good treatment of the planet than it has of ensuring good treatment of fellow humans, and that kind of performance record would probably be the final nail in the coffin for Earth's ecosystems.

There is a cosmic irony (literally) in the fact that it required that we see the world piecemeal to develop the technology and economy to get us to outer space where for the first time we could see life whole—from God's perspective. The image in the rear view mirror was the most important dividend of the space program, allowing us to appreciate Earth as something more than a bunch of parts.

From a scientific perspective, geophysicists and oceanographers could see linkages that spanned continents and hemispheres. The view from space practically begs the scientist to explore the question of linkages between El Niño and the North Atlantic oscillation, for instance, as well as countless other connections and feedbacks that integrate planetary winds, ocean currents, the placement and topography of landmasses, ice sheets, the great forests and deserts, flora and fauna, into something tough and stable but also—and what scientist could possibly ignore this—miraculously beautiful.

Gaia thus provides a framework that helps scientists in tangible, testable ways, but it also melds seamlessly into the intangible. Should good scientists stick to the tangible and ignore the intangible—the beautiful and spiritual aspects of Gaia? Does it necessarily corrupt a scientist, if he or she has an emotional/religious investment in the belief in the transcendence of nature? It is an amusing indicator of the power of conventional wisdom that, as a society, we are likely to celebrate scientists for their piety if their religious beliefs spring from the established religion, even if those beliefs contradict the underpinnings of science. On the other hand, if a scientist professes belief in a nontraditional religion because it is in accord with his or her science, that scientist courts being viewed with suspicion.

Every scientist has a cosmology of some sort, and it is by no means settled that reality favors the materialist/determinist perspective. Among other things, the empirical method still has difficulty digesting quantum mechanics, which operates in a world of probabilities and indeterminateness. If, for instance, a subatomic particle can be influenced by events elsewhere in the universe as can happen under certain circumstances according to the rules of quantum mechanics, then how can an experimenter ever have confidence that he or she has controlled for every variable?

Albert Einstein, whose ideas gave rise to quantum mechanics, expressed dissatisfaction with this unwanted offspring of his in part because the implications of this collection of equations collided with his beliefs about the nature of the universe. I doubt many people would call Einstein a bad scientist for unabashedly identifying himself as a deeply religious man. Actually the religious landscape described by Einstein is not all that different than the nature-centric cosmology implied by Gaia. In a telegraph to a rabbi in New York Einstein (1929) wrote: "I believe in Spinoza's God, who reveals himself in the harmony of all being, not in a God who concerns himself with the fate and actions of men."[1] At other times, Einstein envisioned God as a creative force, revealed through the wonders of nature.

Gaia offers a suggestion of how this creativity can lie in the self-organizing nature of the system itself. It's worth noting that the god implied by Gaia is not some New Age father figure stroking bunny rabbits in a meadow, but more like a vengeful Jehovah of the Old Testament, or perhaps even more aptly, like Shiva, the Hindu deity of destruction and creation. With no particular commitment to any particular creature, Gaia, as Lovelock has argued, might react to us disruptive humans as an infectious agent that is destabilizing the system. As in human physiology, fever (in this case global warming) might be the self-protective reaction to rid the system of a pathogen.

Admittedly, such speculations are premature. Whether the theory/belief/paradigm bundled up in the word Gaia continues to gain adherents depends to some degree on the normal workings of science and the explanatory genius of scientists. If at some point new discoveries about the relationship of the organic and inorganic reveal show-stopping inadequacies in its scientific utility, Gaia will be consigned to the purgatory of the marginal and the outmoded, perhaps still useful in some limited way, but in such circumstances it will never reach center stage as a standard model to be tested and extended. Or possibly some new Lovelock will come along with a model that encompasses and extends Gaia, just as Gaia encompasses and extends the work of a long line of ecologically oriented thinkers including James Hutton (1794), John Muir (1916) and Aldo Leopold (1949), all of whom either explicitly or implicitly viewed Earth as a superorganism.

If Gaia is as productive a theory as I suspect, someone or perhaps many people will take it further. Just as the concept of evolution was in the air in the nineteenth century even before Darwin gave it a form that the scientific world and the public could rally around, so too Gaia was in the air before Lovelock first went to press in 1972. Wherever Gaia goes from here, Western thought owes James Lovelock and his like-minded peers a debt of gratitude, whether or not Lovelock remains the best articulator of Gaia.

The history of science has been that originators have not always been willing to embrace the implications of their own ideas. Darwin resisted the implication that evolution was not goal directed, and as previously mentioned, Einstein balked at the design of the universe implicit in his own work. Early on, Lovelock's confidence in the robustness of Gaia caused him to underestimate humanity's ability to upset the applecart. Gaia, however, has taken on a life of its own, and Lovelock has no more ability to determine where Gaia goes from here than did Einstein have control over the development of quantum mechanics, or Darwin the evolution of evolution. That's as it should be.

Still, Lovelock launched a magnificent ship. As James Parks Morton once remarked, Lovelock has given scientists and religious thinkers the opportunity to address the same reality with a common language. He has offered up a marvelous and enticing explanation/metaphor of life on Earth that allows the scientist to venture from the strictly utilitarian to the spiritual without requiring the adoption of multiple personalities. Perhaps in Gaia is the first sketch of a new religious paradigm.

I humbly count myself as one of those grateful for Lovelock's intuitive genius. I was a pantheist by inclination long before I first encountered Gaia, but he helped give me confidence that I was on the right track, not to mention that he described the tracks that I was on.

## Note

1. *New York Times*, April 25, 1929. p. 60.

## References

Herbert, N. 1985. *Quantum Reality*. New York: Random House.

Herbert, N. 1988. *Faster Than Light: Superluminal Loopholes in Physics*. New York: Penguin Books.

Hutton, J. [1794] 1999. *An Investigation of the Principles of Knowledge, and of the Progress of Reason: From Sense to Science and Philosophy*. Bristol: Thoemmes Press.

Kirchner, J. W. 2002. The Gaia hypothesis: Fact, theory, and wishful thinking. *Climatic Change* 52: 391–408.

Leopold, A. [1949] 1987. *A Sand County Almanac, and Sketches Here and There*. New York: Oxford University Press.

Lovelock, J. [1979] 2000. *Gaia: A New Look at Life on Earth*. Oxford: Oxford University Press.

Muir, J. 1916. *A Thousand Mile Walk to the Gulf*. New York: Houghton Mifflin.

Toynbee, A. 1972. The religious background of the present environmental crisis. *International Journal of Environmental Studies* 3 (1/4): 141–46.

# About the Contributors

**David Abram** is an ecologist, anthropologist, and philosopher, known for bringing the philosophical school of phenomenology to bear on environmental questions. He is director of the Alliance for Wild Ethics and author of *The Spell of the Sensuous* and *Becoming Animal* among other writings.

**Donald W. Aitken** is a principal of Donald Aitken Associates, and an affiliate faculty member at the Frank Lloyd Wright School of Architecture. He is recognized as an international expert on renewable energy.

**Connie Barlow** is the author of four popular science books in the fields of evolution and ecology. Living on the road, she calls herself an "at-large inhabitant" of the continent of North America. She is the founder and webmaster of www.TorreyaGuardians.org and www.TheGreatStory.org.

**J. Baird Callicott** is Regents Professor of Philosophy and chair of the Department of Philosophy and Religion Studies at the University of North Texas. He is the co-editor in chief of the *Encyclopedia of Environmental Ethics and Philosophy* and author or editor of numerous books and dozens of articles and papers in environmental philosophy and ethics.

**Bruce Clarke** is professor of Literature and Science at Texas Tech University. He is author of *Posthuman Metamorphosis: Narrative and Systems*, *Energy Forms: Allegory and Science in the Era of Classical Thermodynamics*, and other books and edited collections.

**Eileen Crist** is associate professor of Science and Technology in Society at Virginia Tech. She is author of *Images of Animals: Anthropomorphism and Animal Mind* and co-editor of *Scientists Debate Gaia: The Next Century*.

**Tim Foresman** is an environmental scientist and president of the Center for Remote Sensing Education. He has served as NASA's Digital Earth manager and as director of the United Nations Environment Programme's Division of Early Warning and Assessment. His work focuses on the use of technology for international environmental protection and sustainability.

**Stephan Harding** is a Gaia theorist, deep ecologist, and ecological scientist at Schumacher College in Devon, UK. His recent work, *Animate Earth*, integrates rational scientific analysis with emotional and intuitive intelligence as a pathway for understanding the Earth system and our environmental predicament.

**Barbara Harwood**   is an adjunct professor of Sustainability at the Frank Lloyd Wright School of Architecture and a LEED-accredited green-building designer. She is an author, designer of energy efficient homes, and environmental consultant.

**Timothy M. Lenton**   is professor of Earth System Science at the University of East Anglia. He is author of numerous publications on how Gaia functions, especially through the development and use of Earth system models.

**Eugene Linden**   is a writer and science journalist. He is author of *Winds of Change: Climate, Weather, and the Destruction of Civilizations*, *The Future in Plain Sight*, and numerous other books about the relationship of humanity with the natural world. He has written for a wide range of publications including *TIME*, *Fortune*, *National Geographic*, and the *New York Times*.

**Karen Litfin**   is associate professor in the Department of Political Science at the University of Washington in Seattle. She is author of *Ozone Discourses: Science and Politics in International Environmental Cooperation* and other publications. She specializes in the intersection of science and international environmental politics and is currently working on the global ecovillage movement.

**James Lovelock**   is an independent scientist, inventor, author, and environmentalist. He originated the Gaia hypothesis in the 1970s and has worked on developing Gaia theory ever since. He is author of *The Revenge of Gaia*, *Homage to Gaia*, and *The Ages of Gaia*, among many other books and publications. His latest book, *The Vanishing Face of Gaia*, was published in 2009.

**Lynn Margulis**   is a biologist and Gaian scientist, best known for championing the theory of endosymbiosis for the origin of eukaryotic cells. Along with James Lovelock, she is one of two original proponents of the Gaia hypothesis. She is author and co-author of many scientific and popular works, including *Symbiotic Planet*, *Microcosmos*, and *What Is Life?* She is Distinguished University Professor in the Department of Geosciences at the University of Massachusetts in Amherst.

**Bill McKibben**   is an environmentalist, educator, activist, and writer who has authored numerous works on global warming, alternative energy, and living lightly on Earth. His books include *Deep Economy*, *Maybe One*, and *The End of Nature*, and his articles have been published in *The New York Times*, *The Atlantic Monthly*, *Harper's*, *Orion Magazine*, and *Rolling Stone*. He is a scholar-in-residence at Middlebury College.

**Martin Ogle**   is chief naturalist of the Northern Virginia Regional Park Authority Potomac Overlook Regional Park in Arlington, Virginia. He organized the October 2006 conference "The Gaia Theory: Model and Metaphor for the 21st Century."

**H. Bruce Rinker**   is the science chair at North Cross School in Roanoke, Virginia. Previously the director of the Pinellas County Environmental Lands

Division in Tarpon Springs, Florida, he is an ecologist, educator, and explorer whose publications include *Forest Canopies*. He studies the links between canopies and soils as an important step toward the long-term conservation of forest resources.

**Mitchell Thomashow**   is president of Unity College in Maine. He is author of *Ecological Identity: Becoming a Reflective Environmentalist*, *Bringing the Biosphere Home: Learning to Perceive Global Environmental Change*, and other works.

**Tyler Volk**   is science director of Environmental Studies and associate professor of Biology at New York University. He is the author of *$CO_2$ Rising: The World's Greatest Environmental Challenge*, *Gaia's Body: Toward a Physiology of the Earth*, and other books.

**Hywel T. P. Williams**   is in the Earth System Modeling Group at the University of East Anglia. He studies the interaction between biological evolution and the physical environment, using simulation models to understand the adaptive dynamics of microbial ecosystems.

# Index

Complacency, and Gaia, 199
Complexity theory, 304
Computer games, educational value
  of, 265
Computer models. *See also* Modeling
  of Gaia perspective
Daisyworld, 9, 326–27 (*see also*
  Daisyworld model)
exceeded by temperature-driven
  changes, 130
limitations of, 81
Computer technology, in
  environmental learning, 258, 259,
  265
Concrete, and water cycle, 55
"Conference of the Parties,"
  fourteenth (COP14), 136
"Conference of the Parties,"
  thirteenth (COPI3), 135, 138
Consciousness, 222–23. *See also*
  Awareness
  and Leopold's earth ethic, 187–88
Conservation, 15
  vs. efficiency, 284
  as moral issue (Leopold), 185, 186
  and previous ecosystems, 288
Conservation ethic, 285–86
"Conservation Ethic, The" (Leopold),
  180
Consumer behaviors, EOS as help
  with, 250
Corporations, and freshwater supply,
  152
Correll, Robert, 129
Costa Rica, golden road of, 328
CO₂. *See* Carbon dioxide
Coweeta Hydrologic Laboratory, 88
Crisis, planetary, 200–201. *See also*
  Mass extinctions
Crutzen, Paul J., 23
Culture, and water, 55–57
Cybernetics, 294–95
  vs. complexity of biosphere, 326
  dangers of applying to Earth, 323–26
  first-order vs. second-order, 295,
    302, 306, 308

first-order, 295, 298, 301, 302, 308
second-order, 295–97, 300–303,
  304
Cycle, carbon. *See* Carbon cycle
Cycle, hydrological, 44–52
  with biodiversity, 119–21
Cycles, in environmental learning,
  269

Daisyworld model, 7, 9, 61, 63, 64–
  65, 305–306, 310n.1, 326–27
Darwin, Charles, 3, 9, 41, 244, 262–
  63, 341
  on earthworm feedback, 319
  on "entangled bank," 320
  on morality, 181, 182
Darwinian evolutionary biology
  and land ethic, 179, 180–81
  neo-Darwinian, 178 (*see also* Neo-
    Darwinian evolutionary theory)
Darwinian natural selection. *See also*
  Evolution
  cooperation vs. competition in, 183,
    205
  cultural appropriation of, 337
  Gaian addition to (environment-
    enhancing effects), 10–11, 22, 78
  as limitation, 41
  and Mendelian genetics (Modern
    Synthesis), 182
  as obvious, 17
  and planetary evolution, xi
  three parts of, 63
Dawkins, Richard, 9, 37
"Dead zones." in Pacific Ocean, 133
*Deep Economy* (McKibben), 287
Deep-time lags, and imperative for
  rewilding, 169–72
"Deep-time" perspective, 166, 168–
  69, 171–72
Democracies, Gaian, 206
  for ecological restoration programs,
    250
Descartes, René, 298
*Descent of Man, The* (Darwin), 181,
  182